Electronics for Guitarists

Denton J. Dailey

Electronics for Guitarists

Third Edition

 Springer

Denton J. Dailey
Butler County Community College
Butler, PA, USA

ISBN 978-3-031-10760-3   ISBN 978-3-031-10758-0   (eBook)
https://doi.org/10.1007/978-3-031-10758-0

This Springer imprint is published by the registered company Springer Nature Switzerland AG
The registered company address is: Gewerbestrasse 11, 6330 Cham, Switzerland

This book is dedicated to my son, Wayne.

Preface

From an electrical engineering standpoint, the guitar is simply a signal source. Of course, the guitarist knows that it's really so much more than that. The guitar is a conduit to the soul of the guitar player. On the other hand, most guitarists know that the electronic circuits used to process and amplify this signal are more than just a collection of transistors and tubes. These circuits and the guitar signal are related in complex and fascinating ways. There are few areas of art and engineering that combine with such dynamic synergy.

If you know absolutely nothing about electronics, starting into this book might be somewhat like showing up for swimming lessons and being thrown into the deep end of the pool. You may panic and struggle to tread water at first, but in the end I'm certain you will come out with some significant knowledge, and maybe even a sweet tube amplifier or effects box you designed and built yourself—and actually understanding how it works!

If you already understand basic circuit analysis, and transistor/linear IC circuit analysis, then this book may serve to give you some insight into the basic principles of various effects and signal processing circuits. If you understand electronics but have never studied vacuum tube circuits, which probably includes most of you who were born in the 1960s or later, I think you will find Chaps. 6 and 7 the most interesting.

If you are already an old pro at tube circuit design/analysis/troubleshooting, then you may find my approach to tube circuit design to be somewhat unconventional. I learned tube theory a long time after I learned transistor theory, and I tend to approach tube circuit design and analysis in a way that is quite different from the typical tube era texts that you might have seen. I'm not saying that my approach is better or worse, just different.

Who This Book Is Written For

If you would like to know how transistor- and vacuum tube–based amplifiers and how various effects circuits work, then this book is for you. In many ways, this book should be of interest to any musician or recording engineer interested in analog signal processing circuit operation. This book should also be useful to electronics hobbyists, technologists, and engineers interested in general audio signal processing and related applications.

This is the book I wish I had when I was in high school—or even junior high school. Would I have understood everything covered here? No way! But there are plenty of basic circuits to experiment with, and qualitative descriptions of circuits and concepts that make some quite advanced material accessible to any curious, technically oriented person.

Analog Rules!

As with previous editions, the main thrust of this book is old-school analog circuitry—lots of coverage of discrete transistors and diodes, classical filter circuits, and of course vacuum tube–based amplifiers. I have added just a bit more in terms of digital circuits this time around. It was decided that going deeply into digital signal processing and software-related areas would just make the book too big. As before, we are sticking mainly to analog circuits.

About the Math

The main obstacle most often associated with understanding electronics is math. It is not necessary to understand the differential equations describing the dynamics of the guitar in order to become a virtuoso. In fact, it is not really necessary to know any mathematics at all to be a great guitar player. However, it is necessary to understand some basic mathematical concepts to gain even an elementary understanding of electronics.

At first glance you'll see that there is a lot of math in this book. While in the strictest sense this is true (there are tons of equations here!), most of the equations are simply analysis formulas that I present without derivation. If you have a basic understanding of algebra and trigonometry, you are in great shape to digest about 90% of the math presented. Familiarity with logarithms, exponentials, and complex numbers is also helpful. Whenever possible, I have tried to explain the principles and motivation behind the equations and circuit theory as clearly and succinctly as possible. A few unavoidable derivatives are used here and there, but no integrals were harmed in the making of this book.

Building the Circuits

Math is very important—in fact absolutely essential—if you want to learn to design your own circuits. But if you only want to build and experiment, you can skip most of the math and still learn a lot. All of the circuits presented here have at least been prototyped and are great starting points for further experimentation. With the exception of the vacuum tube–based circuits, all of the circuits presented operate at relatively safe, low voltages. The battery-powered circuits are especially suitable for beginners.

Most of the transistor- and op amp–based circuits in the book can be built for under $20.00. Often, the most expensive parts of the effects box–type projects will be the case, or perhaps the switches. Should you decide to use an online printed circuit board service to produce your own PCBs, you can expect to pay about $50.00 and up, but then it's easy to make multiple copies which would make nice gifts for your friends and family, or a good way to start a new business.

Vacuum Tubes

Chances are good that one of the main reasons you are reading this book is to learn something about vacuum tube amplifiers. I don't think you will be disappointed. Unless you have studied electronics or it has been your hobby for a while, I recommend that you work through the chapters leading up to vacuum tube ampli- fiers. It's a good idea to learn to build safe, low-voltage circuits before tackling a scratch-built tube amp. You also won't feel too bad if you burn up a few 25-cent transistors as you climb the electronics learning curve.

There is no way around the fact that vacuum tube amplifiers are very expensive to build. Even the smallest tube amplifier will probably cost about $200.00 to build if you order all new parts. A moderately powerful, scratch-built tube amplifier will probably cost $400.00 or more. Even so, there is something that is very cool about seeing the warm glow of those tubes as you play through this amp that you built from the ground up. You might also end up being the amp guru of your neighborhood someday, which is not a bad thing either.

The Third Edition

As with the second edition, I have made corrections, cleaned up some schematics, and clarified a few concepts and explanations. Based on valuable feedback from readers of the previous editions, I have added a number of new circuits and design examples, including noise gates, analog multipliers, some basic digital signal processing concepts, effects loops, and additional tube amp design examples.

Safety

Any circuit that derives power from the 120 V AC line can be dangerous, and common-sense precautions should be taken to prevent shock hazards. This is especially true of vacuum tube circuits which use power supplies of over 500 V in some cases. These voltages can be lethal, and extra caution should be exercised if you decide to build any type of vacuum tube circuit. It is recommended that you consult with a knowledgeable technician or hobbyist if you are inexperienced with high-voltage circuitry.

Disclaimer

The information and the circuits in this book are provided as is without any express or implied warranties. While every effort has been taken to ensure the accuracy of the information contained in this text, the author assumes no responsibility for errors or omissions, or for damages resulting from the use of the information contained herein.

Acknowledgments

As before, I would like to thank noted tube amp guru, and guitarist for the band X, Billy Zoom for engaging in very helpful and informative discussions with me relating to this work. I would also like to thank Gearmanndude for posting an entertaining and informative review of the first edition on his channel http://www.youtube.com/user/gearmanndude.

Contact Information

Although my personal email is easy enough to find, I have created an official Facebook page https://www.facebook.com/ElectronicsForGuitarists/ where I answer questions, post occasional updates, new circuit designs, designs submitted by readers, and even the occasional demo of my amps and effects. Please feel free to check it out and say hello.

Butler, PA, USA   Denton J. Dailey

Internet Links and Descriptions Used in Text

Popular guitar-effects review YouTube channel with good review of the first edition.
http://www.youtube.com/user/gearmanndude
Facebook page for Electronics for Guitarists
https://www.facebook.com/ElectronicsForGuitarists/
Source for electronics components.
www.jameco.com
NSF Guitar technology website.
www.guitarbuilding.org
Supplier specializing in guitar-related electronic components.
www.smallbearelec.com
Supplier specializing in guitar-related electronic components.
www.buildyourownclone.com
Supplier specializing in guitar-related electronic components.
www.coolaudio.com
Supplier specializing in vacuum tube and vintage audio supplies.
www.tubesandmore.com
Supplier specializing in vacuum tube and vintage audio supplies.
www.tubedepot.com
Supplier of transformers for vacuum tube applications.
www.hammondmfg.com
Online repository of vacuum tube data sheets.
http://frank.pocnet.net/

Contents

List of Figures

Chapter 1
Power Supplies

Introduction

All electronic circuits require a power supply of some sort. Many times the power supply for a circuit will simply consist of a 9 V battery, which is especially true for guitar effects boxes (stomp boxes) and pedals. In this chapter, we will examine the operation of linear, single-polarity, and bipolar power supplies, as well as basic vacuum tube diode-based power supply circuits.

A Simple Power Supply Circuit

The circuit in Fig. 1.1a is probably the most common power supply design used in low- to medium-current applications. The component values shown here will provide an output voltage of about 16.5 V_{DC}. We will now discuss some of the design variables of the circuit.

Power is switched via single-pole, single-throw (SPST) switch S_1. The power switch should always be wired in series with the hot line of the AC mains to help prevent shock hazards. If a fuse is used, it should also be wired in series with the hot line as well. Sometimes, for additional safety, both the hot and neutral lines are switched using a double-pole, single-throw (DPST) switch, as shown in Fig. 1.1b.

The terminals of the most common North American 120 V, 60 Hz residential AC line outlets are identified in Fig. 1.2. These are NEMA (National Electrical Manufacturers Association)-type 5–15 and 5–20 sockets, rated for 15 and 20 amp service, respectively.

© The Author(s), under exclusive license to Springer Nature Switzerland AG 2022
D. J. Dailey, *Electronics for Guitarists*,
https://doi.org/10.1007/978-3-031-10758-0_1

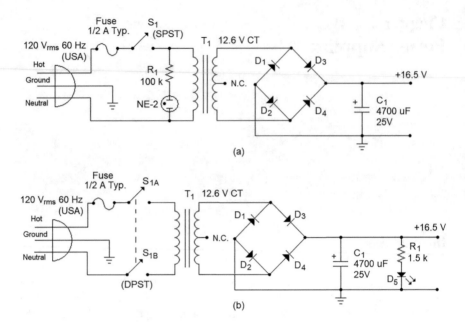

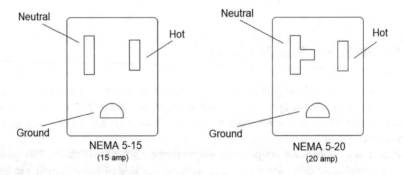

Fig. 1.1 Basic power supply with optional power indicators

Fig. 1.2 North American standard AC line sockets

The Transformer

In this example, transformer T_1 steps the incoming AC line voltage from 120 V_{rms} on the primary down to 12.6 V_{rms} on the secondary. The transformer also isolates the secondary side circuitry from the AC line, which is a desirable safety feature. Secondary voltage ratings of 12.6 V_{rms} and 6.3 V_{rms} are very common and are a throwback to the days of vacuum tubes, where typical tube filaments or heaters operated from these voltages. Such transformers are still commonly called *filament transformers*. We will talk more about this application later in the book. Note that the transformer has a center tap on the secondary winding that is unused in this circuit.

Transformers rated for a secondary current of 1 or 2 amps are pretty common, but if you are only going to power one or two effects circuits, a transformer rated for about 0.5 amp should be sufficient. If you are planning to power a 50 W amplifier or a similar high power load, you will definitely need a much larger transformer.

A final note about transformers: generally transformer secondary voltages are rated at full-load conditions. If you are only lightly loading your transformer, the output voltage will typically be a few volts greater than the indicated value. We will delve deeper into the characteristics of transformers when we examine high-power tube amplifiers in Chap. 7.

The Rectifier

Most power supplies use a full-wave rectifier to convert incoming AC into a pulsating DC waveform. A full-wave bridge rectifier and typical input and output voltage waveforms are shown in Fig. 1.3.

Bridge rectifiers are available as modular units, or they may be built using discrete diodes. Type 1N4001 rectifier diodes are suitable here. Diodes in the 1N400x series are rated to carry 1 A forward current. It is generally ok to substitute diodes with higher voltage and current ratings in a given application. The 1N400x series ratings are given in Table 1.1, where V_{BR} is the rated reverse breakdown voltage.

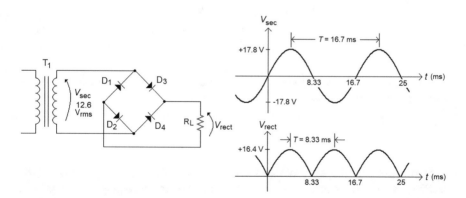

Fig. 1.3 Full-wave rectifier and associated waveforms

Table 1.1 1N400x voltage ratings

Diode	1N4001	1N4002	1N4003	1N4004	1N4005	1N4006	1N4007
V_{BR}	50 V	100 V	200 V	400 V	600 V	800 V	1000 V

Analysis of the Rectifier

In Fig. 1.3, we are assuming the transformer has $V_{sec} = 12.6\ V_{rms}$, which is a common secondary voltage rating. We can convert between peak and rms values of sinusoidal waveforms using the following formulas:

$$V_P = V_{rms}\sqrt{2} \text{ or } V_P \cong 1.414 V_{rms} \tag{1.1}$$

$$V_{rms} = V_P/\sqrt{2} \text{ or } V_{rms} \cong 0.707 V_P \tag{1.2}$$

Using (1.1), the peak secondary voltage works out to be about $V_P \cong 17.8$ V.

During the positive swings of the secondary voltage, diodes D_2 and D_3 conduct (they are forward biased), while D_1 and D_4 act as open circuits (reverse bias). On negative-going voltage swings, D_2 and D_3 are reverse biased, while D_1 and D_4 conduct. This causes current to flow in the same direction through the load for both positive and negative AC voltage swings producing the pulsating DC voltage shown at the bottom of Fig. 1.3.

A typical forward-biased silicon diode will drop about $V_F = 0.7$ V (often called the *barrier potential*). Since the load current must flow through two diodes at any time, we lose approximately $2 \times 0.7\ V = 1.4$ V across the bridge rectifier. So, the output voltage reaches a peak value of

$$\begin{aligned} V_{rect(pk)} &= V_{sec(pk)} - 2V_F \\ &= 17.8\ V - (2 \times 0.7\ V) \\ &= 16.4\ V \end{aligned} \tag{1.3}$$

The frequency f of a periodic waveform is the reciprocal of its period T. As an equation, this is written as

$$f = \frac{1}{T} \tag{1.4}$$

If you examine the waveforms of Fig. 1.3, you will see that the period of the rectified waveform is half as long as the period of the incoming sinusoidal AC waveform. This means that the fundamental frequency of the rectified waveform is twice the AC line frequency.

$$\begin{aligned} f_{rect} &= 2f_{line} \\ &= 2 \times 60\,Hz \\ &= 120\,Hz \end{aligned} \tag{1.5}$$

The Frequency Domain

An oscilloscope allows us to view a signal (usually a voltage) in the time domain, that is, a voltage as a function of time $v(t)$. This is the most familiar and usually the most convenient way to visualize signals. An alternative way to view signals is in the *frequency domain*. The concept of the frequency domain is a bit abstract, but since it is extremely useful and is referenced many times in later chapters, this is a good time for its introduction.

In the frequency domain, we view a signal as a function of frequency $V(f)$. This representation is often called a *frequency spectrum*, or simply a spectrum. It turns out that any periodic waveform that is not precisely sinusoidal in shape actually consists of the sum of (possibly infinitely many) sinusoids of different frequencies. The lowest non-zero frequency in the spectrum is called the *fundamental* frequency. As a point of interest, the mathematical technique used to derive frequency spectrum information is the *Fourier transform*.

Looking at the input and output waveforms for the full-wave rectifier circuit in both the time and frequency domains, we obtain the graphs of Fig. 1.4. Notice that the rectified signal consists of even multiples of the original 60 Hz frequency. Integer multiples of the fundamental frequency are called *harmonics*. The harmonics present in the output of the rectifier decrease in amplitude as frequency goes up, and in principle, they extend to infinite frequency.

By changing the shape of the signal, the rectifier causes energy that originally existed at one frequency (60 Hz here) to be distributed over a range of different frequencies, including DC (0 Hertz). For the purposes of power supply design, the desired function performed by the rectifier is to create a large DC (zero frequency) component on its output. The higher harmonics are undesired and are heavily attenuated by the filter.

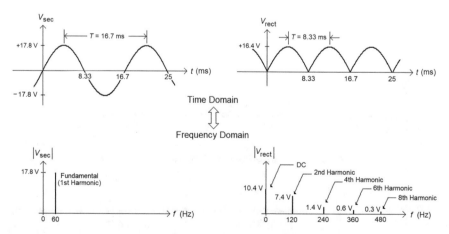

Fig. 1.4 Time and frequency domain representations of circuit waveforms

The Filter

Our power supply has an especially simple filter that is formed by capacitor C_1. The capacitor charges to the peak voltage output from the rectifier and approximately holds this voltage between peaks. Because the transformer/rectifier combination has a low equivalent internal resistance, the capacitor charges very quickly. However, assuming that the load on the power supply is not too heavy, the capacitor discharges much more slowly, holding the output voltage very close to the peak value.

One way of thinking about how the filter works is by observing that the capacitor stores energy supplied by the transformer/rectifier at the positive voltage peaks and then delivers that stored energy to the load, filling in the gaps between pulses. This is shown in Fig. 1.5.

Ripple Voltage

The slight variation of the filtered output voltage is called *ripple*. No filter is perfect, so there will always be at least a little bit of ripple voltage present, but in general, the larger the filter capacitor, the lower the ripple in the output voltage will be. We want to reduce ripple voltage so that we do not hear 120 Hz hum when powering amplifiers and other effects circuits from the power supply.

There are many ways to calculate filter capacitor values, but here's a general rule of thumb for unregulated power supplies: to limit ripple to less than 1% of the total output, use about 1000 μF for each 10 mA of load current. As an equation, this is

$$C(\text{ in } \mu F) \geq 100 I_L(\text{ in mA}) \tag{1.6}$$

For example, if your power supply must deliver 25 mA to an effects pedal, a reasonable filter capacitor value to use would be

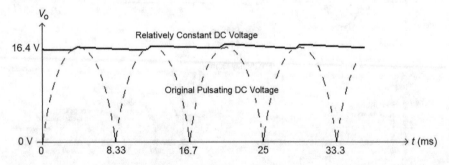

Fig. 1.5 Filtered output voltage

$$C \geq 100 \times 25$$
$$\geq 2500 \ \mu F$$

Several smaller capacitors may be connected in parallel to obtain a higher capacitance value. Total capacitance is $C_T = C_1 + C_2 + \ldots + C_n$ for n capacitors connected in parallel.

The capacitance calculated in the previous example is quite a large value considering the light loading of the supply. It will be shown later in this chapter that when a voltage regulator is used in the design of the power supply, a much smaller filter capacitor may be used because the regulator will provide additional ripple reduction.

In amplifier applications, the effects of supply ripple are most severe under no- and low-signal conditions when output hum is most audible. This is especially true of class A power amplifiers, which operate at relatively high supply current levels under no-signal conditions. As will be discussed in greater detail in later chapters, class B and class AB amplifiers draw relatively low supply current under low-signal conditions, which allows us to get away with using smaller supply filter capacitors.

Operational amplifiers, which are widely used in both amplifier and signal processing applications, tend to be very insensitive to supply voltage ripple. Op amps are discussed in greater detail in Chaps. 3 and 5.

Filter Analysis: The Frequency Domain

In simplest terms, a filter is a network or circuit that has frequency selective characteristics. We will talk much more about filters at various places throughout the book, but for now, it is sufficient to briefly examine a first-order, low-pass filter and its response curve, which are shown in Fig. 1.6.

The *corner frequency* f_C (also called the *critical frequency* or *break frequency*) divides the response into the *passband* and *stopband*. All frequencies less than f_C are in the passband of the low-pass filter, while frequencies greater than f_C are in the stopband. The corner frequency is given by (1.7), which we will use many times in upcoming chapters.

$$f_C = \frac{1}{2\pi RC} \tag{1.7}$$

The actual filter response is traced by the smooth, continuous curve in the graph. The response of the filter drops by 3 dB at the corner frequency f_C. This is the frequency at which power output from the filter has dropped to 50% of maximum. It is a standard practice to plot filter response in decibels (dB) vs. $\log f$, where log is the base-10 logarithm.

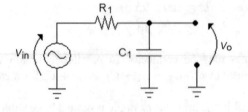

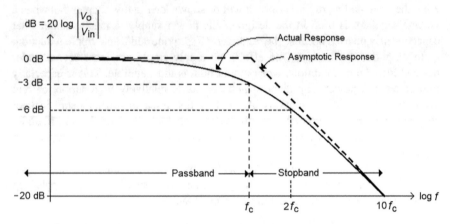

Fig. 1.6 First-order, low-pass filter and response curve

You may be curious where (1.7) comes from. This is the frequency at which the reactance X_C of capacitor C_1 equals resistance R_1 in the filter. Equation (1.7) is derived by setting $X_{C1} = R_1$ and solving for f, where $X_C = 1/(2\pi f C)$. I will leave this as a homework assignment.

Response Curves, Decades, and Octaves

The straight-line approximation of the filter response is called the *asymptotic response*. The asymptotic response is the response that the filter approaches as we move away from f_C in either direction. When using a log-log graph, these asymptotes are straight lines. This is one reason why we use dB vs. log f scales on frequency response graphs. Asymptotic response plots are often called *Bode plots* (pronounced as Boe-Dee), after Hendrik Bode of Harvard University, who is credited with their first use.

In the response plot of Fig. 1.6, starting at f_C, notice that as we move to the left and approach zero frequency (zero is infinitely far to the left on a logarithmically scaled graph), the response becomes more horizontal. As we move to the right from f_C, frequency increases, and the response begins to roll off. The higher the frequency, the more closely the roll-off rate approaches -20 dB/decade. A roll-off rate of -20 dB/decade is equivalent to -6 dB/octave.

A decade is a tenfold increase or decrease in frequency, and an octave is a doubling or halving of frequency. Intervals of one octave and one decade are denoted as $2f_C$ and $10f_C$, respectively, in Fig. 1.6.

Power Indicators

Power indicators aren't absolutely necessary, but they add a nice touch to most projects. Two optional forms of power indicators are shown in the supplies of Fig. 1.1. If you prefer the old-school approach, you can use a neon lamp on the primary side of the transformer. A light-emitting diode (LED) indicator can be added to the low-voltage side of the supply.

Neon Lamps

The typical neon lamp used as a power indicator is the NE-2 lamp, which produces a nice, warm orange glow when operating. The glow is caused by current flow through ionized neon gas inside the bulb. It takes about 70–80 V to ionize the neon, which is why the NE-2 indicator is connected across the primary of the transformer. The value of series resistor R_1 is not critical, with typical values ranging from 100 kΩ and higher. Neon indicator lamps are also available with a built-in series resistor. NE-2 lamps dissipate very little power and operate at very low current levels, typically less than 1 mA.

Light-Emitting Diodes

If you prefer a more modern power indicator, you can use a light-emitting diode (LED). A very common LED package is shown in Fig. 1.7.

Most LEDs will operate well with a forward current of 5–10 mA. Refer back to the power supply schematic of Fig. 1.1. Let's assume that D_5 is a red LED with an approximate forward voltage drop of $VF \cong 1.5$ V. The LED current is given by

$$
\begin{aligned}
I_{\text{LED}} &= \frac{V_O - V_F}{R_2} \\
&= \frac{16.5 \text{ V} - 1.5 \text{ V}}{1.5 \text{ k}\Omega} \\
&= 10 \text{ mA}
\end{aligned}
\tag{1.8}
$$

Unlike silicon rectifier diodes, which typically exhibit a forward voltage drop of about 0.7 volts, different LEDs may have rather widely varying forward voltage

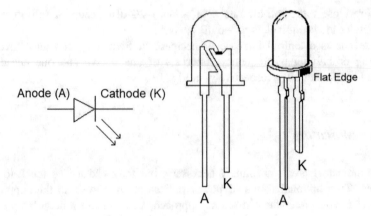

Fig. 1.7 Typical LED schematic symbol, package, and lead identification

drops. In general, the shorter the wavelength of the LED light, the higher its barrier potential will be. Red LEDs tend to have the lowest forward drop (usually around 1.5–2 V), while blue LEDs have the highest (around 3–4 V). White LEDs are normally fabricated by coating a blue LED chip with a phosphor that emits a yellowish light when the LED is turned on. The combination of the phosphor and blue LED emission produces the familiar bluish-white light that is characteristic of these LEDs.

LED Current Limiting Resistor Calculation

Before we leave this topic, let's work out another LED indicator example. Assume we want to add an LED power indicator to a 9 V DC power supply or perhaps a 9 volt battery-powered device. We are going to use a green LED designed to operate with $V_F = 2.5$ V at $I_F = 5$ mA. We solve (1.8) for R_2 and plug in the given values

$$
\begin{aligned}
R_2 &= \frac{V_O - V_F}{I_F} \\
&= \frac{9 \text{ V} - 2.5 \text{ V}}{5 \text{ mA}} \\
&= 1.3 \text{ k}\Omega
\end{aligned}
\tag{1.9}
$$

In practice, we can use a close standard resistor value. Using 1.2 kΩ would be acceptable, causing the LED to be slightly brighter. Using a 1.5 kΩ resistor would result in reduced brightness, but lower current draw.

Incandescent Lamps

Another alternative power indicator is the incandescent pilot lamp. One of the most commonly available indicators is the #47 lamp, shown on the left side of Fig. 1.8. The #47 lamp has a bayonet-style base and is designed to operate from the 6.3 V_{rms} secondary of a filament transformer, drawing a current of about 150 mA_{rms}. Because the lamp draws a fairly heavy current, it is best to connect the lamp before the rectifier, as shown in Fig. 1.8. This prevents the lamp from loading down the rectifier/filter, which would increase output ripple voltage. Of the power indicators discussed, incandescent lamps are by far the most inefficient.

An assortment of neon, incandescent, and LED indicators is shown in Fig. 1.9. Of these indicators, the small LED types will usually be the least expensive. Although

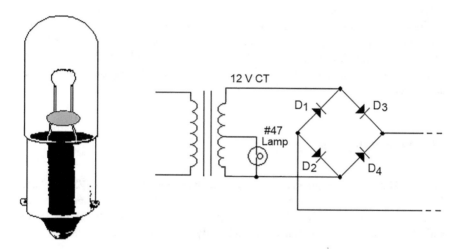

Fig. 1.8 Typical incandescent pilot lamp and connection as a power indicator

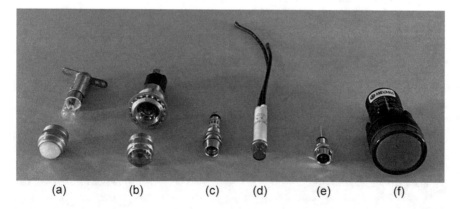

Fig. 1.9 Various power indicators. (**a, b**) Incandescent lamps. (**c, d**) Neon lamps with built-in resistor. (**e, f**) LED

incandescent lamps themselves are rather inexpensive, lamp holders with large colored jewels are much more expensive, costing around $5.00 new. The LED on the far right side of the photo is a large industrial LED indicator, designed to operate directly from the 120 V AC line. These indicators look very cool and are available in a wide range of colors, costing about $5.00 new.

A Basic Regulated Power Supply

Even if it is well-filtered, the output of an unregulated power supply will vary with loading conditions or if the AC line voltage changes. The use of a voltage regulator can usually reduce these effects to negligible levels, as well as providing a large reduction in output ripple voltage.

Figure 1.10a shows the schematic for a regulated, single-polarity power supply. The regulated power supply is identical to those shown in Fig. 1.1 with the addition of a three-terminal linear voltage regulator U_1 (the 78xx) and capacitors C_2 and C_3. Note that two optional power indicators are shown. Either or both may be eliminated.

The 78xx Voltage Regulator

Figure 1.10b shows the pin designations for a 78xx regulator in a TO-220 package. The xx section of the part number designates the rated output voltage. For example, the 5 volt version is the 7805. Available output voltages are 5, 6, 8, 9, 10, 12, 15, 18, and 24 V. The tab of the regulator is electrically connected to the ground terminal.

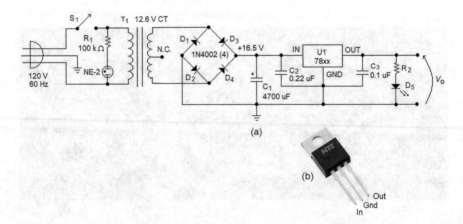

Fig 1.10 (a) Regulated single-polarity supply. (b) Pinout for the 78xx regulator

The 78xx regulators are rated for a maximum load current of 1 A and maximum power dissipation of 15 W; however, you will need to mount the regulator on a heat sink to dissipate more than about 1 watt. Fortunately, the 78xx is a very rugged device featuring short-circuit and thermal overload protection. In fact, the Fairchild Semiconductor data sheet states that the 78xx is "essentially indestructible." Having taught electronics for over 35 years, I have seen more than a few presumably indestructible devices go up in flames, but nevertheless, the 78xx is a pretty tough regulator.

Dropout Voltage

In order for the regulator to work properly, the input voltage must be greater than the output voltage. The minimum voltage required to be dropped across the regulator is called the *dropout voltage*. Typical dropout voltage for the 78xx series is about 2.5 V. As an equation, this is written as

$$V_{in} \geq V_0 + 2.5V \tag{1.10}$$

In the circuit of Fig. 1.10a, since the unregulated voltage is 16.5 V, we could use any regulator from the 7805 up to the 7812 without a problem. There is not quite enough headroom to guarantee that a 7815 would work in this circuit. There are low-dropout (LDO) voltage regulators available that will work with input voltages that are only 0.5 V greater than the regulated output voltage.

The maximum unregulated input voltage should not exceed the rated output plus 15 V. We can state the voltage requirements of the 78xx in a single inequality as

$$(V_0 + 2.5V) \leq V_{in} \leq (V_0 + 15V) \tag{1.11}$$

Power Dissipation

Using a higher than necessary input voltage results in greater power dissipation by the regulator. Consider the circuits shown in Fig. 1.11a, b, where we are using a 7810 to supply 10 V to a load that draws 100 mA.

Applying an unregulated input of 15 V, as shown in Fig. 1.11a, the power dissipation of regulator U1 is

$$\begin{aligned} P_{D(U1)} &= (V_{in} - V_o)I_L \\ &= 5\,V \times 100\ mA \\ &= 0.5\ W \end{aligned} \tag{1.12}$$

Using the maximum allowable input voltage $V_{in} = 25$ V as shown in Fig. 1.11b results in the power dissipation of U2 to be

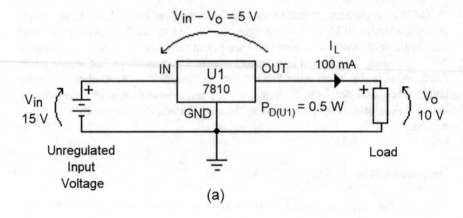

(a)

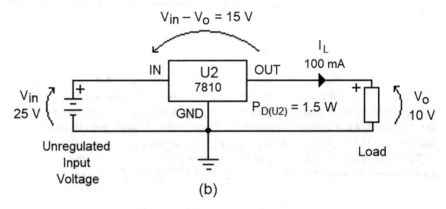

(b)

Fig. 1.11 Operating a 7810 regulator with (**a**) 15 V and (**b**) 25 V input voltages. $I_L = 100$ mA

$$P_{D(U2)} = (V_{in} - V_o)I_L$$
$$= 15 \text{ V} \times 100 \text{ mA}$$
$$= 1.5 \text{ W}$$

Regulator U2 is dissipating three times more power than regulator U1. Considering that 78xx regulators are rated for 15 W max power dissipation, 1.5 W may not seem like much power, but this would cause the regulator to run at about 98 °C (208 °F) without a heat sink. You wouldn't want to touch it.

Internally, the 78xx regulators are fairly complex (16 transistors), with a large portion of the circuit forming a high-gain amplifier. In order to prevent possible oscillation and overheating of the regulator, it is a good practice to mount small decoupling capacitors (C_2 and C_3 in Fig. 1.10) close to the regulator. This is especially true if there are long wires connecting the regulator input to the unregulated supply. The closer you can mount these capacitors to the regulator, the better.

Voltage regulators provide two types of regulation: *load regulation* and *line regulation*. Load regulation is the ability to maintain constant output voltage under varying load currents. Line regulation is the ability to maintain constant output voltage for input line voltage variation.

In addition to providing line and load regulation, the 78xx also reduces output ripple voltage. Typically, a 78xx will reduce ripple voltage by a factor of about 4000 (72 dB). This means that we can use much less filter capacitance to achieve virtually ripple-free output. Usually, the cost saved using smaller filter capacitors more than compensates for the small cost of using a voltage regulator.

Bipolar Power Supplies

A bipolar power supply provides output voltages that are positive and negative with respect to ground. Many circuits that contain operational amplifiers (op amps) require a bipolar power supply.

Using Batteries for Bipolar Power

Two 9 V batteries, connected in series, can be used to form a ±9 V power supply as shown in Fig. 1.12. The junction between the two batteries is designated as being the ground or zero-volt reference node.

Fig. 1.12 Using batteries to form a bipolar power supply

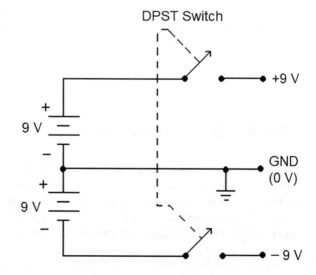

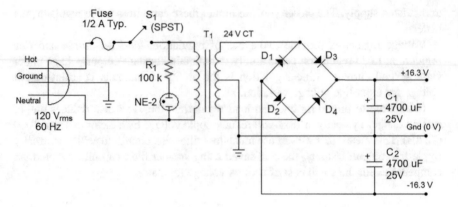

Fig. 1.13 Simple bipolar power supply

If you are using two batteries as a bipolar supply, both batteries must be disconnected from the circuit when turning off power. This requires the use of a double-pole, single-throw (DPST) switch.

A Typical Bipolar Power Supply

An unregulated bipolar power supply is shown in Fig. 1.13. The transformer used here has a 24 V_{rms}, center-tapped secondary, providing DC outputs of approximately ±16 volts. These are usually called the *supply rails*.

Assuming equal loading of both sides of the power supply, we are effectively using half of the secondary winding to derive each output voltage. Because this reduces the secondary voltage to half its normal value, the overall secondary voltage must be doubled to compensate.

A bipolar power supply constructed using point-to-point wiring on perf board is shown in Fig. 1.14. The rectifier is a modular bridge, and each side of the supply is filtered with three 3300 µF capacitors connected in parallel providing 9900 µF per side.

A Regulated Bipolar Power Supply

A regulated bipolar power supply is shown in Fig. 1.15. The positive side of the supply uses a 78xx regulator, while the negative side uses a 79xx regulator. Negative regulators work the same way as positive regulators except for reversal of polarity.

Remember, appropriate values for C_1 and C_2 depend on how much current you expect to draw from the supply (limited to ±1 A using the regulators shown) and

Fig. 1.14 Practical bipolar power supply

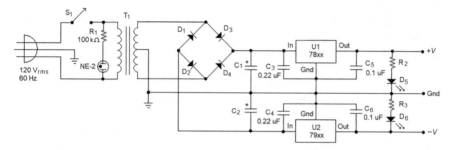

Fig. 1.15 Regulated bipolar power supply

how much ripple voltage you can live with. Because of the excellent ripple rejection of the voltage regulators, 4700 μF would probably work well in most applications.

Three optional power indicator options are shown, the neon lamp and LEDs that may be connected on either supply rail. Resistors R_2 and R_3 would typically range from about 470 Ω for $V_o = \pm 5$ V to about 2.2 kΩ for $V_o = \pm 15$ V. Both LED power indicators could be used, but normally only one would be used.

Two-Diode, Full-Wave Rectifier

An alternative full-wave rectifier, using only two diodes, is shown in Fig. 1.16. This circuit requires the transformer to have a center-tapped secondary winding, even though it has a single-polarity output. The rectifier works as follows:

As the AC line alternates between positive and negative values, the polarity of the transformer secondary winding reverses on every half-cycle of the AC line voltage. When the top of the winding is positive with respect to the bottom, D_1 is forward biased, while D_2 is reverse biased. Current will flow through the circuit in the

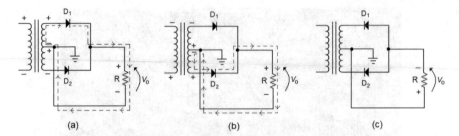

Fig. 1.16 Center-tap full-wave rectifier. (**a, b**) Positive output voltage. (**c**) Negative polarity output version

direction shown in Fig. 1.16a. Note that only the upper half of the secondary is being used during this half-cycle.

When the polarity of the secondary reverses, the bottom of the secondary is positive with respect to the top, causing diode D_1 to be reverse biased, while D_2 will be forward biased. Current flows in the direction indicated in Fig. 1.16b. The important thing to notice is that even though the polarity of the secondary has reversed, the load current flows in the same direction as for the previous half-cycle. The output voltage here is positive with respect to the center tap.

The output produces the familiar pulsating full-wave rectified voltage waveform we discussed previously. If you look closely at the full-wave bridge in Fig. 1.13, you can see that diodes D_1 and D_2 correspond exactly with diodes D_3 and D_4 in Fig. 1.16.

As shown in the schematic, the center tap of the transformer is almost always designated as ground. Also, notice that only half of the secondary winding operates on each half-cycle. Because of this, the secondary voltage must be twice as high as for an equivalent full-wave bridge to obtain a given output voltage.

A negative polarity output is obtained by reversing the direction of the diodes, as shown in Fig. 1.16c. The operation of this section of the rectifier works the same way as described previously, except that the polarity of the output voltage is negative with respect to the center tap. Again, if we refer back to Fig. 1.13, diodes D_1 and D_2 of the bridge rectifier are connected as shown in Fig. 1.16c.

The two-diode, full-wave rectifier is not used very often in solid-state power supply designs. However, this approach is almost always used in the design of full-wave rectified power supplies that are built using vacuum tube diodes, which are discussed next.

Basic Vacuum Tube Diode Power Supplies

Later, in Chaps. 6 and 7, we will be studying the design and operation of vacuum tube-based amplifiers. These amplifiers are really the only application covered in this book in which vacuum tube diodes are used.

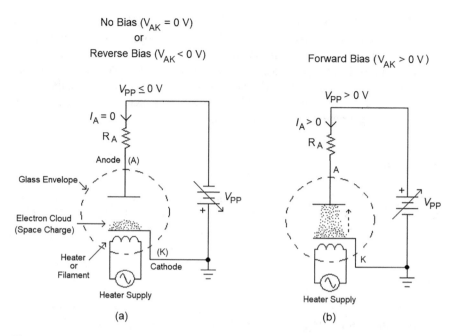

Fig. 1.17 Operation of a vacuum tube diode. (**a**) Reverse or no bias. (**b**) Forward bias

Compared to solid-state diodes, vacuum tube diodes are large, expensive, fragile, and inefficient. There are significant differences in the operating characteristics and design philosophies of solid-state rectified power supplies and their vacuum tube rectified counterparts as well. Those differences help contribute to the unique sound produced by tube amplifiers.

Vacuum Tube Diodes

The vacuum tubes (or *valves* as they are sometimes called) we will consider are *thermionic* devices, that is, their operation is based on the emission of free electrons due to heating of a metal emitter. The basic idea is illustrated in Fig. 1.17 where the cathode of the tube is indirectly heated by a fine tungsten wire filament that is powered by a low-voltage source. An AC heater supply is shown, but a DC voltage source may be used as well. Filament currents usually range from around 450 mA to 3 A, depending on the size of the cathode and the maximum anode current the tube can handle.

The filament heats up the cathode, which is typically made of a nickel alloy, coated with a material that enhances the electron emissivity of the surface. Thermal energy causes free electrons to form a cloud surrounding the cathode. This is sometimes called the *space charge region.*[1] As the name implies, a vacuum tube has all (or nearly all) air removed from inside the glass envelope. This vacuum prevents the heater filament from burning up and the cathode and anode surfaces from oxidizing. Also, if any gas is present in the tube, electrons flowing through the tube would ionize the gas molecules. This would reduce the efficiency of the diode. Occasionally a vacuum tube will develop a slow leak, which will result in the tube giving off a slight blue glow when operating.

Reverse Bias

Applying a voltage V_{PP} that causes the anode to be negative with respect to the cathode ($V_{AK} < 0$ V) reverse biases the diode. In reverse bias, the negatively charged electrons in the space charge cloud are repelled by the anode, and no current flows.

Forward Bias

If the battery is flipped over, V_{PP} is made positive. The anode is now positive with respect to the cathode ($V_{AK} > 0$ V), and the diode will be forward biased. In forward bias, the electrons surrounding the cathode are attracted to the positive anode (often called the *plate* in other types of tubes), and current flows. Ideally, the anode current in a forward-biased vacuum tube diode is described by a relationship called the *Child-Langmuir law,* named after the physicists that derived it. This law is also sometimes called the "three-halves-power law." In simplified form, this is written as

$$I_P = kV_P^{3/2} \qquad (1.13)$$

The constant k is referred to as the *perveance* of the diode. Because the transconductance of the vacuum tube diode is described by a power function, the vacuum tube diode exhibits a much more gradual turn-on characteristic than the solid-state PN junction diode, which has exponential transconductance behavior.

The 5AR4, 5U4-GB, and 5Y3-GT Diodes

The schematic symbols for three very common vacuum tube diodes, the 5AR4, 5U4-GB, and 5Y3-GT, are shown in Fig. 1.18. The 5AR4 has an indirect-heated cathode, like the one shown in Fig. 1.17. The 5U4-GB and 5Y3-GT are direct-

---

[1] Technically, the space charge region occupies the entire space between the cathode and the anode, but unless a forward bias is applied to the tube, virtually all of the electrons remain very close to the cathode surface.

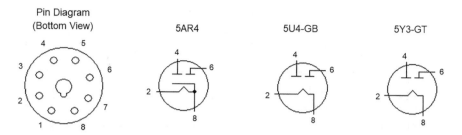

Fig. 1.18 Schematic symbols for 5AR4, 5U4-GB, and 5Y3-GT dual diodes and pin identifications

Table 1.2 Perveance values for common diode tubes

Tube type	Perveance, k
5AR4	0.003
5U4-GB	0.0008
5Y3-GT	0.00035

heated, filamentary cathode tubes. In this type of tube, the heater filament itself acts as the cathode. In an indirect-heated cathode tube, like the 5AR4, the filament heats a metal emitter that serves as the cathode. All three tubes are dual diodes with a common cathode terminal, designed for use as full-wave rectifiers.

As far as I can determine, the value for k in the Child-Langmuir equation is not given on tube data sheets. However, using a little algebra and the published curves for the 5AR4, the 5U4-GB, and the 5Y3-GT, I determined their constants which are presented in Table 1.2.

Using the perveance values of Table 1.2 and Eq. (1.13), I plotted the transconductance curves shown in Fig. 1.19. These curves are almost identical to the published curves.

As you can see, the 5AR4 drops the least voltage at all forward current levels, while the 5Y3-GT has the highest voltage drop. Table 1.3 lists a few of the major parameters for these tubes.

Typical Vacuum Tube Diode Power Supplies

An example of a power supply that can use any of the 5AR4, 5U4-GB, or 5Y3-GT diodes as a full-wave rectifier is shown in Fig. 1.20. For reasons of cost saving and convenience, the center-tap rectifier configuration is almost always used in vacuum tube diode supplies.

Note that the 5 V filament winding of the transformer is at the same potential as the high-voltage V_{PP} output terminal. This is a very dangerously high voltage, which requires great care when testing and troubleshooting is being performed.

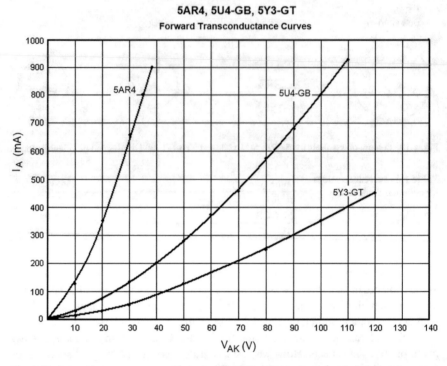

Fig. 1.19 Forward current versus voltage plots for 5AR4, 5U4-GB, and 5Y3-GT diodes

Table 1.3 5AR4, 5U4-GB, and 5Y3-GT parameters		Diode type		
		5AR4	5U4-GB	5Y3-GT
	V_{heater}	5 V_{rms}	5 V_{rms}	5 V_{rms}
	I_{heater}	1.9 A	3.0 A	2.0 A
	$V_{R(max)}$	1700 V	1550 V	1400 V
	I_F (max peak)	3.7 A	4.6 A	2.5 A
	I_F (max DC)	825 mA	1000 mA	440 mA

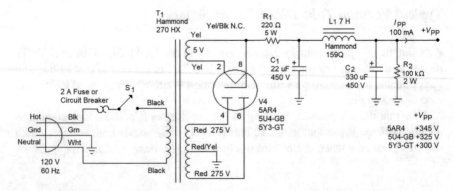

Fig. 1.20 High-voltage power supply using dual-diode, full-wave rectifier and LC pi filter

Resistor R_1 (220 Ω, 5 W) in series with the output of the rectifier limits peak output current. It also can serve as a fuse, preventing damage to the rectifier or power transformer should too much current be drawn from the supply.

Resistor R_2 is called a *bleeder resistor*. The function of the bleeder resistor is to discharge the filter capacitors when the power supply is turned off. If the bleeder was not used, the filter caps could hold a high voltage charge for several hours after power was removed, which could be dangerous. Using the values shown, the bleeder resistor effectively discharges the filter caps in about 3 min. I always use bleeder resistors in high-voltage power supply designs.

The LC, Pi Filter

The filter used here is a capacitor-input, LC, pi filter. The 22 μF capacitor C_1 provides initial peak filtering much the same as in the solid-state power supplies discussed earlier. The 7 H inductor L_1 is referred to as a *smoothing choke*. The choke has a relatively low DC resistance, and so does not cause a large drop in output voltage under load. However, the choke appears as a very high impedance at 120 Hz and higher harmonics, which helps to reduce output ripple. The 330 μF capacitor C_2 shunts any remaining ripple voltage to ground and is typically the same or higher in value than capacitor C_1. Both C_1 and C_2 are electrolytic capacitors, with voltage ratings of at least 450 V in this example.

If you are familiar with solid-state power supplies, but you have never worked with tube circuits before, one thing that you will notice right away is that much lower-value filter capacitors are used in tube power supplies. This is because, in general, tube circuits are high-voltage, low-current circuits, while transistor circuits are usually low-voltage, high-current circuits. If you calculate the product of capacitance and output voltage V_oC_T, you will find roughly equal magnitudes for power supplies of the same power rating.

The output voltages shown in this circuit are based on a total load current I_{PP} of about 100 mA and the choke having winding resistance $R_{DC} \cong 100\ \Omega$. The 5AR4 will only drop about 8 V at this load current. If a 5U4-GB was used, the output voltage would be about 330 V because of the greater voltage drop across the rectifier. The 5Y3-GT has the highest voltage drop of the three diode tubes, resulting in an output voltage of about 310 V under these load conditions.

I should mention that in this book I have generally adopted more modern notation in the vacuum tube circuits presented. For example, in Fig. 1.20, the supply output voltage is labeled V_{PP} which corresponds to the modern transistor supply rail notation V_{CC}. In older references, this supply rail would usually be labeled *B+*.

RC Pi Filter

It's not usually necessary to use a choke in a pi filter. Because power supply chokes are large, heavy, and expensive, it is common to find vacuum tube power supplies

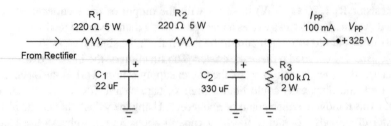

Fig. 1.21 RC pi filter

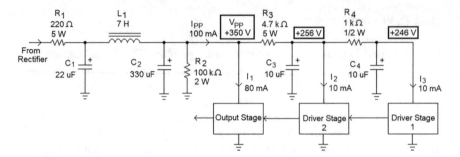

Fig. 1.22 Supplying reduced voltages to various stages

that use pi filters like that shown in Fig. 1.21. Here, the choke is simply replaced with resistor R_2. The main advantage of the RC pi filter is reduced cost. However, this filter is not as effective as the LC pi filter at reducing ripple voltage.

In addition to providing overcurrent protection for the rectifier and power transformer, we can choose values of R_1 and R_2 to reduce output voltage V_{PP} to a desired value. This is a very common practice in vacuum tube amplifiers.

Supply Voltage Distribution

Most tube amplifier designs have multiple stages. The output stage normally operates at the full supply voltage V_{PP}. However, the low-power stages preceding the output stage often operate at lower voltages. In order to supply reduced voltages to these stages, resistors are placed in series with the power supply rail, as shown in Fig. 1.22 (R_3 and R_4).

The values of these resistors are determined based on the current drawn by a given section of the amplifier. Here, R_3 carries 20 mA for both driver stages and has a voltage drop of 94 V. Resistor R_4 carries only 10 mA for the first driver stage and produces a voltage drop of 10 V. Note that these dropping resistors sometimes dissipate significant power. In these cases, wirewound power resistors are usually used.

It is common to use additional filter capacitors (C_3 and C_4) at these points on the supply rail. These capacitors provide additional ripple reduction, which is very important because the early stages of the amplifier are very sensitive to ripple, which is heard as hum on the output of the amp.

Final Comments

Most of the circuits that are presented in the following chapters can be operated from a single 9 V battery, but high-power consumption circuits such as power amplifiers require the use of a power supply. Now that you understand the basics of solid-state power supplies, you should be able to incorporate them into equipment to replace batteries. In addition to the practical material covered, many of the theoretical concepts introduced in this chapter will be used over and over again when we examine the operation of circuits such as tone controls, distortion effects, and envelope followers.

We won't be revisiting vacuum tube circuits for quite a while, but you now have a bit of a taste of what you can look forward to in Chaps. 6 and 7. In my opinion, vacuum tube circuits are some of the most interesting circuits, and I'm glad that we guitar players have helped keep them alive and well.

Summary of Equations

RMS/Peak Voltage Conversion for Sinusoidal Waveforms

$$V_P = V_{rms}\sqrt{2} \text{ or } V_P = 1.414V_{rms} \tag{1.1}$$

$$V_{rms} = V_P/\sqrt{2} \text{ or } V_{rms} = 0.707V_P \tag{1.2}$$

Full-Wave Bridge Rectifier Peak Output Voltage

$$V_{rect(pk)} = V_{sec(pk)} - 2V_F \tag{1.3}$$

Frequency/Period Relationship

$$f = 1/T \tag{1.4}$$

$$T = 1/f \qquad \text{[Reciprocal of (1.4)]}$$

Full-Wave Rectifier Output Frequency

$$f_{\text{rect}} = 2f_{\text{line}} \qquad (1.5)$$

Filter Capacitor for Ripple Voltage < 1%, 120 Hz Ripple (Unregulated Supply)

$$C \text{ (in } \mu\text{F)} \cong 100 I_{\text{L}} \text{(in mA)} \qquad (1.6)$$

Corner Frequency of First-Order, RC Filter

$$f_C = \frac{1}{2\pi RC} \qquad (1.7)$$

LED Current Equation for Fig. 1.9

$$I_{\text{LED}} = \frac{V_{\text{o}} - V_{\text{F}}}{R_2} \qquad (1.8)$$

LED Series Current Limiting Resistor Equation for Fig. 1.10

$$R_2 = \frac{V_{\text{o}} - V_{\text{F}}}{I_{\text{F}}} \qquad (1.9)$$

Input/Output Voltage Limits for 78xx Regulator

$$(V_0 + 2.5\,\text{V}) \leq V_{\text{in}} \leq (V_0 + 15\,\text{V}) \qquad (1.11)$$

Power Dissipation of 78xx Regulator

$$P_{78\text{xx}} = (V_{\text{in}} - V_0) I_{\text{L}} \qquad (1.12)$$

Child-Langmuir Law

$$I_{\text{P}} = k V_{\text{P}}^{3/2} \qquad (1.13)$$

Chapter 2
Pickups and Volume and Tone Controls

Introduction

A pickup is an example of what an engineer would call a *transducer*. A transducer is a device that transforms a non-electrical quantity (temperature, strain, pressure, velocity, etc.) into an electrical signal. Of course, here we are interested in converting the vibration of a string into a corresponding electrical signal.

Volume controls are variable attenuators used to vary the amplitude of the signal produced by a guitar. Tone controls are adjustable filters that modify the frequency response of the instrument pickups. This chapter covers the basic theory of pickups, volume controls, and tone controls and their interaction.

Single-Coil Magnetic Pickups

If a conductor moves through a magnetic field, a voltage will be induced in that conductor. This is shown in Fig. 2.1, where the wire moves from left to right. Assuming a uniform field, the amplitude of the voltage is proportional to the speed at which the wire cuts through the lines of flux and the number of turns of wire. The equation for the induced voltage is called *Lenz's law*, which is

$$V = N\frac{d\Phi}{dt} \tag{2.1}$$

where $d\varphi/dt$ is the rate of change of magnetic flux, which is directly proportional to the speed at which the wire cuts through the lines of flux, and N is the number of turns of wire.

If we were to move the wire back the opposite direction through the field, the polarity of the induced voltage would be reversed. Flipping the magnet over so that the south pole is facing up will also reverse the polarity of the induced voltage, relative to the direction of motion.

Imagine that the wire in Fig. 2.1 is a steel guitar string that was plucked and is vibrating back and forth over the pole of the magnet. A time-varying, AC voltage would be induced in the string, and in principle, we could use the signal induced in the string itself as the output. The induced voltage would be highest as the string moved past the center of the pole, where both flux density and speed are maximum.

In an actual magnetic pickup, many (usually several thousand) turns of wire are wrapped around cylindrical magnetic pole pieces located below each guitar string. The pole pieces couple their magnetic fields to the strings. Because the string is made of steel, as it vibrates and moves back and forth over the pole, the magnetic field is distorted. This movement of the field induces a voltage in the coil, which serves as the guitar signal. In the typical single-coil magnetic pickup, there is one magnetic pole piece for each string, and the coil is wound around all magnets as shown in Fig. 2.2.

Waveforms produced using a Fender Stratocaster with standard single-coil pickups are shown in Fig. 2.3. The upper scope trace shows the waveform for the open A string, using the neck pickup, with volume and tone controls set for maximum (full clockwise). There are three major signal characteristics that we are interested in here: amplitude, fundamental frequency, and overall waveform appearance.

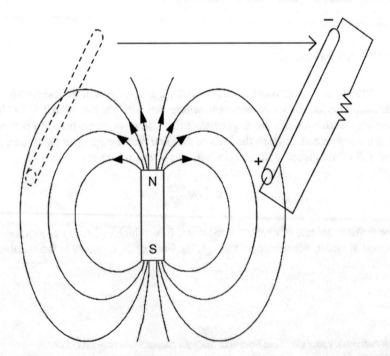

Fig. 2.1 Relative movement of a conductor and magnetic field induces a voltage in the conductor

Fig. 2.2 Typical single-coil
magnetic pickup

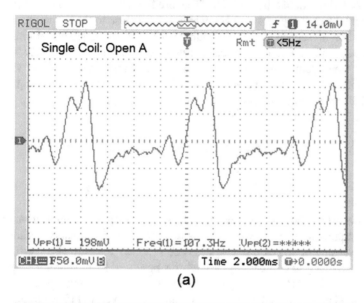

(a)

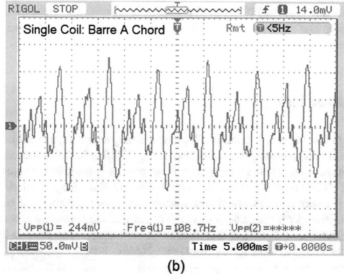

(b)

Fig. 2.3 Single-coil pickup waveforms. (**a**) Open A 110 Hz. (**b**) Barre chord A

As indicated on the scope readout, the amplitude of the waveform is about 198 mV$_{P-P}$. Signal amplitude varies tremendously with the force used to pick or strum the strings and with pickup height. The time I took between hitting the note and capturing the scope display also greatly affected the amplitude of the signal, which decayed rather quickly from an initial spike in amplitude. Based on experimental measurements, this particular guitar is capable of producing a peak signal voltage somewhat greater than 400 mV$_{P-P}$.

Notice that the signal is complex, consisting of the superposition of many different frequency components. The fundamental frequency (or pitch if you prefer) of the A string should be 110 Hz. The oscilloscope is accurately indicating a fundamental frequency of 107.3 Hz, which is very close to the desired tuning.

Digital oscilloscopes often have trouble interpreting complex waveforms such as those produced by speech or musical instruments. If you are getting crazy frequency readings from your scope, you can perform a sanity check on the values by determining the period of the signal the old-fashioned way by measuring the time interval between peaks of the signal. Doing this with Fig. 2.3a gives period and fundamental frequency values of

$$T = 4.6\text{div} \times\ 2\,\text{ms/div} \qquad f = 1/T$$
$$= 9.2\,\text{ms} \qquad\qquad = 1/9.2\,\text{ms}$$
$$= 109\,\text{Hz}$$

The digital oscilloscope derives period and frequency information from the time between triggering events, which means that the scope is probably giving a more accurate reading in this case (`Freq(1)=107.3 Hz`) than my visual estimate.

It is interesting to compare the waveforms produced by plucking a single string versus strumming a chord. Using the same guitar and pickup settings as before, and the output signal produced by strumming the barre chord A (110 Hz fundamental), we get the waveform of Fig. 2.3b. The amplitude is somewhat greater than for the single string, and the waveform is more complex, consisting of additional harmonically related frequency components.

Humbucker Pickups

Any conductor that carries current will produce a magnetic field, and a time-varying current will radiate electromagnetic (EM) energy. Wiring in the home may radiate enough 60 Hz energy that a significant voltage may be induced in a magnetic guitar pickup. This is a common source of annoying 60 Hz hum. Single-coil pickups are especially sensitive to this interference, where the pickup is acting like an antenna that is sensitive to the magnetic component of radiated electromagnetic energy. It is precisely because of this effect that there are some radio receiver designs that use a coil wound around a ferrite rod as an antenna. Humbucker (or humbucking) pickups

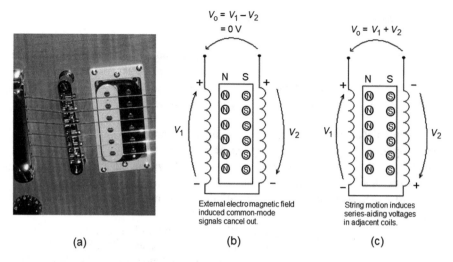

Fig. 2.4 Humbucker pickup and response to external magnetic field vs. string-induced signal

are designed to reduce the effects of stray electromagnetic fields. An example of a guitar with humbucker pickups is shown in Fig. 2.4a.

A humbucking pickup consists of two coils located side by side that are connected such that they form series-opposing voltage sources when coupled to a stray external electromagnetic field. The stray EM field-induced voltages are 180° out of phase and should cancel completely. This is an example of what engineers call *common-mode rejection*, a phenomenon we will see again later in the book. The orientation of the magnetic poles has no effect on coupling to external electromagnetic radiation.

Although voltages induced by external EM fields cancel, when a string vibrates over a pair of humbucker pole pieces, the coils produce voltages that add constructively, generating a large output signal. This occurs because the pole pieces in adjacent coils have opposite magnetic polarity. Figure 2.4b, c illustrates the relative polarities of the windings of the humbucker pickup for electromagnetic field response and string response, respectively.

Often, all four wires of a humbucker are available, and for all practical purposes, you have two separate pickups sitting side by side. It is the close proximity of the pickups to one another that helps to ensure that stray EM fields induce a common-mode signal. However, it is important that the coils be connected properly; otherwise, it is possible to cause string vibration signals to cancel, while stray EM field-induced signals will add constructively. Check the data sheet for your particular pickups for recommended connections, color coding, etc.

Peak and Average Output Voltages

Because there are two coils in series per pickup, humbuckers tend to produce higher amplitude output signals on average than single-coil pickups. As a general rule of thumb, the initial peak amplitude of a typical single-coil pickup will usually be in the range of 200–500 mV, while a humbucker will probably range from 400 to 1000 mV or more.

Although the initial amplitude of the signal produced by the pickup may be relatively large, the average voltage is likely to be about 20–25% of the peak value. Also, the output will drop off quickly after the initial pluck of the string.

The oscilloscope traces shown in Fig. 2.5 were produced using a Gibson Les Paul Standard guitar, using the neck pickup (490R) and volume and tone controls set to maximum (full clockwise).

Comparing the signals in Fig. 2.5 with those shown in Fig. 2.3, we find that for the open A string, the single-coil pickup gives us $v_o = 198$ mV$_{P-P}$, while the humbucker produces $v_o = 488$ mV$_{P-P}$, about 146% greater voltage. The humbucker also appears to generate a signal that has a more pronounced second harmonic content. We can't really make generalizations based on this single comparison, but this does help explain the differences in tonal quality between humbuckers and single-coil pickups. The difference between the single-coil pickup and humbucker for the barre chord A is even more pronounced, as seen when we compare Fig. 2.3b with 2.5b.

The physical height of a pickup is usually adjustable, with the pickup suspended in the guitar body by springs and screws. Pole pieces are also sometimes threaded so that individual spacing from strings may be varied. Moving a pole closer to a string results in a larger output signal, but since the poles are magnetic, if a string is located too close, it could be pulled into that pole.

The type of magnet used in the construction of a pickup will also influence the amplitude of the output signal. Generally, the stronger the magnets used, the greater the output amplitude will be. Traditionally, pickup magnets were made from ceramic or Alnico, but some newer high-output pickups use more powerful rare-earth magnets.

It is important to keep in mind that guitar signals are extremely dynamic, and they are dependent on hard-to-control variables such as fret finger pressure, pick stiffness, pick force, finger-picking vs. strumming, etc., but all things being equal, humbuckers will produce larger output signals than single-coil pickups.

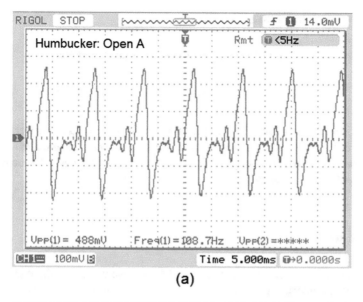

(a)

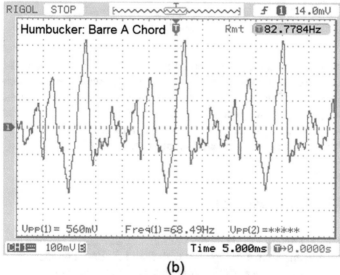

(b)

Fig. 2.5 Waveforms from a Les Paul Standard with humbuckers. (**a**) Open A string. (**b**) Barre chord A

More Magnetic Pickup Analysis

Let's take a look at Eq. (2.1) again. In case you forgot it, here it is:

$$V = N \frac{d\Phi}{dt} \qquad (2.1)$$

The ideal magnetic pickup is simply an inductor. In practice, however, because the coil consists of thousands of turns of thin wire, there is significant winding resistance. Compared to a single-coil pickup, a humbucker has a larger N value (more turns of wire) and so produces a greater output voltage.

Inductance

The inductance L in henrys, of a pickup coil (which we will approximate as being a solenoid or cylinder), is given by

$$L = \frac{\mu N^2 A}{\ell} \qquad (2.2)$$

where μ is the permeability of the pole pieces (H/m), N is the number of turns of wire, A is the cross-sectional area of a pole piece (m^2), and ℓ is the total length of the six pole pieces (m). This formula can be used to give a rough estimate of inductance, should you decide to wind your own pickups from scratch.

Inductance values for single-coil pickups typically range from 1 to 5 H. Humbuckers average a little higher, ranging from about 4 to 15 H.

A Pickup Winding Example

Let's use (2.2) to determine the number of turns of wire needed to produce a single-coil pickup like that shown in Fig. 2.2, with $L = 5$ H. Based on measurements of a Stratocaster pickup coil, we get the following physical dimensions:

$$\text{Pole piece radius, } r = 2.184 \text{ mm } (0.002184 \text{ m})$$
$$\text{Total pole piece length, } \ell = 10 \text{ cm } (0.1 \text{ m})$$

The cross-sectional area of a given pole piece is

$$
\begin{aligned}
A &= \pi r^2 \\
&= 3.141590 \times 002184^2 \\
&= 1.498 \times 10^{-5} \text{m}^2
\end{aligned}
$$

Let's assume the pole piece magnets have approximately the same permeability as electrical steel, which is

$$\mu = 8.75 \times 10^{-4} \text{H/m}$$

Now, we solve (2.2) for N and plug in the various numbers, which gives us

$$N = \sqrt{\frac{\ell \cdot \text{L}}{\mu \cdot \text{A}}}$$
$$= \sqrt{\frac{0.1 \times 5}{0.000875 \times 0.00001498}}$$
$$= 6176 \text{ turns}$$

The length of the wire can be estimated as follows. Approximating the pickup as being a rectangle measuring 0.5 in. × 2.5 in., the length of wire per turn (d) is the perimeter of the rectangle

$$d = 0.5 + 0.5 + 2.5 + 2.5$$
$$= 6 \text{ in./turn}$$

So, the total length d_{total} of the wire is approximately

$$d_{\text{total}} \cong (6\text{in./turn})(6176\text{turns})$$
$$= 37,056 \text{ in.}$$
$$= 3088 \text{ ft.}$$

Winding Resistance

There are practical limits to how much wire we can wrap around a pickup. To get more turns, we must use thinner wire, which is more fragile and has higher resistance per unit length. The diameter and resistance data for some common pickup wire gauges is given in Table 2.1. For comparison, consider that a typical human hair is about 0.003 inches in diameter.

Table 2.1 Common copper pickup wire gauge data

Gauge	Ω/ft.	Diameter (in.)
40	1.08	0.0031
41	1.32	0.0028
42	1.66	0.0025
43	2.14	0.0022

We can use this data to determine the winding resistance of the pickup we just designed. Assuming that we used 43 gauge wire, the resistance of the pickup is

$$R = 2.14 \ \Omega/\text{ft} \times 3088 \ \text{ft}$$
$$= 6.6 \ \text{k}\Omega$$

Typical single-coil pickups have winding resistances in the 5–7 kΩ range, while humbuckers typically range from 6 to 20 kΩ.

Winding Capacitance

Any time conductors are separated by a dielectric (insulating) material, a capacitor is formed. The many turns of wire in a pickup, separated by the thin enamel insulation, will result in capacitance that is distributed through the coil. The more turns of wire, the greater this interwinding capacitance will be. Interwinding capacitance is difficult to predict, but measurements made in the lab for several different single-coil and humbucker pickups averaged about 100 pF for single coils and 200 pF for humbuckers.

Approximate Circuit Model for a Magnetic Pickup

All of the magnetic pickup parameters discussed previously can be combined to form the model shown in Fig. 2.6. The pickup model turns out to be a second-order, low-pass filter.

It's interesting to analyze this pickup model using representative values for single-coil and humbucker pickups. Using PSpice to simulate the pickup circuit using the various R, L, and C values, we obtain the frequency response curves in Fig. 2.7. The circuit component values used and peak frequency parameters are

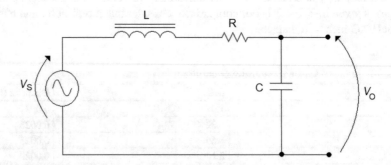

Fig. 2.6 Approximate model for magnetic pickup

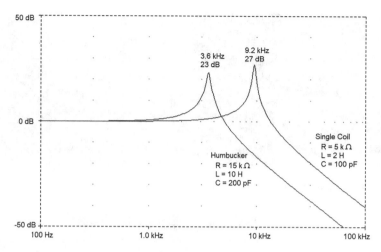

Fig. 2.7 Frequency response for representative single-coil and humbucker pickups

$$\text{Single-Coil}: \quad R = 5\,\text{k}\Omega, L = 2\,\text{H}, C = 100\,\text{pF}$$
$$f_{\text{pk}} = 9.2\,\text{kHz}, \ A_{\text{pk}} = 27\,\text{dB}$$
$$\text{Humbucker}: \quad R = 15\,\text{k}\Omega, L = 10\,H, \ C = 200\,\text{pF}$$
$$f_{\text{pk}} = 3.6\,\text{kHz}, \ A_{\text{pk}} = 23\,\text{dB}$$

These response curves have very large peaks, which occur at the resonant frequency of the circuit. Low-pass (and high-pass) filters that exhibit peaking are said to be *underdamped*. Also, because these are second-order LP filters, the response rolls off at −40 dB/decade once we pass the peak frequencies and enter the stopband. In general, the roll-off rate m of an nth-order filter will be given by

$$m = \pm 20n\,\text{dB/decade} \tag{2.3}$$

The sign is positive for HP and negative for LP responses. The response curves of Fig. 2.7 are only valid for these pickups without volume or tone control circuitry and with no external cable or amplifier connected. These factors can alter pickup response dramatically. We will come back to this topic again after the next section.

Piezoelectric Pickups

Piezoelectric pickups convert strain caused by mechanical vibrations into an electrical signal. Most piezoelectric pickups are constructed of a ceramic material such as barium titanate (BaTiO_3), which may be mounted on a thin metal disk that is glued to the guitar soundboard or built into the bridge assembly.

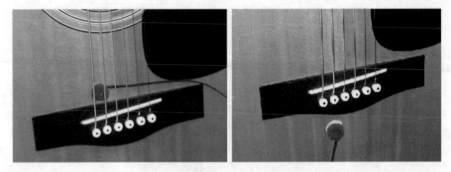

Fig. 2.8 Piezoelectric pickup on an acoustic guitar

Fig. 2.9 Piezoelectric
signal from acoustic guitar.
Open A (110 Hz)

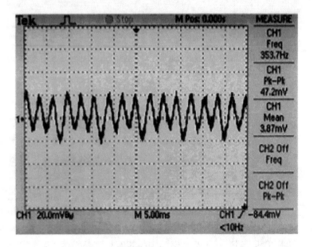

Piezoelectric pickups are best suited for use with acoustic and hollow body electric guitars, where vibration of the body has significantly high amplitude. Normally, the highest signal levels are obtained with the pickup mounted near the bridge. For experimental purposes, a piezo pickup was mounted in several locations on the outside of an acoustic guitar as shown in Fig. 2.8. Internal mounting would be preferred simply to protect the pickup from damage, but the function of the pickup is the same either way. The pickup shown here is a Schatten soundboard transducer.

An oscilloscope trace for the acoustic guitar with the piezo pickup mounted as shown on the left side of Fig. 2.8 is shown in Fig. 2.9. The open A string was picked, and measuring the time interval between the large negative peaks, the period is close to 110 Hz; however, there is very strong third harmonic present as well. Note that the output voltage produced by the piezo pickup (driving a scope probe with 10 MΩ resistance) is about one tenth of that produced by the average magnetic pickup: 47 mV$_{\text{P-P}}$ versus 488 mV$_{\text{P-P}}$ for the humbucker in Fig. 2.5a.

Piezoelectric Pickup Analysis

Whenever two conductors are separated by an insulator, a capacitor is formed. It turns out that the physical structure of a piezoelectric pickup is essentially the same as that of a ceramic capacitor except that, as noted before, the dielectric insulator is made of a material that will generate a voltage when subjected to mechanical strain. This is shown in Fig. 2.10.

A useful approximate circuit model for a piezo pickup is simply a voltage source with a series capacitance. The capacitance will range from around 500 to 1200 pF for typical piezo pickups. The peak output voltage generated by a single piezo pickup will generally be around 20 to 100 mV under typical loading conditions. When the pickup is connected to a load such as the input of an amplifier, a first-order, high-pass (HP) filter is formed. This is shown in Fig. 2.11.

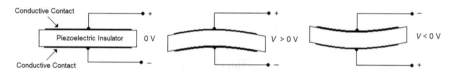

Fig. 2.10 Deformation of piezoelectric material generates a voltage

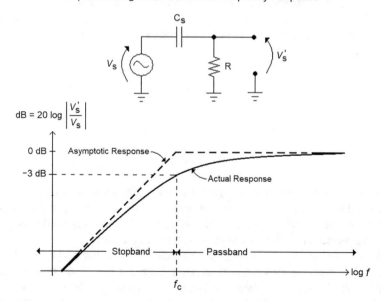

Fig. 2.11 Equivalent circuit for piezo pickup connected to external resistance and frequency response curve

The corner frequency of the equivalent high-pass filter is given by the same equation as for the low-pass filter covered in Chap. 1, which is

$$f_C = \frac{1}{2\pi RC_s} \qquad (2.4)$$

where C_S is the capacitance of the piezo source and R is the resistance being driven by the pickup. It is important that R be large enough to ensure that the lowest frequencies produced by the guitar are within the passband of the filter response. For the HP filter, the higher the value of R, the lower f_c becomes.

Example of Calculation: Input Resistance and Corner Frequency

If we connect a piezo pickup with $C_S = 1000$ pF to an amplifier with input resistance $R_{in} = 100$ kΩ, the lowest frequency we can effectively amplify is

$$
\begin{aligned}
f_C &= \frac{1}{2\pi RC_s} \\
&= \frac{1}{2\pi \times 100 \text{ k}\Omega \times 1000 \text{ pF}} \\
&= 1.6 \text{ kHz}
\end{aligned}
$$

We have a bit of a problem here. The open low E string frequency (E2) is about 81 Hz, which is so far into the stopband of this filter that response would be down about −50 dB. This is clearly not acceptable. The input resistance of the amplifier must be much higher if we want to pass frequencies down to 81 Hz. Solving (2.4) for R lets us calculate the required resistance.

$$R = \frac{1}{2\pi f_c C_s} \qquad (2.5)$$

Using the values $f_c = 81$ Hz and $C_s = 1000$ pF, we get

$$R = 1.9 \text{ M}\Omega$$

This is a very high resistance, higher than the typical input resistance of a guitar amplifier. Further complicating the situation are the effects of amplifier input capacitance and the characteristics of the cable that connects to the amplifier. These factors would have major negative effects on the signal.

Guitar cords are shielded coaxial cables, usually consisting of an outer braided copper shield surrounding a center conductor. The resistance of this type of cable is negligibly small, but the capacitance can be significant. The longer the cord, the greater its capacitance will be. Typical cord capacitances range from about 50 pF/m to 150 pF/m. This would have to be factored into an analysis of the piezo pickup.

There are other factors that further complicate the situation. If we were to add passive volume and tone controls, the piezo pickup would be loaded so heavily that it would be useless. And, even if we left out the tone controls, the inherently high output impedance of the piezo pickup would make the whole system very sensitive to cable microphonics and external noise pickup.

We could perform some more calculations, but that would only confirm what most of you probably already know; piezoelectric pickups require the use of a preamplifier. The preamplifier or just "preamp" serves as a buffer between the pickup and the cord/amplifier. The preamp will normally be located inside the guitar itself and may also have built-in tone/equalizer circuitry. We will take a detailed look at these types of amplifiers in the next chapter.

Sometimes, multiple piezo pickups will be mounted at various locations around the soundboard. If the pickups are connected in series, a much larger output signal will be generated. However, the series connection results in lower equivalent capacitance at the output of the pickups, requiring a higher input resistance for the preamp to get good low-frequency response.

Piezo pickups could also be connected in parallel. This does not increase output voltage, but does increase the equivalent capacitance at the output. This allows lower preamp input resistance to be used while still maintaining good low-frequency response.

Both series and parallel approaches will alter the frequency response of the system. In general, the series connection will raise the lower corner frequency, producing a brighter sound, while the parallel connection will decrease the lower corner frequency, enhancing bass response.

There will be certain positions on the soundboard where vibration is maximized, due to resonance of the guitar at different frequencies. Using different series/parallel connections of multiple pickups at various locations allows response to be tailored to suit individual preferences.

Guitar Volume and Tone Control Circuits

Volume and tone controls generally interact so strongly with one another and with pickups that they can't really be considered as separate components in most guitars. We will start by looking at typical examples of each and then connect them and see how they behave as a whole.

Potentiometers

A potentiometer, often simply called a "pot," is a three-terminal variable resistor. The schematic symbol for a pot and several different variations are shown in Fig. 2.12. The center terminal of the pot is called the *wiper*. The pot of Fig. 2.12b

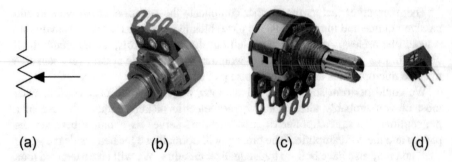

Fig. 2.12 Potentiometer schematic symbol and common packages

is a typical single-turn unit. Figure 2.12c is a dual-gang pot; it is basically two separate pots with a common shaft. These are useful in stereo audio applications. A trim pot is shown in Fig. 2.12d. Trim pots are usually mounted on a printed circuit board (PCB). A standard potentiometer will usually rotate through about 300° from end to end, although multi-turn pots are also available.

The schematic diagram of 2.13a shows a potentiometer connected as a volume control, which is really a variable voltage divider. A pictorial wiring diagram is shown in Fig. 2.13b. In this circuit, as the shaft is rotated clockwise, the output voltage varies from minimum ($V_o = 0$ V) to maximum ($V_o = V_{in}$). The output voltage is given by the *voltage divider equation*

$$V_o = V_{in}\left(\frac{R_B}{R_A + R_B}\right) \tag{2.6}$$

Potentiometer Taper

The *taper* of a potentiometer defines the way its resistance varies as a function of shaft rotation. A *linear taper* potentiometer will produce the characteristic curve of Fig. 2.13c, where the output voltage is a linear function of shaft rotation.

An *audio taper* potentiometer will produce the curve of Fig. 2.13d, where the output voltage is exponentially related to shaft rotation. This is a useful characteristic for volume control applications because the sensitivity of human hearing is approximately logarithmic. The complementary relation between logarithmic hearing sensitivity and the exponential audio taper transfer characteristic results in a perceived linear relationship between pot shaft rotation and loudness of the sound.

There are other potentiometer tapers available, including antilog (also called inverse log) and S-taper, but they are not of particular interest in our applications.

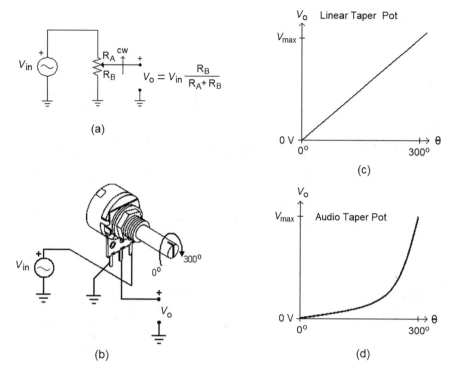

Fig. 2.13 (**a, b**) Potentiometer volume control. (**c**) Linear taper characteristic. (**d**) Audio taper characteristic

The Transfer Function

The output/input characteristic of a network is called its *transfer function*. The concept of the transfer function is not critical to understanding how a volume control works, but I thought it wouldn't be a bad idea to introduce this somewhat abstract concept early on since it will be used frequently in later chapters.

When you come right down to it, a volume control is simply a variable voltage divider. For a voltage divider (and most amplifiers as well), the output and input variables are V_o and V_{in}. Dividing both sides of (2.6) by V_{in} gives us the transfer function of the potentiometer:

$$\frac{V_o}{V_{in}} = \frac{R_B}{R_A + R_B} \tag{2.7}$$

Audio taper potentiometers are best suited for use as volume controls because they make the perceived loudness of the signal change as a linear function of shaft rotation. Before we leave the topic of potentiometers, there is one more common use that we will discuss, that is, using the pot as a simple variable resistor.

Rheostats

A potentiometer can also be used as a simple variable resistor, in which case the pot is functioning as a *rheostat*. Often a rheostat will be drawn schematically as shown in Fig. 2.14.

A pot can serve as a rheostat simply by using the wiper and either end terminal. The unused end may be left open or it may be shorted to the wiper as shown in Fig. 2.15.

Connecting the pot as shown in Fig. 2.15a causes the resistance to increase as the shaft is turned clockwise. For an audio taper pot, the resistance increases exponentially. Connecting the pot as shown in Fig. 2.15b causes the resistance to decrease as the shaft is turned clockwise.

Potentiometers are available with different power dissipation ratings. For the typical volume and tone control applications, inexpensive half-watt potentiometers are more than adequate.

Fig. 2.14 Common symbol used for a variable resistor or rheostat

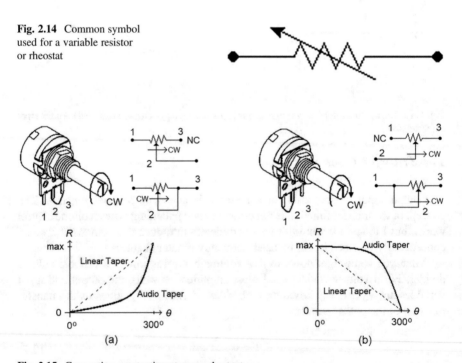

Fig. 2.15 Connecting a potentiometer as a rheostat

Basic Guitar Tone Control Operation

We have already covered a lot of the background information necessary to understand the operation of a tone control. Tone controls can be quite complex, but those found in the guitar itself are usually simple, adjustable low-pass filters.

A very common circuit used to implement volume and tone controls is shown in Fig. 2.16. Resistor R_S is the series winding resistance of the magnetic pickup. This resistance will usually range from 5 kΩ to 10 kΩ for a single-coil pickup and may be up to 20 kΩ or so for series-connected humbuckers. The values of the potentiometers are normally chosen to be at least ten times higher than R_S in order to prevent heavy loading of the pickup. Common potentiometer values used in guitars typically range from 250 kΩ to 1 MΩ, with 500 kΩ being the most common. Notice that the tone control is connected as a rheostat, while the volume control is a true potentiometer (voltage divider). Both pots should have audio taper characteristics.

The input resistance R_{in} of the amplifier to which the guitar is connected will have an effect on the response of the pickup. This is shown connected via the dashed line at the right side of the schematic, through a 0.25 in. *phone jack*. Typical input resistance values for vacuum tube-based amplifiers range from 250 kΩ to 1 MΩ. The effects of cable capacitance and resistance will be neglected here.

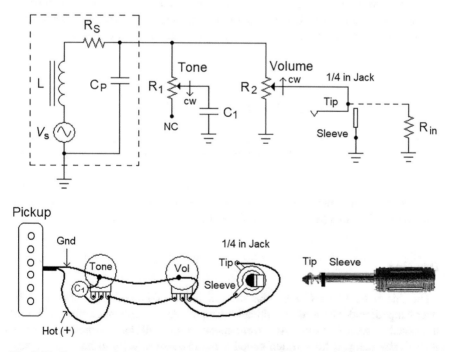

Fig. 2.16 Common volume/tone control schematic and pictorial wiring diagram for single-coil pickup

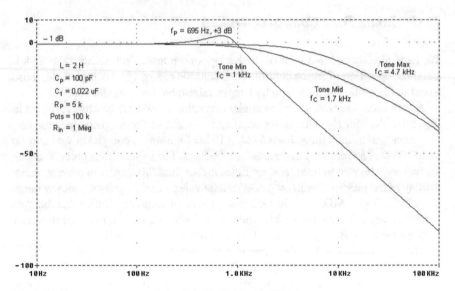

Fig. 2.17 Frequency response of circuit in Fig. 2.16 with 100 kΩ pots

It is possible to perform an analysis of this circuit by hand, but it would fill a few pages or so with phasor algebra, and it wouldn't make for very interesting reading. Instead, the circuit was simulated using PSpice to determine its frequency response. The circuit in Fig. 2.16 uses a single-coil pickup, which was simulated u−ent values:

$$R_S = 5\,\mathrm{k\Omega} \qquad\qquad L = 2\ \mathrm{H}$$
$$R_1 = R_2 = 100\,\Omega \quad C_P = 100\,\mathrm{pF}$$
$$R_{\mathrm{in}} = 1\ \mathrm{M\Omega} \qquad\quad C_1 = 0.022\,\mathrm{\mu F}$$

Examination of the curves in Fig. 2.17 plot indicates that with the tone control set for maximum resistance (full CW rotation), the response is flat at −1 dB up to the corner frequency of 4.7 kHz.

At mid-rotation, the frequency response is flat at −1 dB until the corner frequency at 1.7 kHz. At minimum tone control resistance (full CCW), the network becomes underdamped, with a peak of +3 dB at 695 Hz.

Damping

As a side note, filters can be classified in terms of *damping*. A filter may be overdamped, underdamped, or critically damped. An underdamped filter will have a peaked response curve. An overdamped filter will have droopy response. A critically damped filter (often called a *Butterworth filter*) will have the flattest possible response in its passband. The response curves for the single-coil and

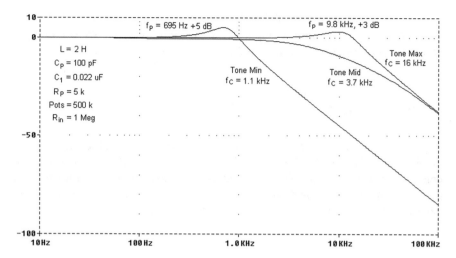

Fig. 2.18 Tone control response with 500 kΩ pots

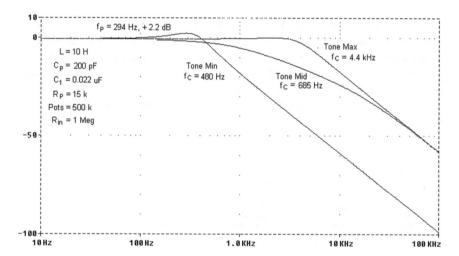

Fig. 2.19 Typical humbucker response

humbucker pickups shown back in Fig. 2.7 are highly underdamped. The addition of tone and volume control circuitry damps the pickups significantly.

Increasing the value of the potentiometers such that $R_1 = R_2 = 500$ kΩ produces the response plot in Fig. 2.18. The increased peaking at both extremes of pot rotation indicate that high frequency response is improved when higher resistance pots are used.

The graph of Fig. 2.19 shows the response for the tone control circuit using $C_1 = 0.022$ μF, 500 kΩ pots, and a humbucking pickup with $L = 10$ H, $R_S = 15$ kΩ, and $C_S = 200$ pF. The higher winding inductance and capacitance of the humbucker results in shifting of the curves toward lower frequencies.

You can experiment with different values of C_1 in the circuit to change the response of the tone control. Using a larger capacitor value, say $C_1 = 0.047$ μF, will lower f_C. A smaller value such as $C_1 = 0.01$ μF will increase f_C.

Multiple Pickups

A guitar with two pickups could be wired as shown in Fig. 2.20. This is really just a duplication of the circuit in Fig. 2.16 where we have independent volume and tone controls for each pickup. With the double-pole, triple-throw (DP3T) switch in position 1, the neck pickup drives the output. Position 2 connects both pickups in parallel, while position 3 connects the bridge pickup to the jack.

Pickup Phasing

You may have wondered why I placed plus signs at the ends of the pickup coils in some of the schematic diagrams. These plus signs simply indicate the relative phase of the coils. Different tone characteristics can be obtained by connecting guitar pickups in and out of phase with each other. The basic two-pickup circuit, with phase reversal, is shown in Fig. 2.21. When phase switch S_2 is in position 1, the pickups are connected in the normal phase relationship. In position 2, the phase of the bridge pickup is reversed.

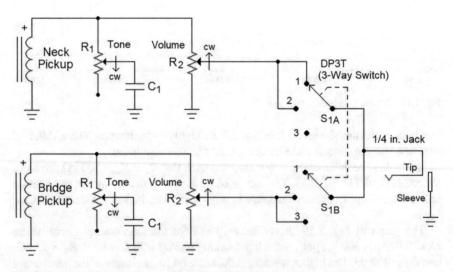

Fig. 2.20 Wiring for two pickups with independent volume and tone controls

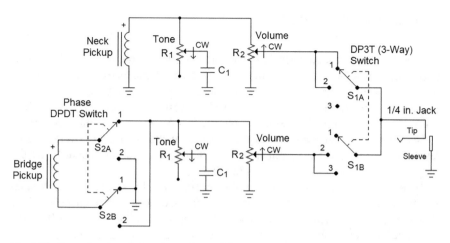

Fig. 2.21 Reversing the relative phase of two pickups

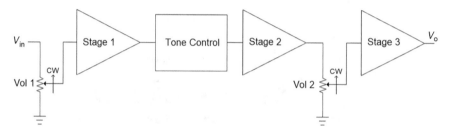

Fig. 2.22 Typical volume and tone control locations in a guitar amplifier

Reversing the phase of a pickup will have an audible effect only when both pickups are used at the same time. This occurs because of the change in constructive and destructive interference relationships between various harmonic components produced in each pickup.

If we add a third pickup, the number of possible wiring configurations increases dramatically. This is especially true if humbuckers are used in nonstandard configurations.

Amplifier Tone Control Circuits

Although amplifiers are the topic of the next chapter, this is as good a place as any to start a discussion on tone control circuits. Practical amplifiers consist of several gain stages and a power output stage. Volume and tone controls are often located at the amplifier input, in between stages, or possibly both as shown in Fig. 2.22.

A Basic Tone Control Circuit

One of the simplest tone controls is the variable filter shown in Fig. 2.23. This is really just a low-pass or treble-cut filter and is the same basic tone control that is used in many guitars, as was shown back in Fig. 2.16. In this circuit, R_o is the output resistance of the driving stage, and R_{in} is the input resistance of the driven stage. The potentiometer is normally chosen to be much greater in value than the output resistance in order to minimize loading, which would reduce the overall signal level. Depending on actual circuit values, insertion loss for this tone control typically ranges from 2 dB to 6 dB (a factor of 0.8–0.5).

When the wiper is at the bottom of the pot (fully clockwise), R_1 is at maximum resistance, and the filter has little effect on the signal, even at high frequencies where $|X_C|$ is very small. When the wiper is set to the upper side of the pot, $R_1 = 0\ \Omega$, and the capacitor tends to shunt higher frequencies to ground, cutting the treble response. The response curves shown in Fig. 2.23 were produced using component values that would be typical for vacuum tube amplifiers.

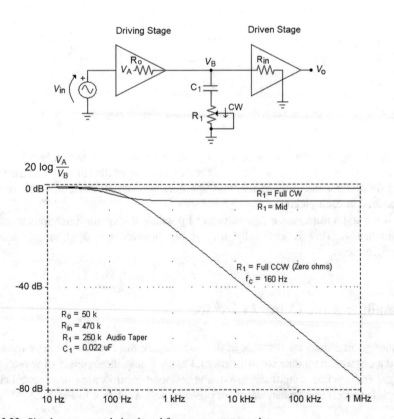

Fig. 2.23 Simple tone control circuit and frequency response plot

It is interesting to note that when the pot is at mid-rotation, treble is down by a constant 6 dB for frequencies above 200 Hz. When the response of a filter drops initially and then levels off, this is called a *shelving response*.

The corner frequency of the filter can be shifted to higher or lower frequencies by changing the value of C_1. If a brighter overall sound is preferred, the corner frequency of the filter can be increased by using a smaller capacitor value. For example, using $C_1 = 0.01\ \mu F$ moves the corner frequency up to $f_C \cong 350$ Hz.

Increasing the capacitor to $C_1 = 0.033\ \mu F$ results in $f_c \cong 100$ Hz. With the pot set for $R_1 = 0\ \Omega$ (max counterclockwise), the corner frequency is given by a modified version of the first-order RC filter equation, which is

$$ f_C = \frac{1}{2\pi(R_o \parallel R_{in})C_1} \tag{2.8} $$

Improved Single-Pot Tone Control

The tone control of Fig. 2.23 is simple and has low insertion loss, but its performance is not very good. The tone control in Fig. 2.24 combines HP and LP filters which allow for adjustment of both bass and treble frequency ranges.

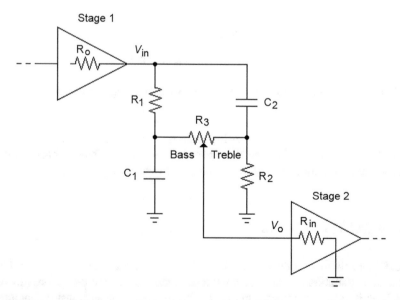

Fig. 2.24 Improved single potentiometer tone control

In this circuit, R_1 and C_1 form a low-pass filter, while R_2 and C_2 form a high-pass filter. Potentiometer R_3 allows variable mixing of the outputs of the low- and high-pass filters, which is applied to the next stage of amplification.

A linear potentiometer would be used in this application. As the wiper of the pot is moved to the left, more low-frequency signal energy is passed to the output, while the high frequency band is attenuated. Moving the wiper to the right attenuates the low-pass output and allows high frequencies to be passed on to the next stage.

This filter is often designed such that the corner frequency of the low-pass section is located near the low end of the guitar frequency range around 100 to 200 Hz, while the high-pass filter corner frequency is typically set to around 600 to 800 Hz. There are no hard and fast rules though, and there is no harm in experimenting with different component values and corner frequencies. Accurate corner frequency equations for this circuit are very complex, but we can calculate the approximate LP and HP section corner frequencies using the usual first-order RC filter equation.

$$f_{C(LP)} \cong \frac{1}{2\pi R_1 C_1}$$

$$f_{C(HP)} \cong \frac{1}{2\pi R_2 C_2}$$

If you would like to experiment with this tone control, some reasonable starting values for components are

$$R_1 = R_2 = 47 \text{ k}\Omega$$
$$C_1 = 0.022 \text{ μF}$$
$$C_2 = 4700 \text{ pF}$$
$$R_3 = 250 \text{ k}\Omega, \text{linear taper pot}$$

A frequency response plot for the circuit, using the component values listed, with $R_o = 50$ kΩ and $R_{in} = 470$ kΩ is shown in Fig. 2.25. At mid-rotation of the pot, response is relatively flat, with an insertion loss of about 12 dB. The response dips to about -24 dB at 250 Hz. An interesting response variation is obtained by swapping the values of C_1 and C_2. Don't be afraid to experiment with different capacitor values to customize the frequency response to suit your own taste.

Baxandall Tone Control

The Baxandall tone control is a classic circuit, named after its inventor Peter Baxandall. Variations of the Baxandall tone control circuit are used more predominantly in hi-fidelity amplifiers, but it can also be used in musical instrument amps. The basic Baxandall circuit is shown in Fig. 2.26.

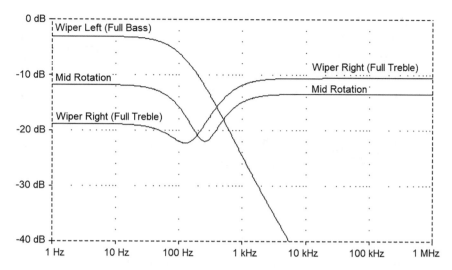

Fig. 2.25 Frequency response of the improved tone circuit

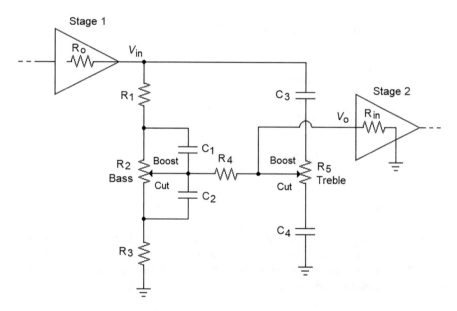

Fig. 2.26 Baxandall tone control circuit

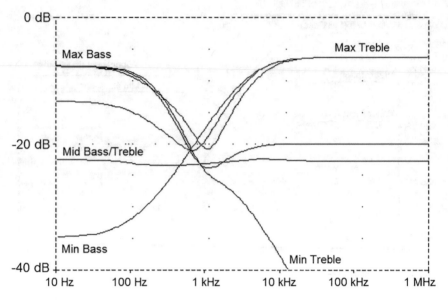

Fig. 2.27 Tone control response for various bass and treble settings

The following component values provide a good starting point for experimentation:

$$R_1 = 100\,\text{k}\Omega$$
$$C_1 = 470\,\text{pF}$$
$$R_2 = 250\,\text{k}\Omega\,(\text{Audio Taper})$$
$$R_3 = 10\,\text{k}\Omega \qquad C_2 = 0.0047\,\mu\text{F}$$
$$R_4 = 150\,\text{k}\Omega \qquad C_3 = 330\,\text{pF}$$
$$R_5 = 250\,\text{k}\Omega\,(\text{Audio Taper}) \quad C_4 = 0.0033\,\mu\text{F}$$

The frequency response of the circuit is shown in Fig. 2.27, for several settings of the bass and treble controls using the component values listed, assuming $R_o = 50\,\text{k}\Omega$ and $R_{in} = 470\,\text{k}\Omega$, which are reasonable ballpark values for vacuum tube amplifiers.

With both bass and treble controls set for maximum, the Baxandall circuit has a response of about -6 dB at low and high frequencies, with a dip to about -20 dB at $f = 935$ Hz. The response is relatively flat at about 23 dB insertion loss with bass and treble controls set to mid-rotation.

Other Tone Control Circuits

Three final examples of passive tone control circuits are shown in Fig. 2.28. The circuit of Fig. 2.28a is typical of the tone controls used in Fender amplifiers, while

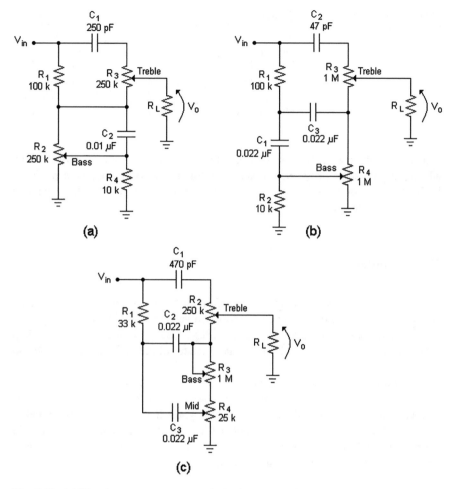

Fig. 2.28 Additional common tone control circuits. (**a**) Fender type, (**b**) Vox type, and (**c**) Marshall type

Fig. 2.28b is typical of Vox amplifiers. Using the component values shown, both of these tone circuits have an average loss of about 18 dB at midpoint adjustment of the potentiometers. Figure 2.28c is representative of Marshall tone controls. This circuit introduces an average loss of about 10 dB at midpoint adjustment.

Like the tone controls presented earlier, the component values given in these schematics are those that would be used in a typical vacuum tube-type amplifier. Generally, the load resistance connected to the tone control will range from 470 kΩ and up.

Tone control circuits can get quite complex, incorporating RLC (resistor-inductor-capacitor) networks, active filters, and specialized integrated circuits. All passive tone controls attenuate the signal to some extent. In general, the more complex the

Table 2.2 Tone control insertion loss

Tone control	Figure	Insertion loss
Basic treble cut	2.23	6 dB (0.5)
Single pot	2.24	12 dB (0.25)
Baxandall	2.26	23 dB (0.0708)
Fender type	2.28a	18 dB (0.126)
Vox type	2.28b	18 dB (0.126)
Marshall type	2.28c	10 dB (0.316)

tone control, the greater the loss will be. This loss can be made up for by using an additional gain stage or by incorporating an amplifier into the tone control circuit itself. The insertion loss values in dB and fractional form for the tone controls presented here are summarized in Table 2.2.

Final Comments

This chapter has presented a somewhat minimalist view of pickup configurations and tone control circuitry. There are literally hundreds of amplifier and guitar pickup and tone control variations that can be used. If you are interested in specific pickup wiring diagrams, you should check out the websites of the major guitar and pickup manufacturers.

The concepts that were presented here will be seen again throughout the following chapters. In Chap. 3, we will use the basic HP filter relationships to determine the frequency response of amplifier coupling and bypass networks. In Chaps. 4 and 7, we will see where tone controls are applied in complete amplifier designs, and in Chap. 5, we will look at some very sophisticated filters when we examine effects circuits such as phase shifters, wah-wahs, and flangers.

Summary of Equations

Lenz's Law

$$V = N\frac{d\Phi}{dt} \tag{2.1}$$

Inductance of a Solenoid Coil

$$L = \frac{\mu N^2 A}{\ell} \tag{2.2}$$

Roll-Off Rate of an *n*th-Order Filter (+ HP, − LP)

$$m = \pm 20n \, \text{dB/decade} \tag{2.3}$$

Corner Frequency of HP or LP, First-Order Filter

$$f_C = \frac{1}{2\pi R C_s} \tag{2.4}$$

Equation 2.4 Solved for *R*

$$R = \frac{1}{2\pi f_C C_s} \tag{2.5}$$

Voltage Divider Equation

$$V_o = V_{in}\left(\frac{R_B}{R_A + R_B}\right) \tag{2.6}$$

Transfer Function (Gain) of a Voltage Divider

$$\frac{V_o}{V_{in}} = \frac{R_B}{R_A + R_B} \tag{2.7}$$

Corner Frequency for Simple Tone Control (Pot = 0 Ω)

$$f_C = \frac{1}{2\pi (R_o \parallel R_{in}) C_1} \tag{2.8}$$

Chapter 3
Small-Signal and Low-Power Amplifiers

Introduction

Amplifiers play a central role in electronics in general and are certainly of great interest to most guitarists. Amplifiers can serve several purposes; usually, the main function of an amplifier is to increase the amplitude of a signal. Other functions of an amplifier might be to buffer or isolate one device or section of a circuit from another or perhaps to provide phase inversion. In this chapter, we will examine some common small-signal amplifier circuits using bipolar and field effect transistors, as well as basic operational amplifier configurations.

Gain

In general, the signals we will be working with are time-varying voltages. The output of a guitar pickup is a perfect example of a time-varying voltage. The voltage gain of an amplifier is defined as the ratio of output voltage to input voltage.

$$A_V = \frac{v_o}{v_{in}} \tag{3.1}$$

Looking at the waveforms shown for the amplifier in Fig. 3.1a, we find

$$A_V = \frac{v_o}{v_{in}} = \frac{4\ V}{2\ V} = 2$$

For Fig. 3.1b, since the output voltage is inverted or phase shifted by 180°, the voltage gain is a negative number. That is,

© The Author(s), under exclusive license to Springer Nature Switzerland AG 2022
D. J. Dailey, *Electronics for Guitarists*,
https://doi.org/10.1007/978-3-031-10758-0_3

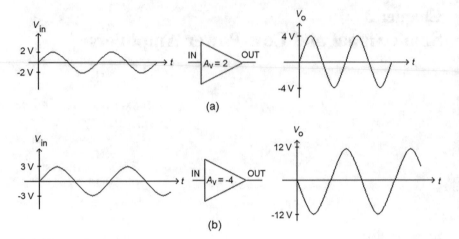

Fig. 3.1 Typical amplifier input and output waveforms. (**a**) Noninverting. (**b**) Inverting

$$A_V = \frac{v_o}{v_{in}} = \frac{-12 \text{ V}}{3 \text{ V}} = -4$$

It is also possible to express the gain of an amplifier in terms of in/out relationships between current and power. Current gain is defined by the equation

$$A_i = \frac{i_o}{i_{in}} \tag{3.2}$$

Similarly, power gain is defined as

$$A_P = \frac{p_o}{p_{in}} \tag{3.3}$$

You may recall the concept of the transfer function, which was presented in the previous chapter. Gain is another expression of this general concept.

Gain may be calculated using ratios of peak, average, or RMS values, as long as we use the same units consistently.

Decibels

We have already seen applications (filter response and voltage regulator ripple rejection) where output/input relationships are expressed in decibels (dB). When working with voltage or current ratios, the equivalent decibel gain is

Table 3.1 Common decibel values

A_V	dB
0.01	-40 dB
0.1	-20 dB
0.5	-6 dB
1.0	0 dB
2.0	6 dB
10	20 dB
100	40 dB

$$A_{V(dB)} = 20 \ \log \left(\frac{v_o}{v_{in}} \right) \quad \text{or} \quad A_{V(dB)} = 20 \ \log A_V \tag{3.4}$$

$$A_{i(dB)} = 20 \ \log \left(\frac{i_o}{i_{in}} \right) \quad \text{or} \quad A_{i(dB)} = 20 \ \log A_i \tag{3.5}$$

To be technically accurate, decibel gain is actually a measure of power gain. Equations (3.4) and (3.5) are only truly accurate when the amplifier input resistance equals the load resistance being driven, that is, $R_{in} = R_L$. In the real world, this almost never happens, but we still use dB voltage gain to give a relative indication of power gain.

In terms of actual input and output power levels, the decibel gain of an amplifier is given by

$$A_{P(dB)} = 10 \ \log \left(\frac{p_o}{p_{in}} \right) \quad \text{or} \quad A_{P(dB)} = 10 \ \log A_P \tag{3.6}$$

In practical situations, it is easiest to measure voltages, so we almost always work with voltage gain. A positive dB value indicates power gain. A negative dB value is a power loss. Table 3.1 lists a few voltage gain/decibel equivalent values.

Most of the amplifiers that we will work with are *voltage amplifiers*, where the primary input and output variables are voltages. We have been designating the voltage gain of an amplifier with the symbol A_V, but sometimes, the Greek letter μ (mu) is used instead.

Other Amplifier Parameters

In Fig. 3.1, the output signals produced by the amplifiers have the exact same sinusoidal shape as the input signals. There is phase inversion in the case of the output of Fig. 3.1b, but the wave shape is still sinusoidal. These are ideal, distortionless amplifiers. A few characteristics of ideal voltage amplifiers are:

Ideal Voltage Amplifier Characteristics

No distortion (or perfect linearity)
Infinite input resistance, $R_{\text{in}} = \infty \ \Omega$
Zero output resistance[1], $R_0 = 0 \ \Omega$
Infinite bandwidth, $\text{BW} = \infty \ \text{Hz}$
Infinite slew rate, $\left.\frac{dv_o}{dt}\right|_{\text{max}} = \infty \ \text{V/s}$

Distortion

Distortion is a fairly complex topic that we will cover in more detail at various points in this book. Whether distortion is a good thing or a bad thing depends on the application. If we were designing high-fidelity audio amplifiers, then we would want our amplifier to be as free of distortion as possible. In the case of a guitar amplifier, stomp box, or some other signal processing circuit, distortion may be exactly the thing we are looking for.

Input Resistance

High input resistance is usually desirable so that the amplifier does not load down the source that is driving it. Figure 3.2a shows a signal source with internal resistance R_S driving an amplifier with input resistance R_{in}. This forms a voltage divider which, as we saw in Chap. 2, will attenuate the input signal.

The actual input voltage that drives the amplifier is given by the familiar equation

$$V_{\text{in}} = V_s \left(\frac{R_{\text{in}}}{R_{\text{in}} + R_s} \right) \tag{3.7}$$

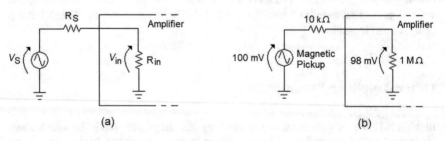

(a)   (b)

Fig. 3.2 Source loading by amplifier input resistance

---

[1]The transconductance amplifier is an exception where ideally $R_0 = \infty$.

You can see the advantage of high input resistance by examining Fig. 3.2b where the actual input voltage applied to the input of the amplifier is

$$V_{in} = V_s \left(\frac{R_{in}}{R_{in} + R_S} \right)$$
$$= 100 \text{ mV} \left(\frac{1 \text{ M}\Omega}{1 \text{ M}\Omega + 10 \text{ k}\Omega} \right)$$
$$= 98 \text{ mV}$$

Here, we see that there is light loading, with only a 2% reduction in signal amplitude.

Output Resistance

Low output resistance is desirable so that the output of an amplifier is not reduced when it is used to drive a low-resistance load. Let's expand the amplifier used previously in Fig. 3.2. Figure 3.3 represents a more complete amplifier model with its output voltage source (the diamond symbol), output resistance ($R_0 = 10 \text{ }\Omega$), and a load resistance ($R_L = 1 \text{ k}\Omega$) included.

If there were no output loading, the output voltage would be 9.8 V. Since the output resistance of the amplifier is not zero ohms, a voltage divider is formed, and the actual output voltage is

$$v_o = A_V v_{in} \left(\frac{R_L}{R_L + R_o} \right)$$
$$= (100)(98 \text{ mV}) \left(\frac{1 \text{ k}\Omega}{1 \text{ k}\Omega + 10 \text{ }\Omega} \right) \tag{3.8}$$
$$= 9.7 \text{ V}$$

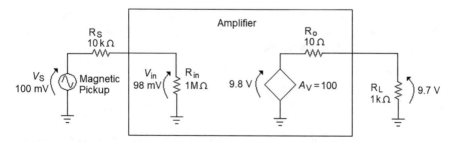

Fig. 3.3 Amplifier model with $A_V = 100$, $R_{in} = 1 \text{ M}\Omega$, and $R_0 = 10 \text{ }\Omega$

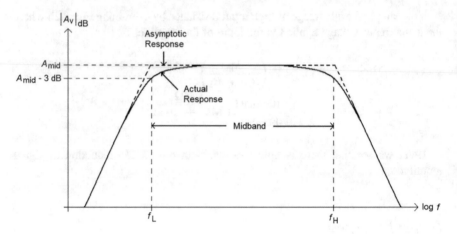

Fig. 3.4 Amplifier bandwidth

Input and output loading effects combine to reduce the effective voltage gain of Fig. 3.3 from an ideal gain of $A_V = 100$ to an actual gain of $A_V = 97$.

Bandwidth

Loosely speaking, bandwidth is the range of frequencies over which the gain of an amplifier remains relatively constant. This range is called the *midband* of the amplifier. A hypothetical amplifier frequency response plot is shown in Fig. 3.4. Note that both the actual and asymptotic response plots are shown. The low and high corner frequencies, designated f_L and f_H, occur where the response drops 3 dB from the midband gain, A_{mid}. The bandwidth of the amplifier is given by the equation

$$BW = f_H - f_L \tag{3.9}$$

The audio frequency range extends from 20 Hz to 20 kHz. A good-quality high-fidelity audio amplifier will typically have a frequency response that exceeds this range by a decade or more, especially on the high side. Many amplifiers have frequency response that extends down to DC.

Slew Rate

The *slew rate* of an amplifier is the maximum possible rate of change of the output voltage. The easiest way to visualize slew rate is to imagine that the input to an amplifier changes suddenly, as shown in Fig. 3.5. The input waveform here is called

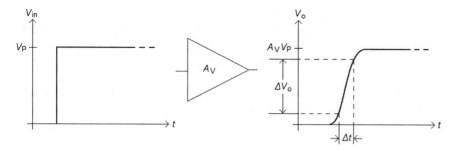

Fig. 3.5 Effect of slew rate on amplifier step response

a *step function*. The output of an ideal amplifier would change instantly in response to the step. A real amplifier takes time Δt to change its output. The slope of the output voltage response is the slew rate.

$$SR \cong \frac{\Delta V_o}{\Delta t} \, (V/s) \tag{3.10}$$

In general, the instantaneous rate of change of the output voltage is given by the *derivative* of v_0 with respect to time (we get the derivative from (3.10) when $\Delta t \to 0$). The maximum rate of change is written as

$$SR = \frac{dv_o}{dt}\bigg|_{max} \, (V/s) \tag{3.11}$$

Slew rate and bandwidth are related. An amplifier with high bandwidth will usually have a high slew rate as well, although the relationship between the two parameters is not always simple. There are even some amplifiers that have different slew rates for positive- and negative-going output voltages (the LM3900, which is covered in this chapter, is an example of such an amplifier).

Amplifier Classifications and Biasing

It is very likely that you have heard of amplifiers being classified as classes A, B, AB, C, and so on. There are even classes D, E, F, G, and H. Of these different classifications, classes A, B, and AB are of particular interest to us. For the sake of completeness, the other amplifier classes will be described briefly.

Class C amplifiers are primarily used in radiofrequency (RF) transmitter and receiver applications. Class D, E, and F amplifiers are different variations of *switching amplifiers*. Switching amplifiers, while more common in high-fidelity applications and public address systems, are not used as often as guitar amplifiers, and so we will not be concerned with them here.

Class G and H amplifiers have dynamically switched power supply voltage rails. At low output levels, the amplifier operates from low-voltage supply rails. When the output voltage exceeds a certain level, the supply rails are switched to a higher voltage. This results in increased amplifier efficiency. Again, this amplifier topology is not used in guitar amp applications very often.

Traditional amplifier classifications A, B, AB, and C are determined on the basis of how the transistor or vacuum tube in the circuit is biased, so before we discuss the details of these amplifier classifications, we must cover a little background on transistor operation and biasing.

Bipolar Junction Transistors

It's likely that anyone reading this book has some familiarity with transistors. While we can't go too deeply into their theory of operation, we can get a good qualitative understanding of basic transistor operation in typical applications. This section gives a quick overview of the circuit behavior of *bipolar junction transistors (BJTs)*. Modern BJTs are almost always silicon devices. Germanium transistors are still used occasionally, mainly in effects circuits, and were very common in the early days of semiconductor development. We will use both silicon and germanium transistors in the circuits presented here.

BJTs come in two flavors, NPN and PNP. Schematic symbols for NPN and PNP devices are shown in Fig. 3.6a. Both types of transistor serve the same functions but have opposite polarities for terminal voltages and currents. In digital applications, the transistor is used as an electronic switch; it is either on or off. In analog applications, the transistor is usually used as a continuously variable current source. In both cases, most of the time, the base terminal is used to control the operation of the transistor.

The circuit in Fig. 3.6b will be used to examine the basic operation of the BJT. The variable voltage source V_{BB} connected to the base controls the transistor. A second voltage source labeled $+V_{CC}$ is the main power supply for the circuit. There are three main areas of operation for a BJT: the *active, saturation*, and *cutoff region*. Simplified circuit models that approximate the collector-to-emitter characteristics for each area of operation are shown in Fig. 3.6c–e. In these models, we are not showing the base terminal, which is where the control signal is applied.

The Active Region

In the active region, the transistor behaves like a *current-controlled current source* (the diamond-shaped symbol in Fig. 3.6c). For our purposes, the active region is the most relevant area of operation in the design of amplifiers. When the base bias voltage V_{BB} is increased, current will begin to flow into the base terminal. This

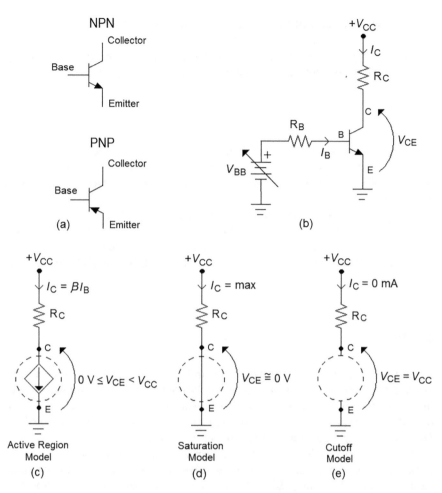

Fig. 3.6 (a) Transistor schematic symbols. (b) Test circuit and transistor behavior in (c) active, (d) saturation, and (e) cutoff regions of operation

causes a proportional collector current to flow. The base current controls the collector current and we have $I_C = \beta I_B$.

Beta β (or alternatively h_{FE}) is the primary gain parameter for the bipolar transistor. It tells us how effectively the base current controls the collector current. Beta varies quite a bit from one transistor to another, typically ranging from about 50 to 300.

Saturation

If the base voltage is increased enough, the transistor will saturate. This is where the transistor is acting like a closed switch. In *saturation*, the transistor is turned fully on, behaving approximately like a short circuit from collector to emitter. In saturation, collector current is at its maximum, while collector-to-emitter voltage is minimum. In circuits typical of those we will be studying, $V_{CE(sat)} = 0.2$ V, which is generally small enough to be ignored, so we will usually assume $V_{CE(sat)} \cong 0$ V.

Cutoff

If the base voltage is reduced to approximately 0 V, the transistor will enter cutoff. In cutoff, the transistor is turned off and behaves like an open circuit from collector to emitter. Collector current is reduced to zero, while collector-to-emitter voltage is maximum.

There are many similarities between bipolar transistors, vacuum tubes (triodes, tetrodes, and pentodes), and field effect transistors (FETs). Each has similar modes of operation, but device physics differs significantly between them. We could have used any of these devices in this discussion, but we will stick to BJTs for the time being.

Biasing

If you have ever read anything about vacuum tube-based amplifiers, especially on online discussion forums, you have probably heard the term *biasing* thrown around. In a nutshell, biasing is the establishment of the no-signal operating conditions of the tube(s) or transistor(s) in a circuit. This no-signal operating point is often called the *Q-point*, where Q stands for *quiescent* (quiet, i.e., no signal).

In the context of this chapter, we connect external components (resistors, DC voltage sources, diodes, and even other transistors) to bias the transistor for some desired values of V_{CE} and I_C. Since these are quiescent or no-signal values, we add a Q to the subscripts: V_{CEQ} and I_{CQ}.

There is a lot of mystique surrounding the concept of biasing, especially when it comes to tube amplifiers. However, there is really nothing mysterious about biasing, and hopefully, this book will help to demystify it.

Class A

In a class A amplifier, the transistor is biased up into the active region. Unless the amplifier is overdriven and clipping occurs, the transistor will remain in the active region at all times. When this is the case, the transistor has a *conduction angle* of 360°. In other words, the transistor operates in the active region for the entire 360° period of a sinusoidal driving signal.

Class A amplifiers have very low efficiency. For a capacitively coupled, class A amplifier, the maximum theoretical efficiency is $\eta = 25\%$. In practice, you are likely to get an efficiency of 10% or lower. Transformer coupling, which is rarely used in transistor circuits, but used often in vacuum tube circuit designs, allows a maximum theoretical class A efficiency of 50%. Because class A amplifiers have very low efficiency, they are used most often in the low-power sections of transistor amplifiers. However, there are many vacuum tube guitar and some hi-fi amplifiers that use class A biasing throughout, including the high-power output stage.

An interesting and possibly counterintuitive characteristic of class A amplifiers is that power dissipation of the transistor is maximum under no-signal conditions. The average power dissipation of the transistor actually decreases as the output signal amplitude increases. This will be explained soon, when the concept of the load line is discussed.

Class B

Under no-signal conditions, a transistor that is operating class B will be in cutoff. When an input signal is applied, the transistor will turn on and enter the active region, where it is capable of providing amplification. An NPN transistor will turn on when the input signal goes positive sufficiently to overcome the base-emitter barrier potential ($V_{BE} \cong 0.7$ V). Conversely, a PNP transistor will turn on when the input goes sufficiently negative ($V_{BE} \cong -0.7$ V). Thus, for a given transistor, the conduction angle for class B is somewhat less than 180°. Theoretical maximum efficiency for class B is $\eta = 78.5\%$. Class B biasing is discussed in more detail in Chap. 4.

Class AB

As you have probably guessed, class AB biasing is a compromise between class A and class B biasing. In class AB biasing, the transistor is just slightly biased into the active region, just enough to overcome the base-emitter barrier potential, V_{BE}. As with class B, class AB is used very often in solid-state power amplifier designs, which is covered in Chap. 4.

The Load Line

A graph of collector characteristic curves (I_C vs. V_{CE}) for a typical NPN transistor such as a 2N3904 is shown in Fig. 3.7. Each curve is generated by setting the base current I_B to a fixed value and then sweeping collector-to-emitter voltage V_{CE} from zero to an upper value (8 V is used in this example). The collector current I_C is plotted for each sweep of V_{CE}, resulting in a family of curves that characterize a particular transistor. These curves contain a lot of information, but we don't want to get too bogged down in the details at this time.

We are most interested in the heavy diagonal line added to the graph, which is called a *load line*. The load line is a plot of all possible operating points for a

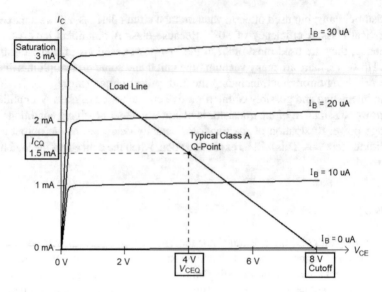

Fig. 3.7 Possible load line and collector curves, showing regions of transistor operation

transistor in a given circuit, spanning from saturation to cutoff. The load line is determined by the various components connected to the transistor in a circuit. The circuit designer chooses the circuit topology and power supply voltage and resistor values to obtain a desired load line.

Clipping

When an input signal such as a sine wave is applied to the amplifier, the transistor base current will vary with the signal. This causes the collector current to vary, moving the Q-point back and forth on the load line. Small input signals move the Q-point a small amount. A large enough input signal can cause the Q-point to move far enough that the transistor can hit saturation or cutoff. When this happens, the amplifier is said to be driven into *clipping*. As the term implies, clipping causes the top and bottom of the output signal to be flattened off.

Locating the Q-point at the center of the load line causes symmetrical clipping of the top and bottom of the output signal and gives the maximum possible output voltage swing before clipping occurs. When the Q-point is not centered, the output will clip asymmetrically. If you would like to see an example of asymmetrical clipping, you can look ahead to the oscilloscope display in Fig. 3.16c.

DC and AC Load Lines

The load line of Fig. 3.7 describes transistor operating limits for DC conditions only and so is technically a *DC load line*. Under AC signal conditions, the circuit may behave differently due to the presence of coupling and bypass capacitors and possibly inductive components like transformers. Because of this, we can also construct an *AC load line* for the amplifier. The AC load line is really the most important of the two, and it tells us how the amplifier output voltage and current will vary when a load is coupled to the output. Additional details on the construction of DC and AC load line analysis will be covered when we study vacuum tube amplifiers in Chaps. 6 and 7.

Class A Power Dissipation Characteristics

Let's take a closer look at the DC load line of Fig. 3.7. At the Q-point, the transistor power dissipation is

$$P_{DQ} = V_{CEQ}I_{CQ}$$
$$= 4 \text{ V} \times 1.5 \text{ mA}$$
$$= 6 \text{ mW}$$

Under no-signal conditions, the transistor simply sits at the Q-point dissipating power, but producing no output.

In saturation, ideally the transistor power dissipation is zero.

$$P_{D(\text{sat})} = V_{CE(\text{sat})}I_{C(\text{sat})}$$
$$= 0 \text{ V} \times 3 \text{ mA}$$
$$= 0 \text{ mW}$$

Similarly, in cutoff, the transistor power dissipation is zero as well.

$$P_{D(\text{cut})} = V_{CE(\text{cut})}I_{C(\text{cut})}$$
$$= 8 \text{ V} \times 0 \text{ mA}$$
$$= 0 \text{ mW}$$

If we drive the amplifier such that the Q-point moves continuously back and forth from saturation to cutoff, say with a sine wave, the average power dissipation of the transistor is much lower than under no-signal conditions. In fact, if the Q-point is centered on the load line and a sinusoidal signal drives the amp to maximum unclipped output, the average power dissipation of the transistor is $P_{ave} = P_{DQ}/2$.

The transistor in a class A amplifier actually runs cooler when it is driven to maximum output signal level.

Very few solid-state power amplifiers are class A designs. A good clue that your home or car audio power amp is *not* class A is the fact that as you turn up the volume, the amplifier runs hotter, not cooler. In Chap. 7, you will see many examples of class A power amplifiers, when vacuum tube amp designs are presented.

The Common Emitter Configuration

The basic topology of a common emitter (CE) amplifier stripped of all external components except for the input signal source and load is shown in Fig. 3.8. Notice that the output signal is inverted relative to the input. Phase inversion is a major characteristic of the CE configuration. The term *common emitter* is derived from the fact that the emitter terminal of the transistor is connected to common ground. Keep in mind that the circuit of Fig. 3.8 would require additional components for biasing in order to function.

A few of the general characteristics of the common emitter amplifier are listed below. Recall that β relates the base and collector currents through $I_C = \beta I_B$. Typically, $\beta \cong 100$.

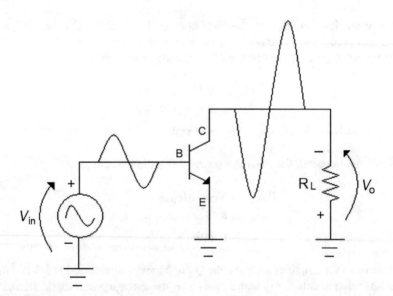

Fig. 3.8 Common emitter (CE) configuration. Output is inverted (180° phase) with respect to input

Common Emitter Characteristics

$\Phi = 180$ The CE amplifier exhibits phase inversion (A_V and A_i are negative numbers).

$|A_V| \gg 1$ The CE amplifier is capable of high voltage gain.

$1 < |A_i| < \beta$ The current gain of the CE amplifier may be significant, but is less than beta.

$A_P \gg \beta$ Power gain is higher for the CE configuration than for other configurations.

The common emitter is the most frequently used configuration in small-signal, low-power, audio frequency (AF) amplifier applications.

The Emitter Follower (Common Collector) Configuration

The emitter follower can be recognized by noting that the input signal is applied to the base terminal, while the output signal is taken from the emitter, as shown in Fig. 3.9.

Some of the major characteristics of the emitter follower are summarized below.

Emitter Follower Characteristics

$\Phi = 0°$ v_0 is in-phase with v_{in} (A_V and A_i are positive numbers).

$0 < A_V < 1$ The EF has voltage gain less than unity. Often, we assume that $A_V \cong 1$.

$1 < A_i < \beta$ The current gain of the EF amplifier may be significant, but is less than beta.

$A_P < \beta$ Power gain is always less than the beta of the transistor.

Low R_o Often, output resistance can be quite low ($R_o < 1 \, \Omega$).

Fig. 3.9 Emitter follower (EF) or common collector (CC) configuration

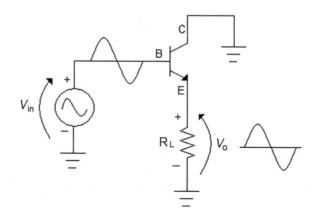

The emitter follower configuration is not used as often as the common emitter in class A, small-signal audio applications, except as a buffer or for impedance matching purposes. The emitter follower is used most often in class B and AB power amplifier applications.

The Common Base Configuration

Of the three basic configurations, common base is generally the least used in audio applications. The common base configuration is shown in Fig. 3.10.

The main characteristics of the common base configuration are listed below.

Common Base Characteristics

$\Phi = 0°$	v_o is in-phase with v_{in} (A_V and A_i are positive numbers).
$A_V \gg 1$	Voltage gain can be high (as high as for the CE amplifier).
$A_i < 1$	The current gain of the CB amplifier is always less than 1.
$A_P < \beta$	Power gain is always less than the beta of the transistor.
Low R_{in}	Lowest input resistance of the three configurations.
High BW	Best high-frequency performance of the three configurations.

Because of the excellent high-frequency performance of the common base, this configuration is used often in radiofrequency (RF) amplifier and oscillator applications.

Fig. 3.10 Common base (CB) configuration

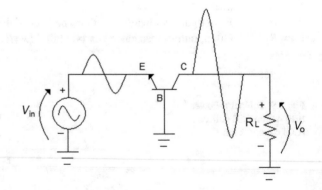

Field Effect Transistors

Functionally, field effect transistors (FETs) serve in the same general applications as bipolar transistors. There are certain applications, however, where FETs have definite advantages over bipolar transistors. In particular, junction FETs (*JFETs*) are most suitable for input stage designs where high input resistance is required. A preamplifier for a piezoelectric pickup is an excellent example of this sort of application.

The second area where FETs are often used is in the output stage of a high-power amplifier. In this case, power *MOSFETs* (metal-oxide-semiconductor FETs), and in rare cases power JFETs, would be used in place of BJT power transistors.

Most of the concepts covered previously for BJTs apply directly to FETs. FETs come in two polarities, N-channel and P-channel, similar to NPN and PNP bipolar transistors. Also like BJTs, FETs (and vacuum tubes too) can be biased for class A, B, and AB operation. The three fundamental transistor configurations have direct equivalents for FETs, which are shown in Fig. 3.11, using N-channel JFETs and MOSFETs. The common source (CS) is equivalent to the common emitter, the source follower (SF) is equivalent to the emitter follower, and the common gate (CG) is equivalent to the common base.

When comparing JFETs with normal bipolar transistors, there are a few significant differences to consider. First, if you are after high input resistance, FETs are superior to BJTs. It is not uncommon to have $R_{in} \geq 1$ MΩ for a JFET or MOSFET common source amplifier. The equivalent BJT CE amplifier will typically have an input resistance $R_{in} \leq 10$ kΩ.

If high voltage gain is important, the BJT is far superior to the JFET. A common source amplifier built using a very popular N-channel JFET, the MPF102, will typically have a voltage gain of $|A_V| \leq 10$. An equivalent common emitter amplifier built using a 2N3904 NPN transistor may easily have $|A_V| > 100$.

In terms of voltage gain, MOSFETs typically fall someplace in between BJTs and JFETs. A low-power N-channel MOSFET such as the BS170, operated in the common source configuration, will typically yield $|A_V| \cong 30$ or so. Like JFETs, MOSFETs have extremely high input resistance, but they are highly susceptible to damage from static electricity and should be handled with care. BJTs and JFETs are not very sensitive to static electricity and generally do not require special handling. We will revisit JFETs and MOSFETs again later in this chapter.

Bipolar Transistor Specifications

The typical transistor data sheet lists an intimidating array of electrical parameters, characteristic curves, model parameters for Spice simulation, and other information about a given device. Fortunately, we usually only need to know a few basic

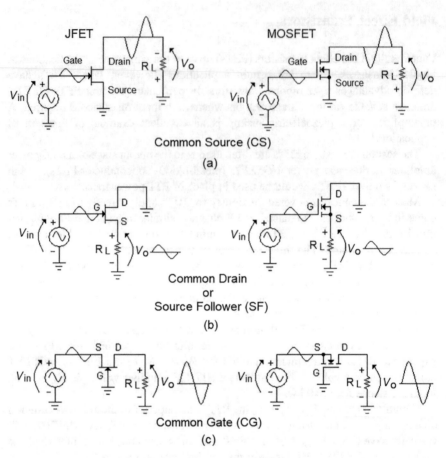

Fig. 3.11 FET amplifier configurations. (**a**) Common source. (**b**) Common drain or source follower. (**c**) Common gate

transistor parameters for most design and analysis applications. The circuits presented in this chapter have been designed to work with nearly any general-purpose NPN silicon transistor you can get your hands on. For example, the 2N2222, 2N3904, and MPSA42 are inexpensive general-purpose, low-power NPN transistors that can be used interchangeably in the BJT-based circuits presented in this chapter.

Important differences between transistors that must be considered are case styles (which determine power dissipation limits) and pin assignments (you can't mix up the C, B, and E terminals). Depending on the application, maximum voltage and current ratings, maximum frequency limits, and typical beta spread may also be important. Table 3.2 lists some of the major parameters for a sampling of commonly available transistors.

Notice that the silicon low-power, general-purpose transistors have very good high-frequency performance. High-power transistors tend to have worse high-

Table 3.2 Sample bipolar transistor data

Part no.	Type	Case	$I_{C(max)}$	BV_{CE}	$P_{D(max)}$	β	f_T
2N109	AF PNP Ge	TO-5	10 mA	30 V	125 mW	30	300 kHz
2N404	AF PNP Ge	TO-5	100 mA	25 V	150 mW	135	4 MHz
2N1639	AF PNP Ge	TO-1	50 mA	35 V	120 mW	100	4 MHz
2N2222A	GP NPN Si	TO-92	1 A	40 V	625 mW	50–300	300 MHz
2N3904	GP NPN Si	TO-92	200 mA	40 V	625 mW	70–300	300 MHz
2N3906	GP PNP Si	TO-92	200 mA	40 V	625 mW	80–300	250 MHz
MPSA42	GP NPN Si	TO-92	500 mA	300 V	625 mW	40–200	50 MHz
TIP29C	Power NPN Si	TO-220	1 A	100 V	30 W	40–75	3 MHz
TIP30C	Power PNP Si	TO-220	1 A	100 V	30 W	40–75	3MHz
TIP102	Power Darlington NPN Si	TO-220	8 A	100 V	80 W	1000–20,000	–
TIP106	Power Darlington PNP Si	TO-220	8 A	100 V	80 W	1000–20,000	–
2N3055	Power NPN Si	TO-3	15 A	60 V	115 W	20–70	2.5 MHz
MJ2955	Power PNP Si	TO-3	15 A	60 V	115 W	20–70	2.5 MHz

frequency performance because in order to carry high current and dissipate high power, the transistor must have a physically larger silicon chip area. This causes power transistors to have higher junction capacitances. Junction capacitances form parasitic low-pass filters that degrade high-frequency performance.

Generally, there is also a trade-off between breakdown voltage rating BV_{CE} and β. High-beta transistors tend to have lower breakdown voltage. The TIP102 and TIP106 seem to defy this statement, but these are Darlington transistors, which are actually a composite device made from two BJTs combined to form a very high-beta device. While Darlington transistors have the advantage of very high beta, they also have the disadvantage of very poor high-frequency performance and slow switching speeds, compared to standard BJTs with similar current handling capabilities.

The first three transistors listed in Table 3.2, the 2N109, 2N404, and 2N1639, are obsolete germanium, audio frequency PNP devices. Germanium transistors were available before silicon devices and were used in early solid-state circuit designs. One of the main differences between germanium and silicon transistors is the forward-biased base-emitter barrier potential. Forward-biased Ge and Si barrier potentials are approximately

$$V_{BE(Ge)} \cong 0.3\,\text{V (Germanium)}$$
$$V_{BE(Si)} \cong 0.7\,\text{V (Silicon)}$$

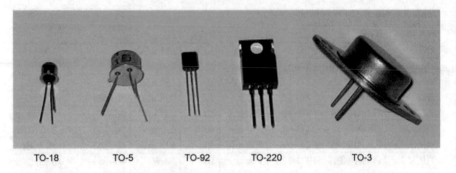

TO-18   TO-5   TO-92   TO-220   TO-3

Fig. 3.12   Common transistor case styles

Germanium semiconductors also have much higher leakage currents (often thousands of times higher) than silicon devices of the same junction area.

There are a number of other differences between vintage germanium and modern silicon transistors of the same general current/power handling capability as well. Vintage germanium transistors, due mainly to older fabrication methods, have relatively poor high-frequency characteristics and low breakdown voltage ratings. Substitution of a silicon transistor for a germanium transistor (and vice versa) usually requires modification of the biasing components used in the circuit.

The high-power transistors listed in the lower half of Table 3.2 are presented in complementary NPN/PNP pairs. Complementary transistors have nearly identical parameter values but are opposite in polarity. The complementary pairs listed here are the TIP29C/TIP30C, TIP102/TIP106, and 2N3055/MJ2955. Complementary transistor pairs are often used in power amplifier designs, which are covered in Chap. 4.

The various transistor case styles listed in Table 3.1 are shown in the photo of Fig. 3.12. These case styles are extremely common in current production transistors, including FETs, but you may come across any of a number of obsolete transistor case styles in older equipment.

Basic BJT Amplifier Operation

This section will walk you through some basic transistor circuit analysis techniques that are very useful and even sometimes interesting. Even if you don't care to study the equations, you can just skim this section and still gain some insight into how a simple transistor amplifier works.

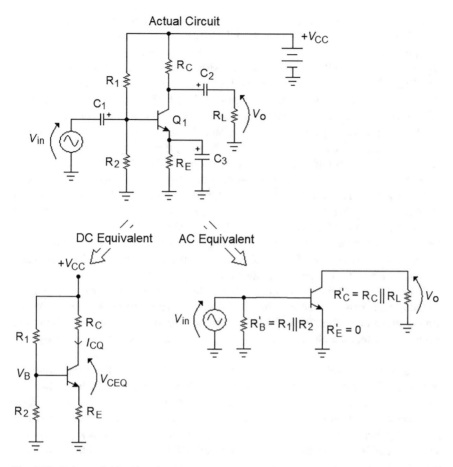

Fig. 3.13 Voltage divider biased, common emitter amplifier

Voltage Divider Biased CE Amplifier

The circuit of Fig. 3.13 is capable of good performance as a general-purpose, low-power amplifier or even as a distortion circuit as you will soon discover. Here is a brief explanation of the functions of the various components in the circuit.

Resistors R_1 and R_2 form a voltage divider which is used to produce a low voltage that serves to bias transistor Q_1 on. The emitter resistor R_E sets the quiescent current I_{CQ} of the transistor, which for low-power amplifiers is typically a few milliamps or so. The emitter current and collector current only differ by about 1% or less, so usually we assume $I_E \cong I_C$.

Collector resistor R_C is chosen to set V_{CE} such that the transistor Q-point is close to the center of the active region (or the center of the load line, if you prefer). This voltage is generally set to $V_{CEQ} \cong V_{CC}/2$.

Capacitors C_1 and C_2 are called *coupling capacitors*, and in simplest terms, they serve to pass AC signals while blocking DC. The values of the coupling capacitors are chosen such that, at the lowest frequency to be amplified, the capacitive reactance X_C is negligibly small compared to the resistance of the circuit to which we are coupling. Capacitor C_3 is the *emitter bypass capacitor*; it bypasses AC signals around the emitter resistor, directly to ground. Emitter bypassing increases voltage gain at the expense of decreased input resistance and increased distortion.

Notice that, in the upper schematic diagram, the power supply voltage source V_{CC} is shown explicitly as a battery. Most of the time, the battery symbol will not be drawn in order to simplify the schematic, as is done in the DC equivalent schematic.

Because of the presence of the coupling and bypass caps, the circuit behaves differently under DC and AC conditions. If the DC power supply V_{CC} is active and the AC source v_{in} is turned down to zero volts, after a short time, the capacitors will have charged to their final values. Under these conditions, the circuit is treated as the DC equivalent shown in the lower part of Fig. 3.13 where all capacitors behave like open circuits. When we analyze an amplifier, we always start with the DC equivalent circuit, which is used to determine the Q-point.

Once the DC Q-point analysis is done, the AC source is activated, and the coupling and bypass capacitors are treated as short circuits ($X_C \cong 0\ \Omega$). Here, the emitter bypass cap effectively shorts the emitter to ground, so in the AC equivalent circuit, the external emitter resistance is $R'_E = 0\ \Omega$ as shown in the lower right of Fig. 3.13.

Ideally, the internal resistance of a voltage source is zero ohms; therefore, the power supply voltage source V_{CC} provides a low-resistance path to ground for signals. This effectively places R_C in parallel with R_L. This gives us $R'_C = R_C \parallel R_L$. Also, an AC input signal "sees" two parallel paths to ground via biasing resistors R_1 and R_2; therefore, the equivalent resistance from base to ground is $R'_B = R_1 \parallel R_2$. The convention we are using is to denote effective resistance values in the AC equivalent circuit with primes, i.e., R'_B, R'_C, R'_E, and so on.

This is a good place to review the formula for two resistors in parallel, $R_1 \parallel R_2$, which has two equivalent forms

$$R_{eq} = \frac{1}{\frac{1}{R_1} + \frac{1}{R_2}} \tag{3.12a}$$

Or equivalently

$$R_{eq} = \frac{R_1 R_2}{R_1 + R_2} \tag{3.12b}$$

DC Q-Point Analysis Equations for Fig. 3.13

The following equations will be used to determine the Q-point of the transistor in Fig. 3.13:

$$R'_B = R_1 \parallel R_2 \tag{3.13}$$

$$V_{Th} = V_{CC}\left(\frac{R_2}{R_1 + R_2}\right) \tag{3.14}$$

$$I_{CQ} = \frac{V_{Th} - V_{BE}}{\frac{R'_B}{\beta} + R_E} \tag{3.15}$$

$$V_{CEQ} = V_{CC} - I_{CQ}(R_C + R_E) \tag{3.16}$$

$$r_e = \frac{26\,\text{mV}}{I_{CQ}} \tag{3.17}$$

We won't go into the details here, but (3.13) through (3.16) are derived using Ohm's law, Kirchhoff's laws, and the basic transistor relationship $I_C = \beta I_B$. Equation (3.17) tells us the *dynamic resistance* r_e of the forward-biased B–E junction of the transistor, which is important in the upcoming AC analysis.

The convention we are using here is to represent parameters that are internal to the transistor with lowercase, italic letters, so r_e is an equivalent dynamic resistance that is a property of the transistor itself. This notational convention will be applied in upcoming JFET and vacuum tube circuit analysis as well.

Beta Independence

In a well-designed circuit, the term R'_B/β in (3.15) should be small compared to R_E. This ensures that the Q-point location is essentially independent of β. If you take a look back at Table 3.1, it is clear that beta varies widely from transistor to transistor. If you measured the beta of a randomly chosen 2N3904, you might get anything from $\beta = 70$ to 300. If our design required a specific value of beta to operate correctly, we would have to handpick every transistor, which would be inconvenient. As a general rule of thumb, a circuit will have a stable, β-independent Q-point if the following inequality is satisfied:

$$\beta R_E \gg R'_B \tag{3.18}$$

The symbol $\gg$ means "much greater than," which is at least a 10-to-1 ratio. Satisfying (3.18) allows us to calculate the collector current using the approximate equation

$$I_{CQ} \cong \frac{V_{Th} - V_{BE}}{R_E} \tag{3.19}$$

You might be asking yourself how (3.18) can be used if we don't know β ahead of time. The important thing to know is that even though beta is fairly unpredictable, it is generally around 100 or more. With this in mind unless there is some compelling reason to do otherwise, we will assume that beta is approximately 100. For emphasis, we'll make this a numbered equation.

$$\beta \cong 100 \tag{3.20}$$

As noted earlier, a second assumption we will make is that a forward-biased silicon PN junction, such as the B–E junction of a silicon transistor, will always drop approximately 0.7 V. For emphasis again, let's make this a numbered equation as well.

$$V_{BE} \cong 0.7\,V \tag{3.21}$$

AC Analysis Equations for Fig. 3.13

The following equations will be used in the AC analysis of the amplifier in Fig. 3.13:

$$R'_C = R_C \parallel R_L \tag{3.22}$$

$$R_{in} = R'_B \parallel \beta(r_e + R'_E) \tag{3.23}$$

$$R_o = R_C \tag{3.24}$$

$$A_V = \frac{-R'_C}{r_e + R'_E} \tag{3.25}$$

$$V_{o(max)} = I_{CQ}R'_C \tag{3.26}$$

$$V_{o(min)} = -V_{CEQ} \tag{3.27}$$

Equations (3.26) and (3.27) tell us the maximum unclipped output voltage of the amplifier. For an amplifier with a Q-point that is perfectly centered on the AC load line, clipping will be symmetrical with $V_{0(max)} = |V_{0(min)}|$ or equivalently $I_{CQ}R'_C = V_{CEQ}$.

We will use modified versions of these equations in the AC analysis of FET and vacuum tube-based amplifiers as well.

Common Emitter Amplifier Analysis Example

The circuit in Fig. 3.14 is suitable for use as a preamplifier or overdrive circuit. Although the transistor specified in this example is a 2N3904, nearly any general-purpose, silicon, NPN transistor could be used in this circuit (2N2222, MPSA42, BC107, etc.). We will perform an analysis of this amplifier using the equations

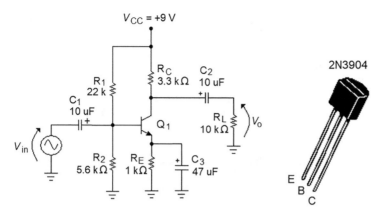

Fig. 3.14 Practical CE, class A amplifier circuit and pin designations for 2N3904

presented in the previous section and compare the theoretical predictions with measurements made on an actual circuit. We will assume $\beta = 100$.

DC Q-Point Analysis

We start the DC analysis by calculating the equivalent external base resistance R'_B and base biasing voltage V_{Th}.

$$R'_B = R_1 \| R_2 \qquad\qquad V_{Th} = V_{CC}\left(\frac{R_2}{R_1 + R_2}\right)$$

$$= 22\text{ k}\Omega \| 5.6\text{ k}\Omega \qquad\qquad = 9V\left(\frac{5.6\text{ k}\Omega}{22\text{ k}\Omega + 5.6\text{ k}\Omega}\right)$$

$$= 4.5\text{ k}\Omega \qquad\qquad\qquad\qquad = 1.8\text{ V}$$

We will calculate the collector current I_{CQ} using both Eqs. (3.15) and (3.19) to see if the approximation is valid.

$$I_{CQ} = \frac{V_{Th} - V_{BE}}{\dfrac{R'_B}{\beta} + R_E}$$

$$= \frac{1.8\text{ V} - 0.7\text{ V}}{\dfrac{4.5\text{ k}\Omega}{100} + 1\text{ k}\Omega}$$

$$= 1.05\text{ mA}$$

$$I_{CQ} \cong \frac{V_{Th} - V_{BE}}{R_E} \quad \text{(approximate equation)}$$

$$\cong \frac{1.8\text{ V} - 0.7\text{ V}}{1\text{ k}\Omega}$$

$$\cong 1.1\text{ mA} \quad \text{(approximation is accurate enough)}$$

In practice, I would usually just round either result off to $I_{CQ} \cong 1$ mA because we're typically using 5% tolerance resistors, and the battery that powers the circuit may be somewhat less than 9 V as well.

The DC collector-to-emitter voltage is

$$
\begin{aligned}
V_{CEQ} &= V_{CC} - I_{CQ}(R_C + R_E) \\
&= 9\,V - 1.1\,mA(3.3\,k\Omega + 1\,k\Omega) \\
&= 4.3\,V
\end{aligned}
$$

The DC analysis is completed with calculation of the dynamic emitter resistance.

$$
\begin{aligned}
r_e &= \frac{26\,mV}{I_{CQ}} \\
&= \frac{26\,mV}{1.1\,mA} \\
&\cong 24\,\Omega
\end{aligned}
$$

AC Analysis

Now we are ready to perform the AC analysis of the amplifier. Let's assume that we are driving a load resistor $R_L = 10$ kΩ, which is reasonable for this type of circuit.

Since the emitter resistor is completely bypassed by capacitor C_3, the external AC resistance in series with the emitter terminal is

$$
R'_E = 0\,\Omega
$$

The input resistance of the amplifier is

$$
\begin{aligned}
R_{in} &= R'_B \,\|\, \beta(r_e + R'_E) \\
&= 4.5\,k\Omega \,\|\, 100(24\,\Omega + 0\,\Omega) \\
&= 1.6\,k\Omega
\end{aligned}
$$

Equation (3.24) tells us that the output resistance is approximately the same as R_C. This is so because in the active region, the collector terminal has very high (ideally infinite) resistance. Therefore, "looking" back from the perspective of the load resistor, the only path to ground is through the collector resistor R_C.

$$
R_o = R_C = 3.3\,k\Omega
$$

With a 10 kΩ load on the output, the equivalent AC collector resistance is

$$R'_C = R_C \parallel R_L$$
$$= 3.3\,\mathrm{k\Omega} \parallel 10\,\mathrm{k\Omega}$$
$$= 2.5\,\mathrm{k\Omega}$$

Now we can determine the voltage gain of the amplifier. Remember, the negative sign simply indicates that the output is $180°$ out of phase with the input signal.

$$A_V = \frac{-R'_C}{r_e + R'_E}$$
$$= \frac{-2.5\,\mathrm{k\Omega}}{24\,\Omega + 0\,\Omega}$$
$$= -104 \quad (\text{or, equivalently, } 40.3\ \mathrm{dB})$$

The output of the amplifier will clip at the maximum and minimum output voltages, which are

$$V_{o(\max)} = I_{CQ}R'_C \qquad V_{o(\min)} = -V_{CEQ}$$
$$= 1.1\ \mathrm{mA} \times 2.5\ \mathrm{k\Omega} \qquad = -4.3\ \mathrm{V}$$
$$= 2.8\ \mathrm{V}$$

The maximum unclipped peak-to-peak output voltage, which is also called the *output compliance*, is two times the smaller of $V_{o(\max)}$ and $|V_{o(\min)}|$, so in this example we have

$$V_{o(P-P)} = 2V_{o(\max)}$$
$$= 2 \times 2.8\ \mathrm{V}$$
$$= 5.6\ \mathrm{V_{P-P}}$$

Experimental Results

The amplifier of Fig. 3.14 was constructed and tested in the lab. A randomly selected 2N3904 transistor was used, and a 9 V alkaline battery provided power. It is a standard practice to use a 1 kHz sinusoidal input as an audio test signal, but the frequency is not critical.

The actual prototype circuit is shown constructed on solderless breadboards or protoboards in Fig. 3.15b. The protoboards are attached to an aluminum base, and two 0.25 inch jacks have been mounted to provide easy connections to a guitar or amplifier.

Fig. 3.15 (a) Amplifier
circuit with input attenuator.
(b) The actual circuit built
on protoboards

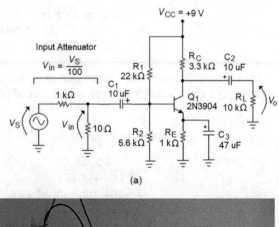

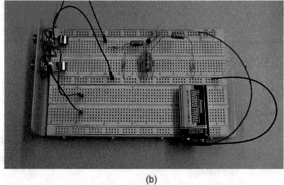

(b)

Some Practical Testing and Measurement Tips

Testing this circuit can be a bit tricky. Because the voltage gain of the amplifier is high, small input signal levels (around 10 mV$_{P-P}$) must be used to prevent output distortion. This is a problem because most oscilloscopes will not give accurate voltage readings at such low signal amplitudes. Using a 10× probe, the maximum vertical sensitivity of the scope used (a Rigol DS1052E) is 20 mV/division. Another problem is that most signal generators are not adjustable down to much less than about 100 mV$_{P-P}$ or so.

Because of these problems, a 100-to-1 attenuator (really, just a simple voltage divider) was connected at the input of the amplifier as shown in Fig. 3.15a. Using the attenuator, a signal generator voltage of $V_S = 1$ V$_{P-P}$ produces $V_{in} = 10$ mV$_{P-P}$.

Remember, the input attenuator would not be used in the actual circuit. It is simply a convenient and accurate way to allow a very small input signal to be applied to the amplifier. A simple sinusoidal audio oscillator is presented at the end of this chapter (see Fig. 3.46) that is suitable for use in amplifier tests such as this.

The Q-point values I_{CQ} and V_{CEQ} were first measured with no signal applied to the amplifier. The measured values are listed along with the theoretical values calculated previously.

Q-point

Parameter	Theory	Measured
I_{CQ}	1.1 mA	1.2 mA
V_{CEQ}	4.3 V	4.2 V

In terms of the AC signal performance of the amplifier, we are usually most interested in determining the voltage gain and how much distortion is introduced in the output signal. Input signals of $V_{in} = 10$, 50, and 100 mV$_{P-P}$ were applied to the amplifier, resulting in the output voltage waveforms and frequency spectra shown in Fig. 3.16. In order to reduce display clutter, the input signal is not shown, but the output signal phase is $180°$ as expected.

(a) $V_{in} = 10$–mV$_{P-P}$, $V_o = 920$ mV$_{P-P}$. Negligible distortion present in output
(b) $V_{in} = 50$ mV$_{P-P}$, $V_o = 4.60$ V$_{P-P}$. Significant second and third harmonics present in output
(c) $V_{in} = 100$ mV$_{P-P}$, $V_o = 6.80$ V$_{P-P}$. Significant even and odd harmonics past 12 kHz present in output

Recall that the expected theoretical voltage gain magnitude of the amplifier was calculated to be $|A_V| = 104$. The analysis equations presented here are based on the assumption that we used an ideal transistor, and coupling/bypass capacitors that are perfect AC short circuits, which yields optimistic predictions.

Because Fig. 3.14 uses full emitter bypassing, transistor parameter variations will have a noticeable effect on voltage gain. In an experimental circuit such as this, you can expect to see voltage gain variations of $\pm 15\%$ from one transistor to another.

$$V_{in} = 10\,mV_{P-P}$$

For $V_{in} = 10$ mV$_{P-P}$ (Fig. 3.16a), the output voltage has no visible distortion with amplitude
$V_o = 920$ mV$_{P-P}$. The measured voltage gain magnitude is

$$|A_V| = V_o/V_{in}$$
$$= 920\ mV_{P-P}/10\ mV_{P-P}$$
$$= 92.0$$

We find that the actual gain is about 12% less than the predicted value of $|A_V| = 104$. This is not unusual.

$$V_{in} = 50\,mV_{P-P}$$

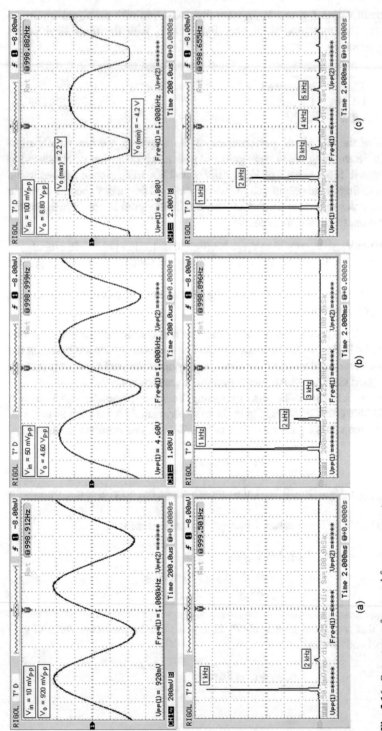

Fig. 3.16 Output waveforms and frequency spectra

With 50 mV$_{P-P}$ applied to the amplifier, the output voltage waveform is noticeably distorted. The positive-going portion of V_0 appears squashed, reaching a peak of about 1.6 V, while the negative-going portion of the output reaches about -2.6 V. The effective overall voltage gain of the amplifier has decreased slightly, but the difference is too small to resolve on the scope.

$$|A_V| = V_o/V_{in}$$
$$= 4.60 \ V_{P-P}/50 \ mV_{P-P}$$
$$= 92.0$$

The asymmetry of the output voltage waveform is caused by nonlinearity of the transistor, which causes harmonics of the original signal frequency to appear in the output. The output signal harmonics are visible in the lower spectrum analyzer displays in Fig. 3.16. The horizontal axis of the spectrum display is scaled 625 Hz/div. The vertical scale of the spectrum display is 200 mV$_{rms}$/div, so if you wish, you can read the amplitudes of the various frequency components directly. For example, the 2 kHz component in 3.16 b has an amplitude of

$$1.4 \ div \times \ 200 \ mV_{rms}/div = 280 \ mV_{rms}$$

We can calculate the peak amplitude of this output component as follows:

$$V_P = \sqrt{2} \ V_{rms} = 1.414 \times 280 \ mV_{rms} = 396 \ mV_{pk}$$
$$\mathbf{V_{in} = 100\,mV_{P-P}}$$

With $V_{in} = 100$ mV$_{P-P}$, the amplifier is overdriven, causing severe clipping of the negative-going portion of the output signal. Because of this distortion, the effective voltage gain of the amplifier has dropped to

$$|A_V| = V_o/V_{in}$$
$$= 6.80 \ V_{P-P}/100 \ mV_{P-P}$$
$$= 68.0$$

The positive-going portion of V_0 is just starting to clip. Comparing the measured clipping levels with the calculated max and min output voltages, we find that they are in reasonably close agreement.

Clipping levels	Theory	Measured
$V_{o(max)}$	2.8 V	2.2 V
$V_{o(min)}$	-4.3 V	-4.2 V

Once clipping occurs, a large number of significant harmonics are created in the output signal. As you can see in the spectrum display of Fig. 3.16c, both even and odd harmonics are popping up.

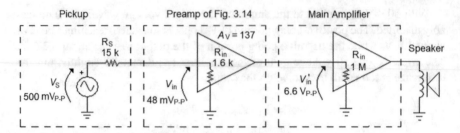

Fig. 3.17 Using the CE amplifier of Fig. 3.14 as a guitar preamp

Amplifying a Guitar Signal

The amplifier in Fig. 3.14 was connected as a preamp between a guitar and an amplifier, as shown in Fig. 3.17. The guitar used was the Les Paul Standard which was used to produce waveforms previously in Chap. 2 (see Fig. 2.5). The main amplifier was a custom-built amp with input resistance $R_{in} = 1$ MΩ.

Since we are no longer driving a 10 kΩ load, but rather the 1 MΩ input resistance of the main amplifier, we will recalculate the gain of the preamp using (3.22) and (3.25). You can verify that this gives us $|A_V| = 137$. Based on experimental measurements, this is probably somewhat optimistic (the actual gain would probably be about 100 or so), but let's assume it is correct.

The increased load resistance not only increases A_V but in this case also increases the positive clipping level as well, giving us more output compliance or *headroom*. Again, you can verify that using (3.26) with $R_L = 1$ MΩ, we get $V_{o(max)} = 3.6$ V. The new peak-to-peak unclipped output voltage is

$$V_{o(P-P)} = 2 \times 3.6 \text{ V}$$
$$= 7.2 \text{ V}_{P-P}$$

Looking back at Fig. 2.5, we find that this guitar normally produces a maximum output voltage in the range from 400 to 600 mV$_{P-P}$. Using an average input of about 500 mV$_{P-P}$, we might expect the preamp to distort severely since the output compliance is 7.2 V$_{P-P}$. However, we still haven't accounted for loading of the guitar pickup. The pickup winding resistance is $R_S = 15$ kΩ, and the actual input voltage applied to the preamp is found by using (3.7).

$$V_{in} = V_S \left(\frac{R_{in}}{R_{in} + R_S} \right)$$
$$= 500 \text{ mV}_{P-P} \left(\frac{1.6 \text{ k}\Omega}{1.6 \text{ k}\Omega + 15 \text{ k}\Omega} \right) \qquad (3.7)$$
$$= 48 \text{ mV}_{P-P}$$

The preamp is really loading down the guitar pickup. Putting all of this together, the amplitude of the signal applied to the main amplifier V'_{in} is

$$V'_{in} = 48 \ \text{mV}_{P-P} \times 137$$
$$= 6.6 \ \text{V}_{P-P}$$

This will not make the preamp clip, but it is easily large enough to overdrive the main amplifier or an effects box. When it comes to audio equipment, the proof is in the listening. The preamp certainly boosted the signal to the main amp; however, there was no audible distortion produced by the preamplifier even at this high output level.

The low input impedance of this amplifier makes it impractical for general use in an effects chain. Some effect circuits might not like driving such a low impedance device. However, this is a great circuit for learning the basics of small-signal amplifiers built with discrete transistors. We will see a number of improved designs that are similar to this circuit that incorporate BJTs, JFETs, MOSFETs, and op amps that have much better performance.

Frequency Response

To complete the analysis of the preamp, we should make sure the bandwidth of the amp meets our needs. The good news here is that the high-frequency performance of Fig. 3.14 is so good that we don't even need to think about it. The high corner frequency of the amplifier extends into the MHz range, and the slew rate will be very fast as well. This is good news because high-frequency analysis is generally a lot trickier than low and audio frequency work.

The low-frequency response of the preamp is determined by the size of the coupling and bypass capacitors. Capacitor C_1 forms a high-pass filter at the input, C_2 forms a high-pass filter at the output, and C_3 forms a high-pass filter on the emitter. If we are processing guitar signals, we should make sure that the corner frequency of each high-pass filter is lower than 82 Hz, which is roughly the open low E string frequency.

Referring to Fig. 3.14, the approximate corner frequencies of the input coupling network $f_{C(in)}$, the output coupling network $f_{C(out)}$, and the emitter bypass network $f_{C(E)}$ are given by applying the familiar first-order RC filter equation:

$$f_{C(in)} \cong \frac{1}{2\pi R_{in} C_1} \tag{3.28}$$

$$f_{C(out)} \cong \frac{1}{2\pi R'_C C_2} \tag{3.29}$$

$$f_{C(E)} \cong \frac{1}{2\pi r_e C_3} \tag{3.30}$$

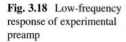

Fig. 3.18 Low-frequency response of experimental preamp

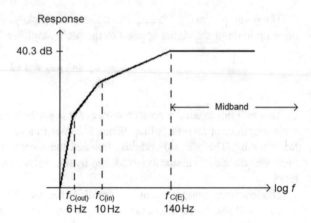

These equations are approximations but they are close enough to give useful results. In BJT-based audio amplifiers, we will almost always be using electrolytic coupling and bypass capacitors, which normally have tolerances of +20% to −5%. There is no point in deriving complex corner frequency equations when our capacitors vary over such a wide range.

Using the values given for the experimental circuit, the corner frequencies are

$$f_{C(in)} \cong \frac{1}{2\pi R_{in} C_1} \qquad f_{C(out)} \cong \frac{1}{2\pi R'_C C_2} \qquad f_{C(E)} \cong \frac{1}{2\pi r_e C_3}$$

$$\cong \frac{1}{2\pi (1.6\,k\Omega)(10\,\mu F)} \qquad \cong \frac{1}{2\pi (2.5\,k\Omega)(10\,\mu F)} \qquad \cong \frac{1}{2\pi (24\,\Omega)(47\,\mu F)}$$

$$\cong 10\,Hz \qquad \cong 6\,Hz \qquad \cong 140\,Hz$$

The low-frequency response of the preamp is dominated by the highest coupling/bypass corner frequency, which in this case is $f_{C(E)} \cong 140$ Hz. Strictly speaking, we should use a bigger emitter bypass capacitor. Using $C_3 = 100\,\mu F$ would give us $f_{C(E)} = 66$ Hz, but in all honesty, you wouldn't even hear the difference. The asymptotic low-frequency response is shown in the graph of Fig. 3.18.

Negative Feedback

In that last section, we really put that preamp circuit through a circuit analysis meat grinder, and you might be wondering how much more could possibly be said about this circuit. I promise we are almost finished, but we really need to briefly touch on the concept of negative feedback.

Fig. 3.19 Amplifier with
negative feedback

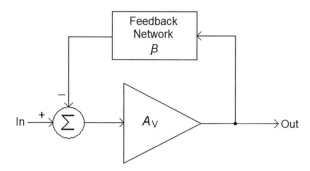

Negative feedback occurs when we sample the output of an amplifier and send some (or possibly all) of it back to the input, 180° out of phase with the input signal. A block diagram showing this concept is given in Fig. 3.19.

In general, negative feedback will:

1. Reduce gain.
2. Reduce distortion.
3. Stabilize amplifier parameters.
4. Increase bandwidth.
5. Increase input resistance.
6. Decrease output resistance.

Basically, negative feedback allows us to trade gain for improved stability, reduced distortion, increased bandwidth, and, depending on the exact details of the circuit, increased input and decreased output resistances.

We will talk more about negative feedback again in this and in later chapters, but for now, let's apply this concept to the experimental common emitter amplifier from Fig. 3.14. The voltage gain and input resistance of this amplifier are given by (3.23) and (3.25) which are repeated for convenience.

$$A_V = \frac{-R'_C}{r_e + R'_E} \qquad R_{in} = R'_B \parallel \beta(r_e + R'_E)$$

The important thing to notice is that these two equations include the external AC equivalent emitter resistance term R'_E.

With the emitter bypassed to ground by C_3, the external AC resistance is $R'_E = 0\,\Omega$. This gives us maximum A_V, but it also accounts for much of the nonlinearity and unpredictability of the voltage gain. The problem is that the dynamic emitter resistance r_e changes directly with collector current, which causes the voltage gain to vary nonlinearly with output voltage swing. This is the reason we saw the squashing/stretching characteristic of the output waveforms back in Fig. 3.16b, c.

If we add a resistor R_X in series with the emitter bypass capacitor as shown in Fig. 3.20, then the external AC emitter resistance is given by

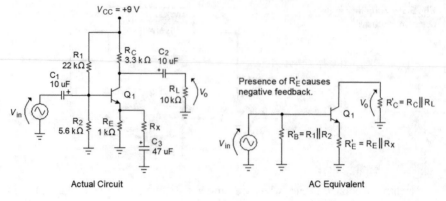

Actual Circuit   AC Equivalent

Fig. 3.20 Adding resistor R_X creates negative feedback in the AC equivalent circuit

$$R'_E = R_E \parallel R_X \tag{3.31}$$

The external AC emitter resistance R'_E produces negative feedback in the following way. Assuming that the input voltage is going positive, the emitter current will increase. This causes a positive-going voltage drop across R'_E, which opposes the input signal that was its original cause. In this way, negative feedback creates a signal that tends to cancel out the original signal. The greater the amount of negative feedback, the less effect the original input signal has on the output. Negative feedback works the same way for negative-going input signals as well.

It doesn't take much external emitter resistance to stabilize the amplifier. As long as $R'_E \geq r_e$, nonlinearity will be drastically reduced. And as an additional bonus, the input resistance of the amplifier will be increased as well. To demonstrate these ideas, let's determine R'_E, A_V, and R_{in} for two different values of R_X (recall that for this circuit, we have $r_e = 24 \ \Omega$).

$R_X = 22 \ \Omega$

$R'_E = R_E \parallel R_X$

$\quad = 1 \ \text{k}\Omega \parallel 22 \ \Omega$

$\quad \cong 22 \ \Omega$

$A_V = \dfrac{-R'_C}{r_e + R'_E}$

$\quad = \dfrac{-2.5 \ \text{k}\Omega}{24 \ \Omega + 22 \ \Omega}$

$\quad \cong -54$

$R_{in} = R'_B \parallel \beta(r_e + R'_E)$

$\quad = 4.5 \ \text{k}\Omega \parallel 100(24 \ \Omega + 22 \ \Omega)$

$\quad \cong 2.3 \ \text{k}\Omega$

$R_X = 100 \ \Omega$

$R'_E = R_E \parallel R_X$

$\quad = 1 \ \text{k}\Omega \parallel 100 \ \Omega$

$\quad \cong 91 \ \Omega$

$A_V = \dfrac{-R'_C}{r_e + R'_E}$

$\quad = \dfrac{-2.5 \ \text{k}\Omega}{24 \ \Omega + 91 \ \Omega}$

$\quad \cong -22$

$R_{in} = R'_B \parallel \beta(r_e + R'_E)$

$\quad = 4.5 \ \text{k}\Omega \parallel 100(91 \ \Omega + 22 \ \Omega)$

$\quad \cong 3.2 \ \text{k}\Omega$

Local and Global Feedback

The negative feedback applied by placing R_X in the circuit of the preceding example is a form of *local feedback*. Local feedback primarily affects the stage to which it is applied. In this particular case, the additional AC equivalent emitter resistance created is sometimes called *swamping resistance*. The term swamping resistance comes from the fact that variations in r_e are swamped out by the presence of R'_E, which remains constant.

In a multiple-stage amplifier, we can apply local feedback to individual stages as done in the last example, but we can also form one or more feedback loops that enclose several amplifier stages. This is called *global feedback*. Global feedback will generally result in an amplifier that has very linear gain, right up to clipping. Global feedback is often used in high-fidelity audio amplifiers, but it's not as common in guitar amplifiers, where distortion is desirable.

A JFET Common Source, Class A Amplifier

Recall from previous discussions that if we were to design a preamplifier that was suitable for use with a piezoelectric pickup, we would want the preamp to have very high input resistance. High input resistance reduces loading of the piezo pickup and ensures that good low-frequency response can be obtained using a reasonably small input coupling capacitor. This application is ideally suited for an amplifier built using a JFET.

JFET Parameters

The typical JFET data sheet will list a large number of device parameters and characteristic curves. There are three main JFET parameters that are of interest to us here: I_{DSS}, V_P, and g_{m0}. Here are brief descriptions of these parameters.

I_{DSS} This is the maximum possible drain current that the JFET can carry. $I_D = I_{DSS}$ when $V_{GS} = 0$ V, and $V_{DS} \geq V_P$. When this occurs, the JFET is said to be saturated.

V_P This is the JFET pinch-off voltage. This is the minimum value of V_{DS} required to operate the JFET as a voltage-controlled current source. V_P is also the magnitude of the gate-to-source voltage V_{GS} required to turn off the JFET, that is, $V_P = -V_{GS(off)}$. I prefer to work with $-V_P$ and V_P for convenience of notation.

g_{m0} The maximum possible transconductance of the JFET is denoted as g_{m0}. Transconductance in general (g_m) is the gain parameter of the JFET. In amplifier applications, the JFET behaves as a voltage-controlled current

Fig. 3.21 JFET drain
characteristic curves

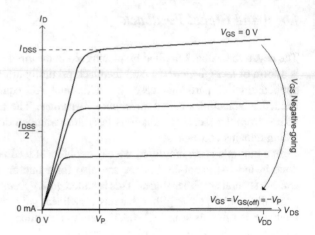

source, and g_m tells us how sensitive the drain current I_D is to the control
voltage V_{GS}. Maximum transconductance g_{m0} occurs when $V_{GS} = 0$ V and
$I_D = I_{DSS}$.

The preceding parameters are related through the family of N-channel JFET drain
curves shown in Fig. 3.21. Note the similarity between these curves and those of the
NPN bipolar transistor shown back in Fig. 3.7.

If you were asked to choose a class A amplifier Q-point location on the JFET
characteristic curve plot, where would you place it? If you thought about this and
decided that a reasonable location would be at $I_{DQ} = I_{DSS}/2$ and $V_{DQ} = V_{DD}/2$
(or perhaps $V_{DQ} = (V_{DD} - V_P)/2$), then you probably have a good handle on the
concepts that we are discussing.

The JFET we will work with in this book is the MPF102, a popular N-channel
device with excellent low noise and high-frequency performance. The data sheet for
the MPF102 states that I_{DSS} may range from 2 to 20 mA. Based on my experience,
these extreme values are very unlikely. Running curves for ten randomly selected
MPF102s using a Tektronix Type 575 transistor curve tracer gave actual I_{DSS} values
that ranged from 8 to 12 mA.

Based on these measurements, the following average values were determined:

MPF102 N-Channel JFET Parameters (Typical Values)
$I_{DSS} = 10$ mA
$V_P = 3$ V
$g_{m0} = 6.7$ mS

At the time of writing, MPF102s are readily available for about $1.50 USD. There
are other N-channel JFETs with similar specifications that may be substituted for the
MPF102, including the J111, 2N5485, and 2N5486. All of these devices have the
same pin designations as the MPF102 (see Fig. 3.23).

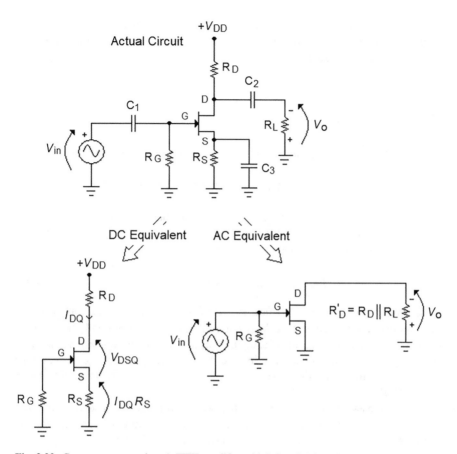

Fig. 3.22 Common source, class A JFET amplifier with DC and AC equivalent circuits

JFET Amplifier Overview

A common source (CS), class A, N-channel JFET amplifier is shown in Fig. 3.22. The JFET CS configuration is analogous to the BJT common emitter (CE), providing significant voltage gain and current gain, as well as output signal phase inversion.

Here is a basic summary of how the JFET amplifier works. Power is supplied by the V_{DD} source, which is equivalent to V_{CC} in a bipolar transistor circuit. Capacitors C_1 and C_2 are coupling capacitors, while C_3 bypasses the source terminal to ground. As in the BJT circuit, these capacitors act as open circuits to DC while acting approximately as short circuits at the AC signal frequency.

Resistor R_S is chosen to set the drain current I_{DQ} to some desired value. Typically, we are after $I_{DQ} = I_{DSS}/2$, which is half the maximum JFET drain current. The JFET in Fig. 3.22 uses what is called *self-bias* or *source feedback bias*. This biasing arrangement can't be used with BJTs or enhancement MOSFETs, but a similar biasing technique is often used in vacuum tube circuit designs. Drain resistor R_D is chosen to set the desired V_{DSQ} value.

JFET Centered Q-Point

It can be shown that to obtain $I_{DQ} \cong I_{DSS}/2$, we can use a source resistor given by

$$R_S \cong 1/g_{m0} \quad \text{(for centered Q-point)} \tag{3.32}$$

The drain resistor R_D is normally chosen to obtain $V_{DSQ} \cong V_{DD}/2$, which approximately centers the amplifier Q-point. Assuming that $I_{DQ} = I_{DSS}/2$, the operating point transconductance g_m is given by

$$g_m \cong 0.75 g_{m0} \quad \text{(when } I_D = I_{DSS}/2\text{)} \tag{3.33}$$

Because the gate-source junction is effectively a reverse-biased PN junction, there is only a very small gate leakage current, which should normally be less than 1 nA. This is small enough to be insignificant, and the effective resistance looking into the gate terminal is practically infinite. Because the gate leakage current is so small, we can make the gate resistor very large. In practice, R_G is usually set to around 1 MΩ.

Source bypass capacitor C_3 shorts the source terminal to ground in the AC equivalent circuit, maximizing voltage gain, much like emitter bypassing for the BJT amplifier. The voltage gain of the bypassed CS amplifier is given by

$$A_V = -g_m R'_D \tag{3.34}$$

Where,

$$R'_D = R_D \parallel R_L \tag{3.35}$$

Finally, the input and output resistances of the amplifier are given by

$$R_{in} = R_G \tag{3.36}$$

$$R_o \cong R_D \tag{3.37}$$

BJT vs. JFET

Unlike bipolar transistors, which are normally-off devices, JFETs are normally-on devices. The maximum collector current of a BJT is limited by external resistors R_C and R_E. As noted previously, the maximum drain current for a JFET is the basic device parameter I_{DSS}. With no bias voltage applied to the gate terminal, the JFET will allow maximum drain current I_{DSS} to flow. The G–S junction must be reverse biased ($V_{GS} < 0$ V for an N-channel JFET) to bias the JFET into the middle of its active region. This can be accomplished by applying a negative DC voltage directly to the gate terminal (gate biasing), but most often source feedback biasing is used instead.

In source feedback biasing, the gate terminal is held at ground potential, while the source terminal is raised to a positive voltage with respect to ground. The voltage that is dropped across the source resistor, $I_{DQ}R_S$, produces the JFET bias voltage. Since the gate is at ground potential ($V_G = 0$ V), the voltage dropped across R_S raises the source terminal above ground potential, which reverse biases the gate-source junction giving the bias voltage $V_{GS} = -I_{DQ}R_S$.

Piezoelectric Pickup Preamplifier

A practical piezo pickup preamplifier, powered by a 9 V battery, is shown in Fig. 3.23. An input jack is not shown here because the amplifier and battery would normally be mounted internally in the acoustic guitar. This is done to avoid long cable runs to the amp, which reduces the chance of noise pickup.

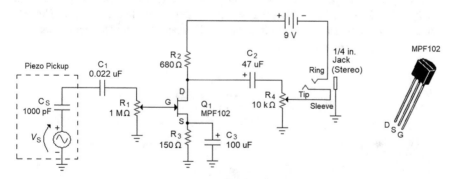

Fig. 3.23 Battery-powered JFET piezo preamplifier with volume control and pin designations for the MPF102

This circuit uses an MPF102 JFET so the device parameters will be approximately

$$I_{DSS} = 10 \text{ mA}, \quad V_P = 3 \text{ V}, \quad g_{m0} = 6.7 \text{ mS}$$

We can use (3.32) to determine if the given value of R_S will bias the JFET for $I_{DSS}/2 = 5$ mA.

$$\begin{aligned} R_S &= 1/g_{m0} \\ &= 1/6.7 \text{ mS} \\ &= 149.3 \, \Omega \end{aligned}$$

Since the circuit has a source resistor $R_3 = 150 \, \Omega$ which is close to the calculated value, the JFET is biased for $I_{DQ} \cong 5$ mA, which is $I_{DSS}/2$. Making R_3 larger will reduce I_{DQ}, and making R_3 smaller will increase I_{DQ}.

Using Ohm's law and Kirchhoff's voltage law, the drain-to-source voltage is given by

$$\begin{aligned} V_{DSQ} &= V_{DD} - I_{DQ}(R_D + R_S) \\ &= 9\text{V} - 5 \text{ mA}(680 \, \Omega + 150 \, \Omega) \\ &\cong 4.9 \text{ V} \end{aligned} \tag{3.38}$$

The operating point transconductance is given by (3.33).

$$\begin{aligned} g_m &= 0.75 \, g_{m0} \\ &= 0.75 \times 6.7 \text{ mS} \\ &= 5.025 \text{ mS} \end{aligned}$$

This preamp is intended to be connected to a typical guitar amplifier with $R_{in} \geq 100$ kΩ, which is so high that we can disregard its loading effect. The AC equivalent drain resistance is found using (3.35), where the 10 kΩ potentiometer R_4 serves as the load resistance R_L.

$$\begin{aligned} R'_D &= R_D \parallel R_L \\ &= 680 \, \Omega \parallel 10 \, k\Omega \\ &= 637 \, \Omega \end{aligned}$$

The voltage gain is

$$\begin{aligned} A_V &= -g_m R'_D \\ &= -5.025 \text{ mS} \times 637 \, \Omega \\ &= -3.2 \end{aligned}$$

As you can see, the voltage gain is not very impressive, but in this application, high input resistance is our priority, and getting a little voltage gain along the way is icing on the cake. If, by chance, the transconductance of your JFET happened to be unusually high, you might get voltage gain around $A_V = -5$ or so.

Recall that the piezoelectric pickup may be modeled as a voltage source V_S in series with a capacitance C_S (see Fig. 2.11). Technically, we do not need an input coupling capacitor in this application, so you could leave it out of the circuit if you wanted to. But just in case the preamp should be used in a different application and connected to source that has a significant DC offset, the coupling capacitor prevents the Q-point from being affected. Coupling capacitor C_1 was chosen to be very large compared to C_S, so it has negligible effect on the input corner frequency. The corner frequency of the HP filter formed at the gate terminal is given by (3.39).

$$
\begin{aligned}
f_{C(in)} &= \frac{1}{2\pi R_G C_S} \\
&= \frac{1}{2\pi(1\,M\Omega)(1000\,pF)} \\
&\cong 159\,Hz
\end{aligned}
\tag{3.39}
$$

For excellent bass response, we would like $f_{C(in)} \leq 82$ Hz. Using a 2.2 MΩ gate resistor would give us a 72 Hz corner, but the difference in response would be very slight.

The output coupling capacitor HP filter has a corner frequency of

$$
\begin{aligned}
f_{C(out)} &= \frac{1}{2\pi R'_D C_2} \\
&= \frac{1}{2\pi(637\,\Omega)(47\,\mu F)} \\
&\cong 5\,Hz
\end{aligned}
\tag{3.40}
$$

Finally, the corner frequency of the source bypass HP network is given by

$$
\begin{aligned}
f_{C(S)} &= \frac{1}{(2\pi)\left(\dfrac{1}{g_m} \,\|\, R_S\right)C_3} \\
&= \frac{1}{2\pi(199\,\Omega \,\|\, 150\,\Omega)(100\,\mu F)} \\
&\cong 19\,Hz
\end{aligned}
\tag{3.41}
$$

The circuit was constructed and tested using a Schatten soundboard transducer mounted on a Yamaha FG-335 acoustic guitar (see Fig. 2.8) and found to give excellent performance, driving a variety of guitar amplifiers with no noticeable distortion.

Phone Jack Power Switching

Notice that in Fig. 3.23 the output jack functions as the power switch. Inserting a mono 0.25 in. plug into a stereo jack shorts the ring terminal to the sleeve, which connects the negative terminal of the battery to ground. This is a common power switching arrangement in many effects stomp boxes. An input jack may be used for power switching as well, as will be shown in the next example.

Increasing Voltage Gain

There are a number of techniques that we can employ to increase the voltage gain of the common emitter and common source amplifiers we just examined. Turning our attention back to the BJT CE amp of Fig. 3.14, we used an emitter bypass capacitor to reduce R'_E to zero in (3.25). Emitter bypassing reduces the voltage gain equation to

$$A_V = \frac{R'_C}{r_e}$$

To further increase the voltage gain of this single-stage amplifier, we have two choices. First, we could reduce r_e by increasing the collector current I_{CQ}, which is feasible but also will increase power dissipation, decrease battery life, decrease R_{in}, and change the Q-point location. Even if we manage to deal with these negative effects, we soon reach the point of diminishing returns when reducing r_e because the internal connecting wires, the contacts, and the inherent resistance of the emitter bulk material itself will add a few ohms of resistance that can't be eliminated.

The second choice is to increase R'_C. It is not usually possible to change the load resistance, so we would have to increase the value of collector resistor R_C. In Fig. 3.20, simply increasing R_C forces the transistor toward saturation, which will cause clipping to occur sooner. The only practical way to use a much higher value collector resistor is to redesign the circuit using a higher voltage for V_{CC} say +12 or 15 V. Alternatively, we might use a bipolar power supply of ±9, 12, or 15 V. The increased supply voltage allows us to drop more voltage across R_C without moving the Q-point too close to saturation. If we want to power the amplifier using a single 9 V battery, these are not practical options, without resorting to a charge pump (covered later in this chapter). The main point to take from this discussion is that as a practical matter, we can only squeeze so much gain from a single transistor.

A JFET-BJT Multiple-Stage Amplifier

Although there are many applications where a single transistor will provide sufficient gain, most practical amplifiers consist of multiple cascaded stages. We can
cascade a slightly modified version of the JFET circuit (Fig. 3.23) with the previous
BJT amplifier (Fig. 3.14) to form a single higher-gain amplifier, which is shown in
Fig. 3.24.

Because the amplifier stages are capacitively coupled by C_2, there is no DC
interaction between them, and the Q-points of the transistors will be the same as
previously calculated. This is a very nice characteristic of capacitive coupling. On
the downside, capacitive coupling does not allow frequency response to extend to
DC (0 Hz), and using coupling capacitors adds extra components in the signal path.

The voltage gain of the BJT output stage will depend on the actual load resistance
that is being driven. If we assume there is no load or a high resistance load (100 kΩ
or greater), the theoretical gain of the second stage is $A_{V(2)} \cong -137$ as previously
calculated (see Fig. 3.17).

The voltage gain of the first stage will be somewhat lower than previously
calculated because the input resistance of stage 2 loads the first stage. Applying
(3.31), we have

$$R'_D = R_D \parallel R_{in(2)}$$
$$= 680 \, \Omega \parallel 1.6 \, k\Omega$$
$$= 477 \, \Omega$$

Now, using the JFET gain equation (3.30), the voltage gain of the first stage is
found to be

$$A_{V(1)} = -g_m R'_D$$
$$= -5025 \, \mu S \times 477 \, \Omega$$
$$\cong -2.4$$

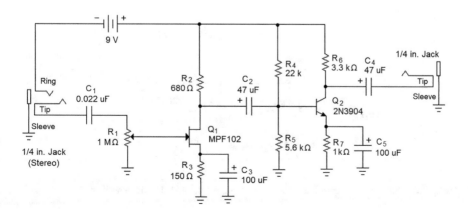

Fig. 3.24 Two-stage, low-power amplifier

The overall voltage gain of the amplifier is the product of the individual stage gains, which is

$$A_V = A_{V(1)}A_{V(2)}$$
$$= -2.4 \times -137 \tag{3.42}$$
$$\cong 329$$

Notice that the overall voltage gain is noninverting. The phase inversions of the two cascaded stages cancel out; the product of two negative numbers is a positive number.

The JFET gate resistor was replaced with a 1 MΩ potentiometer. The input resistance of the amplifier is 1 MΩ regardless of the setting of the pot.

$$R_{in} = R_1 = 1\,M\Omega$$

As the potentiometer is adjusted, the effective gate resistance as seen by the JFET changes from 1 MΩ at full volume to 0 Ω at zero volume. However, because the gate leakage current is so small, this resistance variation has virtually no effect on the Q-point of the JFET.

The advantage of placing the volume control at the input of the preamp is that if we happen to be using a high-amplitude source, we can adjust the volume down so that the preamp output isn't distorted or clipped. Of course, for maximum flexibility, you could always place pots at both the input and output if you want to.

Some Useful Modifications

A modified version of the two-stage amplifier is shown in Fig. 3.25, where output level and tone controls have been added. Adding 100 kΩ potentiometer R_9 to adjust the output level of the amplifier is practical in this application because the load being

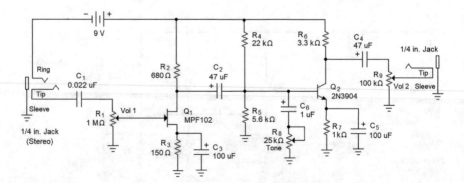

Fig. 3.25 Simple tone control and output level control pot added to the amplifier

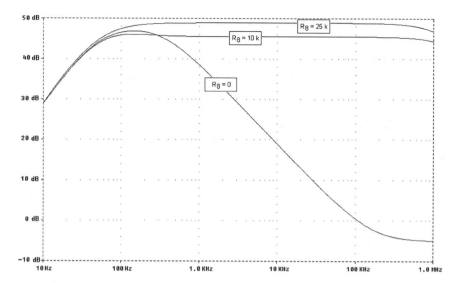

Fig. 3.26 Frequency response for the two-stage preamp with simple tone control

driven will typically be a guitar amplifier or an effect circuit (fuzz, distortion, etc.), which should have a very high input resistance. This output level adjustment technique would not be used at the output of a power amplifier, but it is common in low-power stages. Reasonable potentiometer values for R_9 in Fig. 3.25 range from 10 kΩ to 1 MΩ.

With both input and output level adjustments available, it is possible to adjust the circuit such that the preamp, the guitar amp, or even both may be overdriven. A slightly modified version of this amplifier will be used to create a classic distortion effect in Chap. 5.

The tone control is a very simple circuit consisting of potentiometer R_8 and capacitor C_6. When R_8 is adjusted to 25 kΩ, the tone control has negligible effect on the frequency response of the amplifier, which will be relatively flat from about 50 Hz to over 1 MHz. As the pot is adjusted for lower resistance values, the tone control section provides a low-impedance path to ground for high frequencies at the base of Q_2, mainly providing a treble-cut function in this application. Frequency response curves for several different potentiometer settings are shown in Fig. 3.26.

Note that the tone control is most effective between 0 and 10 kΩ, with only a small effect up to 25 kΩ. The action of the tone control could be made more directly proportional to shaft rotation with the use of an audio taper potentiometer, wired as in Fig. 2.13d.

A Closer Look at Transconductance

We've been talking a lot about transconductance, which is a topic that is important enough to deserve some additional study. We know that transconductance is the gain parameter of the JFET. Transconductance is also the gain parameter for the MOSFET, and it is also a very important gain parameter in the analysis of vacuum tube circuits. The concept of transconductance may even be applied in the analysis of BJT circuits and some specialized operational amplifier circuits as well. For an FET, transconductance is defined as

$$g_m = \frac{dI_D}{dV_{GS}}\bigg|_{V_{DS}=\text{const.}} \tag{3.43}$$

This bit of calculus basically says that transconductance is the rate of change of drain current with respect to gate-to-source voltage, with the drain-to-source voltage held constant. The larger g_m is, the more sensitive the FET is to changes in gate voltage. If we take a look at the transconductance curves for the typical JFET, MOSFET, and BJT, we can get a more intuitive understanding of this concept.

BJT, JFET, and MOSFET transconductance current curves are shown in Fig. 3.27, where the individual device curves are shown on the left. In order to make a more clear comparison between these devices, their transconductance curves have been scaled and shifted to fit nicely on the plot at the right side of the figure. The approximate transconductance of a given device can be obtained from the curves using the following equation:

$$g_m \cong \frac{\Delta I}{\Delta V} \tag{3.44}$$

For each of the curves, the change in control voltage (ΔV) is the same, but notice that the resulting change in current (ΔI) is greatest for the BJT and smallest for the JFET. The transconductance of the MOSFET is somewhere in between. This is the reason that in most situations, the common emitter BJT will produce much higher voltage gain than an equivalent common source JFET or MOSFET amplifier.

While (3.43) and (3.44) are useful in defining the basic concept of transconductance, they are not very useful in practical circuit analysis applications. We used a rule of thumb (3.33) to determine g_m for the JFET, and a (somewhat) practical transconductance equation will be derived for the MOSFET in the next section. In case you are curious, and I'm sure you are, the transconductance of a BJT is given by $g_m = 1/r_e$.

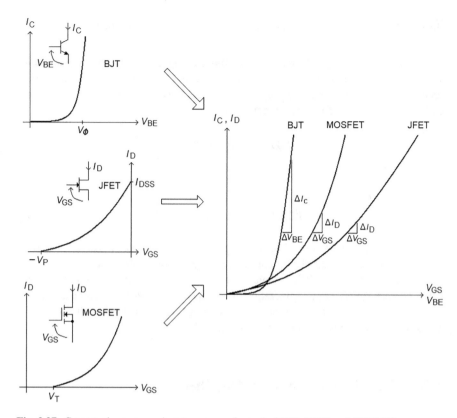

Fig. 3.27 Comparative transconductance curves for typical BJT, JFET, and MOSFETs

You may be wondering where the term transconductance comes from. Recall that the transfer function of a network is the ratio of the output variable to the input variable, output/input. For a voltage amplifier, this is the voltage gain, $A_V = V_0/V_{in}$. The ratio I/V is conductance G, which is measured in Siemens, S.[2] The output/input variables in (3.43) or (3.44) are $\Delta I/\Delta V$. Because these quantities represent the transfer characteristics of a given device, the parameter is called transconductance: a contraction of *transfer function* and *conductance*. We use a lowercase, italic letter to designate transconductance because it is a dynamic, internal property of the device.

You may see g_m designated as g_{fs} on some FET data sheets. This stands for "forward transconductance, common source configuration." We will stick with g_m in this book.

---

[2] Some references, especially older vacuum tube data sheets, use the now obsolete unit "mhos" as the unit of conductance. The mho is represented symbolically by an upside down omega.

BJT, JFET, and MOSFET Transconductance Equations

Device manufacturers produce curves like those in Fig. 3.27 based on measurements made on large numbers of actual transistors. These curves are also described quite accurately by specific mathematical relationships based on device physics for the BJT, JFET, and MOSFET. These equations are:

$$\text{BJT}: \quad I_C = I_s e^{V_{BE}/\eta V_T} \tag{3.45}$$

$$\text{JFET}: \quad I_D = I_{DSS}\left(1 - \frac{V_{GS}}{-V_P}\right)^2 \tag{3.46}$$

$$\text{MOSFET}: \quad I_D = k(V_{GS} - V_T)^2 \tag{3.47}$$

These equations look pretty intimidating, but fortunately, we almost never need to use them for most circuit design and analysis applications. For now, the important thing to notice is that the BJT collector current equation is exponential e^x, while the JFET and MOSFETs have quadratic (or *square-law*) x^2, drain current equations. These equations are very important in the study of device distortion characteristics which we will examine in great detail in Chap. 5.

A MOSFET Common Source Amplifier

Although they are most often used in high-power circuits and digital logic integrated circuits, MOSFETs may also be used in general-purpose, small-signal amplifier applications. The BS170 is an example of a commonly available N-channel, enhancement-mode MOSFET that can be used in a variety of applications. Although the BS170 is designed primarily for use in relatively high current (up to 2 A peak) switching applications such as PWM motor speed control, we will use it to implement a general-purpose, low-power, class A amplifier suitable for use in guitar signal processing applications. The schematic symbol and pin designations for the TO-92 package version of the BS170 are shown in Fig. 3.28.

The diode shown connected from drain to source of the BS170 is built into the transistor. This is what is called a *parasitic diode*; it is a byproduct of the internal construction of the MOSFET chip. In normal operation, this diode is reverse biased and has no effect on circuit operation, but it can actually protect the MOSFET from damage when it is used in certain switching applications.

The schematic symbol for the MOSFET graphically shows us that the gate terminal is insulated from the rest of the device. The insulating layer here is a very thin layer of silicon dioxide (SiO_2) that is formed during fabrication. Because of this construction, the MOSFET has practically infinite gate input resistance ($R_G > 10^9\ \Omega$). But, because the SiO_2 gate insulation layer is so thin, MOSFETs are notoriously susceptible to damage from static discharge during handling.

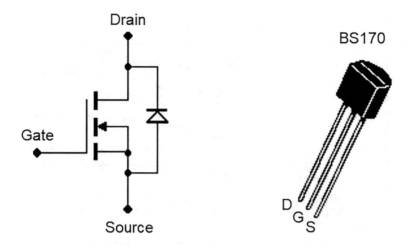

Fig. 3.28 BS170 N-channel MOSFET schematic symbol and TO-92 package

Enhancement-mode MOSFETs have some characteristics in common with JFETs and some with BJTs. Like the JFET, the drain current of the MOSFET is controlled by the gate-to-source voltage through a square-law relationship, as seen in Eqs. (3.46) and (3.47). BJTs are exponential devices, and they are usually modeled as current-controlled current sources.

Similar to BJTs, enhancement-mode MOSFETs like the BS170 are normally-off devices. In order to use the BS170 as a class A amplifier, it must be biased into the active region like a bipolar transistor. A few important parameters for the BS170 are listed below, along with brief explanations.

BS170 N-Channel MOSFET Parameters (Typical Values)
$V_T = 2.1 \text{ V}$
$g_m = 320 \text{ mS}$
$k = 0.08$

V_T This is the threshold or turn-on voltage of the MOSFET. V_T is the minimum gate-to-source voltage V_{GS} required to cause the device to begin carrying drain current. This is analogous to the base-to-emitter barrier potential ($V_{BE} \cong 0.7 \text{ V}$) of a typical BJT in the active region.

g_m Transconductance. The value given here ($g_m = 320 \text{ mS}$) is from the BS170 data sheet which specifies g_m at $I_D = 200 \text{ mA}$. We will be operating at a drain current of around 5 mA, so g_m will be much lower than the published value.

k This is a proportionality constant for the MOSFET, which is used in Eq. (3.47), $I_D = k(V_{GS} - V_T)^2$. This parameter is not given on the BS170 data sheet, but using the published transconductance curves, I derived a value of $k \cong 0.08$.

The drain curves for a typical BS170 are shown in Fig. 3.29. The similarity between these curves and those of the typical JFET and BJT is quite apparent. As

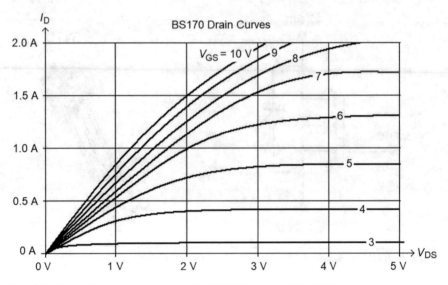

Fig. 3.29 Drain characteristic curves for the BS170 N-channel MOSFET

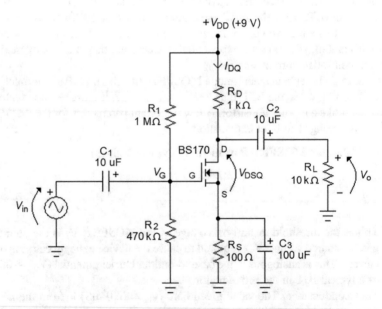

Fig. 3.30 Practical common source, low-power amplifier using a BS170 MOSFET

you can see from the scale of the drain curves, the BS170 is definitely designed for operation at rather high drain current levels. These curves are not very useful to us because we will be operating the MOSFET at a very low drain current (around 5 mA or so).

A simple class A, common source amplifier using the BS170 is shown in Fig. 3.30. Notice that this circuit uses voltage divider biasing much like the BJT amplifier we examined previously. Source resistor R_S provides DC negative feedback which helps to stabilize the Q-point of the MOSFET. Source bypassing ensures that the voltage gain of the amplifier is maximized.

Theoretical Analysis

As usual, the analysis begins with determination of the Q-point. First, we apply the voltage divider equation to the gate terminal network which gives us the gate bias voltage V_G. We require $V_G > V_T$ to ensure that the MOSFET is biased into conduction.

$$V_G = V_{DD}\left(\frac{R_2}{R_1 + R_2}\right)$$
$$= 9 \text{ V}\left(\frac{470 \text{ k}\Omega}{1 \text{ M}\Omega + 470 \text{ k}\Omega}\right)$$
$$= 2.9 \text{ V}$$

The drain current is found using Eq. (3.48) which you may recognize from algebra class as a version of the famous quadratic formula. All quadratic equations have two solutions (often called roots). We use the solution found by subtracting the radical in the numerator. The second solution, which would be found by adding the radical in the numerator, gives a meaningless result (this is called an *extraneous root*).

$$I_{DQ} = \frac{-B - \sqrt{B^2 - 4AC}}{2A} \tag{3.48}$$

The constants in the formula are determined as follows:

$$A = R_S^2, \ B = -\left(2R_S(V_G - V_T) + \frac{1}{k}\right), \ C = (V_G - V_T)^2$$

Plugging the various values into the formulas, with $k = 0.08$, we get the following results:

$$A = 100^2 = 10,000$$
$$B = -\left(2 \times 100\,(2.9 - 2.1) + \frac{1}{0.08}\right) = -172.5$$
$$C = (2.9 - 2.1)^2 = 0.64$$

Note that the negative sign of B cancels when used in the quadratic formula. Evaluating (3.48), the Q-point drain current is

$$I_{DQ} = \frac{172.5 - \sqrt{172.5^2 - (4 \times 10,000 \times 0.64)}}{2 \times 10,000}$$
$$= 5.4 \text{ mA}$$

Applying KVL and Ohm's law just as we did for the BJT and JFET circuits covered earlier, we find the Q-point drain-to-source voltage to be

$$V_{DSQ} = V_{DD} - I_{DQ}(R_D + R_S)$$
$$= 9V - 5.4 \text{ mA}(1 \text{ k}\Omega + 100 \text{ }\Omega)$$
$$= 3.1 \text{ V}$$

We need to determine a reasonably accurate value of g_m to allow us to predict the voltage gain of the amplifier. Recall from the previous section on transconductance that g_m for the MOSFET is found by taking the derivative of I_D with respect to V_{GS}, which is Eq. (3.43): $g_m = dI_D/dV_{GS}$. Applying this operation to (3.47) gives us

$$g_m = 2k(V_{GS} - V_T) \tag{3.49}$$

The transconductance of the MOSFET is determined using a slightly modified version of (3.49) where $V_{GS} = V_G - I_D R_S$ which gives us

$$g_m = 2k(V_G - I_D R_S - V_T)$$
$$= 0.16(2.9 \text{ V} - 0.54 \text{ V} - 2.1 \text{ V})$$
$$= 41.6 \text{ mS}$$

The equivalent AC drain resistance is

$$R_D' = R_D \parallel R_L$$
$$= 1 \text{ k}\Omega \parallel 10 \text{ k}\Omega$$
$$= 909 \text{ }\Omega$$

Because the source resistor is bypassed to ground, the voltage gain is given by the familiar equation

$$A_V = -g_m R_D'$$
$$= -41.6 \text{ mS} \times 909 \text{ }\Omega$$
$$= -37.8$$

The input resistance of the amplifier is

$$R_{in} = R_1 \| R_2$$
$$= 1\,M\Omega \| 470\,k\Omega$$
$$= 319.7\,k\Omega$$

It is interesting to compare the effects of emitter and source bypassing on the operation of BJT, JFET, and MOSFET amplifiers. BJT emitter bypassing results in increased voltage gain but also reduced input resistance. Usually, we want the input resistance of the amp to be as high as possible, so there is a trade-off between A_V and R_{in} in the design of the common emitter amplifier. Source bypassing increases voltage gain but has no effect on the input resistance of either the JFET or MOSFET common source amplifiers at audio frequencies.

The output voltage clipping points can be predicted using slightly modified versions of Eqs. (3.26) and (3.27), which we used with the BJT amplifier of Fig. 3.14.

$$V_{o(max)} = I_{DQ} R'_D \qquad V_{o(min)} = -V_{DSQ}$$
$$= 5.4\,mA \times 909\,\Omega \qquad = -3.1\,V$$
$$= 4.9\,V$$

Based on these calculations, we should be able to get a maximum unclipped output voltage of

$$V_{o(P-P)} = 2|V_{o(min)}|$$
$$= 2 \times 3.1\,V$$
$$= 6.2\,V_{P-P}$$

Experimental Results

Since we are operating this MOSFET at a very low drain current compared to its intended, typical operating level, I decided to test the circuit using three randomly selected BS170s to see if circuit operation is reliable and predictable. A fresh 9 V alkaline battery was used to power the circuit. The results of the theoretical analysis and experimental measurements are given below. As you can see, the circuit produced very consistent results, in good agreement with the theoretical analysis.

Parameter	Theory	Experimental measurements		
		Q_1	Q_2	Q_3
I_{DQ}	5.4 mA	5.1 mA	5.4 mA	5.2 mA
V_{DSQ}	3.1 V	3.3 V	3.1 V	3.3 V
A_V	−37.8	−29	−30	−30
$V_{o(max)}$	4.9 V	5 V	5 V	5 V
$V_{o(min)}$	−3.1 V	−3 V	−3 V	−3 V
$V_{o(P–P)}$	6.2 $V_{P–P}$	6 $V_{P–P}$	6 $V_{P–P}$	6 $V_{P–P}$

Overall, the performance of this amplifier is really quite good, especially considering that the BS170 is not designed to be used as a small-signal, class A amplifier. I will leave it up to you to perform a low-frequency response analysis on this circuit. As with the similar JFET and BJT circuits of this chapter, the high frequency response will extend well into the MHz range.

This amplifier may be used as a building block in other designs and could be substituted for any of the BJT or FET amplifiers presented in this chapter. Sometimes it's just nice to mix things up a bit, and using MOSFETs can put a unique and interesting spin on your project.

Operational Amplifiers

It can be difficult to meet all of the performance requirements of a given amplifier design using a small number of discrete transistors. The operational amplifier or *op amp* is an alternative that can be used in these cases. An op amp is an integrated circuit (IC) that approximates the characteristics of an ideal amplifier. Op amps can easily be configured to realize inverting or noninverting amplifiers, differential amplifiers, and many other functions as well. The schematic symbol for an op amp is shown in Fig. 3.31.

Internally, op amps are fairly complex devices usually consisting of a dozen or more transistors. If you are curious, a brief look at some representative internal circuitry is presented at the end of this section.

Fig. 3.31 Schematic symbol for an op amp

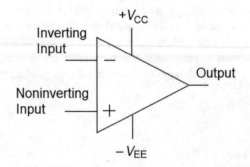

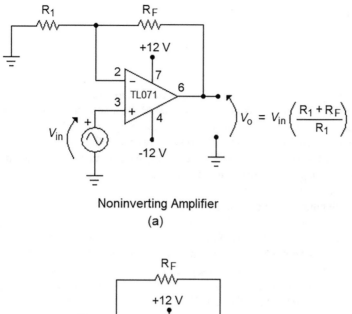

Noninverting Amplifier

(a)

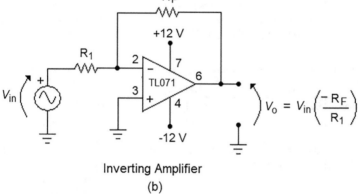

Inverting Amplifier

(b)

Fig. 3.32 Basic op amp configurations. (**a**) Noninverting. (**b**) Inverting

Most op amps are designed to operate from a bipolar power supply. Because of the advanced biasing techniques used in op amp designs, there is a lot of flexibility in the power supply voltages that may be used. Typically, op amps are designed to operate from supply voltages ranging from ± 5 V to ± 22 V. There are also op amps that are designed specifically to operate from single-polarity power supplies.

The main attraction of op amps is the ease with which they can be used. The basic inverting and noninverting configurations are shown in Fig. 3.32. The voltage gain equations for the circuits are

Noninverting Amplifier

$$A_V = \frac{R_F + R_1}{R_1} \quad \text{or} \quad A_V = 1 + \frac{R_F}{R_1} \tag{3.50}$$

Inverting Amplifier

$$A_V = \frac{-R_F}{R_1} \tag{3.51}$$

The simplicity of the voltage gain equations relative to those of the discrete BJT and FET amplifiers is obvious. We can set the gain of the amplifiers simply by choosing two resistor values, which can save a lot of design time and effort. Because the op amps are connected to symmetrical bipolar power supplies, the output terminal is automatically biased for $V_{0Q} \cong 0$ V, so coupling capacitors are not required. This is called *direct coupling*. Because these amplifiers are direct coupled, the frequency response extends down to DC.

There are hundreds of different op amps available, and many are suitable for the applications we will be discussing. Table 3.3 lists several commonly available op amps and a few of their primary specifications. We take a closer look at just a few of these parameters here. Other parameters will be discussed on an as-needed basis.

The application in which an op amp is to be used will determine which specifications are most important. For example, for high-fidelity audio applications, the TL061, the TL071, and the RC5532 are superior in frequency response

Table 3.3 Common op amp specifications (typical values unless otherwise noted)

	LM741C	LM358	TL061	TL071	RC5532
Technology	BJT	BJT	BJT	BJT	BJT
		PNP Darlington input	JFET input	JFET input	
Special notes	Dual and quad equiv. versions available	Dual (LM358)	Dual (TL062)	Dual (TL072)	Dual op amp
		Quad (LM324)	Quad (TL064)	Quad (TL074)	
		Single-supply operation to 3 V[a]	Low power		
A_{OL}	150,000	100,000	6000	200,000	100,000
R_{in}	2 MΩ	5 MΩ	10^{12} Ω	10^{12} Ω	300 kΩ
R_o	50 Ω	–	100 Ω	200 Ω	300 Ω
I_B	80 nA	40 nA	30 pA	65 pA	200 nA
I_{io}	20 nA	15 nA	5 pA	5 pA	10 nA
V_{io}	2 mV	5 mV	3 mV	3 mV	0.5 mV
GBW	1 MHz	1 MHz	1 MHz	3 MHz	10 MHz
SR	0.5 V/μs	0.5 V/μs	3.5 V/μs	13 V/μs	8 V/μs
CMRR	90 dB	85 dB	86 dB	100 dB	100 dB
SVRR	96 dB	100 dB	95 dB	100 dB	100 dB
Freq. comp.	Internal	Internal	Internal	Internal	Internal
Offset comp.	Yes	No	Yes	Yes	No
$\pm I_{o(max)}$	25 mA	50 mA	10 mA	25 mA	38 mA

[a]Input and output may be driven to ground in single-supply operation

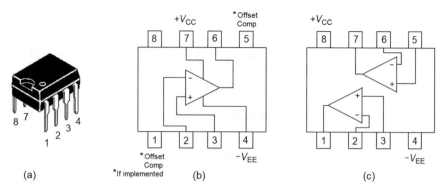

Fig. 3.33 Op amps of Table 3.3. (**a**) 8-pin DIP pin designations. (**b**) Single op amp. (**c**) Dual op amp

characteristics to the LM741 and LM358 and would be preferred. The TL061 has much lower quiescent power dissipation than the TL071, making it preferred in battery-powered circuits, although the TL071 has better high-frequency performance.

Compared to the other op amps listed, the LM358 has the advantage of being designed specifically to operate from a low-voltage, single-polarity power supply. This makes the LM358 ideal for battery-powered applications. In addition, the output of the LM358 can swing to within about 10 mV of the negative supply rail (which is ground with a single-polarity supply) and within 1.5 V of the positive supply rail. True rail-to-rail op amps are available (e.g., the OPA342) that can produce a maximum output signal swing that drives to within about 1 mV of either supply rail.

All of the op amps listed in Table 3.3 have the same pin assignments in the dual op amp package. The 741, TL061, and TL071 are also compatible in single amplifier packages. The most convenient package for our purposes is the 8-pin *DIP* (dual in-line pin). The DIP package and op amp pin designations are shown in Fig. 3.33. Quad devices are also available, such as the LM324 which is a quad LM358.

Basic Noninverting and Inverting Op Amp Equations

Recall that when negative feedback is used to reduce gain, most other amplifier parameters will improve. Resistors R_1 and R_F determine the amount of negative feedback present in the circuit which is usually represented by β (don't confuse this with the beta of the bipolar transistor). The amount of feedback can vary from $\beta = 0$ (no feedback) to $\beta = 1$ (100% feedback). The feedback factors for the op amp configurations, along with equations for some other important amplifier parameters, are given below.

Noninverting Amplifier

$$\beta = \frac{R_1}{R_1 + R_F} \tag{3.52}$$

$$R_{\text{in (F)}} = R_{\text{in}}(1 + \beta A_{\text{OL}}) \tag{3.53}$$

$$R_{o(F)} = \frac{R_o}{1 + \beta A_{\text{OL}}} \tag{3.54}$$

$$\text{BW} = \text{GBW}\left(\frac{R_1}{R_1 + R_F}\right) \tag{3.55}$$

Inverting Amplifier

$$\beta = \frac{R_1}{R_F} \tag{3.56}$$

$$R_{\text{in}(F)} \cong R_1 \tag{3.57}$$

$$R_{o(F)} = \frac{R_o}{1 + \beta A_{\text{OL}}} \tag{3.58}$$

$$\text{BW} = \text{GBW}\left(\frac{R_1}{R_1 + R_F}\right) \tag{3.59}$$

Most of the time, the op amp behaves so much like an ideal amplifier that aside from A_V, and possibly BW, we won't need to worry about calculating these parameters. However, just to help put things into perspective, let's see what effect negative feedback has in typical noninverting and inverting op amp circuits.

Noninverting and Inverting Amplifier Analysis

Figure 3.34 shows noninverting and inverting amplifiers using a 741 op amp. We will determine A_V, β, $R_{\text{in}(F)}$, $R_{o(F)}$, and bandwidth *BW* for each circuit.

Noninverting

$$A_V = \frac{R_F + R_1}{R_1}$$
$$= \frac{20\ \text{k}\Omega + 2\ \text{k}\Omega}{2\text{k}\Omega}$$
$$= 11$$

$$\beta = \frac{R_1}{R_1 + R_F}$$
$$= \frac{2\ \text{k}\Omega}{2\text{k}\Omega + 20\ \text{k}\Omega}$$
$$= 0.091$$

Inverting

$$A_V = \frac{-R_F}{R_1}$$
$$= \frac{-20\text{k}\Omega}{2\ \text{k}\Omega}$$
$$= -10$$

$$\beta = \frac{R_1}{R_F}$$
$$= \frac{2\ \text{k}\Omega}{20\text{k}\Omega}$$
$$= 0.1$$

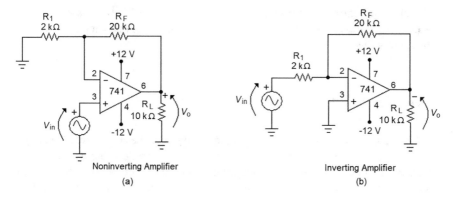

Fig. 3.34 (a) Noninverting and (b) inverting amplifiers

$R_{in(F)} = R_{in}(1 + \beta A_{OL})$

$= 2\,M\Omega(1 + (0.091 \times 150,\ 000))$

$= 27.3\,G\Omega$

$R_{o(F)} = \dfrac{R_o}{1 + \beta A_{OL}}$

$= \dfrac{50\ \Omega}{1 + (0.091 \times 150,000)}$

$= 0.0037\ \Omega$

$BW = GBW\left(\dfrac{R_1}{R_1 + R_F}\right)$

$= 1\ MHz\left(\dfrac{2\ k\Omega}{20\ k\Omega + 2\ k\Omega}\right)$

$= 90.91\ kHz$

$R_{in(F)} = R_1$

$= 2\,k\Omega$

$R_{o(F)} = \dfrac{R_o}{1 + \beta A_{OL}}$

$= \dfrac{50\ \Omega}{1 + (0.1 \times 150,000)}$

$= 0.0033\ \Omega$

$BW = GBW\left(\dfrac{R_1}{R_1 + R_F}\right)$

$= 1\ MHz\left(\dfrac{2\ k\Omega}{20\ k\Omega + 2\ k\Omega}\right)$

$= 90.91\ kHz$

One of the most impressive parameters is the input resistance of the noninverting amplifier, which is over 27 GΩ. The input resistance of the inverting configuration is approximately the same as R_1 which may occasionally be as high as 100 kΩ or so.

The output resistance of both configurations is very low, which means that changes in load resistance will not affect voltage gain. This does not mean that the op amp can effectively drive a low-resistance, high-current load such as an 8 Ω loudspeaker however. If you look at Table 3.2, you will see that, for example, the 741 is only capable of delivering ±25 mA to a load, which, with $R_L = 8\ \Omega$, would cause current limiting and output clipping to occur at $V_o = ±0.2$ V.

As a general rule of thumb, the minimum load resistance that the typical op amp can drive effectively is about 1 kΩ.

Finally, the bandwidth of both circuits is the same at 90.91 kHz. This is more than adequate for audio applications, but the situation is not quite that simple, as the next subsection will show.

Power Bandwidth

In addition to bandwidth, the slew rate of the op amp must be considered. You might remember way back at the beginning of the chapter where slew rate was defined as the maximum rate of change of the output of an amplifier. For the discrete transistor circuits we studied earlier, the slew rate was so high that we didn't need to think about it, but this may not be true for some op amps.

The slew rate and output voltage amplitude are used to determine what is often called the *power bandwidth* or *large-signal bandwidth* of the amplifier. Assuming a sinusoidal signal is being amplified, the power bandwidth is calculated using

$$BW_P = \frac{SR}{2\pi V_{o(max)}} \tag{3.60}$$

where SR is expressed in volts/second and $V_{o(max)}$ is the maximum peak sinusoidal output voltage the op amp is expected to produce. This equation tells us the frequency at which a sinusoidal output signal will just begin to exceed the slew rate of the op amp, when producing maximum output voltage.

As an example, let's determine the power bandwidth for the LM741, the TL061, and the TL071 op amps with $V_{o(max)} = 10$ V.

LM741	TL061	TL071
$BW_P = \dfrac{SR}{2\pi V_{o(max)}}$	$BW_P = \dfrac{SR}{2\pi V_{o(max)}}$	$BW_P = \dfrac{SR}{2\pi V_{o(max)}}$
$= \dfrac{500{,}000 \text{ V/s}}{(2\pi)(10V)}$	$= \dfrac{3{,}500{,}000 \text{ V/s}}{(2\pi)(10 \text{ V})}$	$= \dfrac{13{,}000{,}000 \text{ V/s}}{(2\pi)(10 \text{ V})}$
$= 7.9$ kHz	$= 55.7$ kHz	$= 207$ kHz

So now we see that there are two types of bandwidth, traditional *small-signal* bandwidth (BW) and power bandwidth (BW_P), which is a *large-signal* parameter.

Exceeding the small-signal bandwidth of an amplifier with a sinusoidal input signal results in a sinusoidal output signal that is smaller than expected and delayed in phase. However—and this is important—no new signal frequency components are created.

Exceeding the power bandwidth of an amplifier will cause nonlinear distortion which creates new frequency components that were not in the original signal. Figure 3.35 illustrates the effects of exceeding these limits.

The slew distortion shown in Fig. 3.35b is extreme, but even exceeding the slew rate by a small amount will introduce some distortion. In this case, the slew

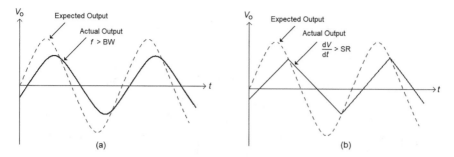

Fig. 3.35 Distortion caused by (**a**) exceeding bandwidth and (**b**) exceeding slew rate

distortion will create odd multiples (harmonics) of the fundamental frequency in the output signal. Distortion is not always a bad thing, though, and there are times when nonlinear distortion is desired, as in the case of effects circuits like fuzz boxes and overdrives. We will talk more about nonlinearity and distortion in Chap. 5.

Single-Polarity Supply Operation

For convenience, it is desirable for stomp-box type effects circuits, external pre-amplifiers, and similar devices to operate from a single 9 V battery. We've already seen that this is not a problem in capacitively coupled discrete transistor circuits. But the op amp circuits we've looked at so far require a bipolar power supply. We could connect two batteries to form a bipolar voltage source as was shown back in Fig 1.12, but a single battery solution is really what we're after.

Noninverting Amplifier

A noninverting op amp configured to work from a single 9 V battery is shown in Fig. 3.36. The LM358 is specified here because it is nearly ideal for use in single-supply circuits, though the other op amps listed in Table 3.3 may be used as well. This circuit looks intimidating, but it's not as complex as it appears. Here's how it works. Resistors R_A and R_B form a voltage divider that establishes a bias voltage equal to half of the supply voltage, in this case $V_B = V_{CC}/2 = 4.5$ V. This bias voltage is applied to the noninverting input of the op amp through R_C. The input bias current of the LM358 is typically very small (around 40 nA), so there is negligible voltage drop across R_C, which is chosen to be large to keep the input resistance of the amplifier high. Capacitor C_3 is used to reduce input noise and acts as an AC bypass around the biasing resistors, preventing the input signal from being coupled to the supply voltage.

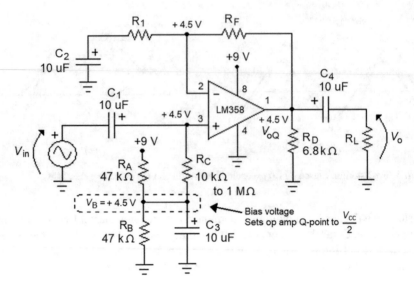

Fig. 3.36 Noninverting op amp using a single-polarity supply

Resistor R_F forms a feedback path around the op amp. At DC, coupling capacitors C_1, C_2, and C_4 act as open circuits, and the noninverting DC gain of the op amp A_V (DC) is unity. Since $A_{V(DC)} = 1$, the output of the op amp will equal the bias voltage ($V_B = +4.5$ V) at the noninverting input. The op amp is now biased so that $V_{oQ} = V_B = +4.5$ V. Essentially, we are biasing up the op amp to the center of its load line, just like we did with the class A amplifiers discussed earlier. Resistor R_D is recommended to reduce output distortion at small-signal voltages and does not otherwise affect operation of the circuit.

Under AC signal conditions, the capacitors are approximately short circuits, and the gain of the op amp is given by (3.50), which is repeated here for convenience.

$$A_V = \frac{R_F + R_1}{R_1} \quad \text{or} \quad A_V = 1 + \frac{R_F}{R_1} \tag{3.50}$$

The circuit in Fig. 3.36 is capable of producing a maximum output voltage of about 7 to 8 V_{P-P} before clipping begins to occur. The input resistance of the amplifier is approximately

$$R_{in} \cong R_C \tag{3.61}$$

The Two Golden Rules of Op Amp Analysis

In the previous discussion, a few subtle but important op amp analysis assumptions were used. These assumptions can be stated as two rules of op amp behavior that can be used to analyze a large number of op amp circuits that employ negative feedback.

Rule 1
No current flows into or out of the op amp input terminals: $I_{(+)} = 0$, $I_{(-)} = 0$

Rule 2
The op amp always forces the inverting input to the same voltage present at the noninverting input: $V_{(-)} = V_{(+)}$

The first rule is normally valid because the input resistance of the typical op amp is very high, and the input bias current is very small. The second rule is a direct consequence of negative feedback and the very high open-loop gain of the op amp. The output terminal will drive the feedback loop (and the load) until the difference between the inverting and noninverting input terminals is almost zero volts. Since the inverting terminal is forced to follow the noninverting input, we say that there is a *virtual short circuit* between these terminals. The two rules of op amp behavior apply regardless of whether single-polarity or bipolar power supplies are used.

Inverting Amplifier

An inverting op amp, operating from a single-polarity supply, is shown in Fig. 3.37. The analysis of the inverting amplifier is similar to that of the noninverting amp. As

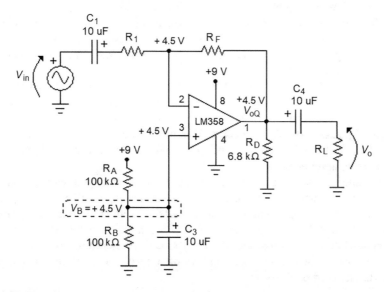

Fig. 3.37 Inverting op amp using a single-polarity supply

before, R_A, R_B, and C_3 establish a low noise bias voltage of 4.5 V at the noninverting input. The virtual short circuit behavior of the op amp forces the output terminal to 4.5 V. Because of Rule 1 and the fact that C_1 blocks DC, no current flows through R_F, and the inverting terminal is held constant at 4.5 V. Total current drain from the battery is about 1.2 mA, which gives a life of about 17 h for a 9 volt alkaline battery.

When an AC signal is applied, it is coupled through C_1 causing a time-varying current to flow through R_1. In order for the op amp to maintain 4.5 V at the inverting input (Rule 2), the output voltage changes such that all current flowing through R_1 is forced to flow through R_F. We can apply Ohm's law and Kirchhoff's laws to show that the voltage gain is once again given by (3.51) which is repeated here.

$$A_V = \frac{-R_F}{R_1} \tag{3.51}$$

The advantages of the single-supply circuit are that only one battery is required and we do not need a DPST power switch (see Fig. 1.12). The main disadvantage of this circuit is increased complexity. Another possible disadvantage is that low-frequency response no longer extends to 0 Hz, but this is normally not a concern in most effect circuit applications.

Parasitic Oscillation

It is tempting to try to use the highest-performance op amp you can get in a given circuit. Most of the time, this approach will work in your favor. However, lower-bandwidth/slew rate op amps like the LM741 and LM358 are less susceptible to instability and undesired parasitic oscillation caused by less than ideal circuit layout and poor power supply decoupling. These problems frequently occur when you are breadboarding and testing prototype circuits and sometimes even when your PC board layout is not perfect.

Parasitic oscillation may occur at frequencies well into the tens or even hundreds of megahertz range. This oscillation is not audible but can usually be easily viewed on an oscilloscope. The oscilloscope capture in Fig. 3.38 is typical of what you might observe if parasitic oscillation occurs. This high-frequency oscillation may cause unusual circuit behavior and overheating of components, especially output transistors in power amplifiers.

If parasitic oscillation is causing problems, try cleaning up your wiring (don't use wires longer than necessary, etc.). You can also try placing 0.01 μF capacitors from the op amp power supply pins to ground, mounting them as close to the op amp as possible. If you still have high-frequency oscillation problems and you are using a high-bandwidth op amp like the TL072, you can try switching to an LM358 or similar lower-bandwidth op amp. This will often resolve the problem, at a cost of somewhat reduced performance.

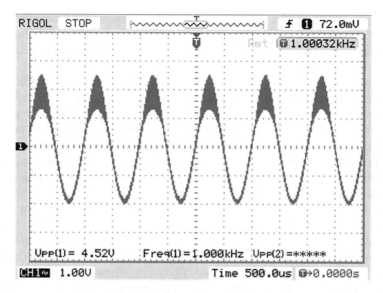

Fig. 3.38 Typical appearance of parasitic oscillation occurring on positive signal peaks

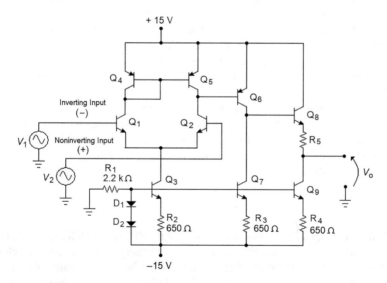

Fig. 3.39 Internal circuitry of a simple op amp

Inside the Op Amp

A schematic for a very simple op amp is shown in Fig. 3.39. The heart of the op amp is a transistor configuration called a *differential pair* (also sometimes called a *long-tailed pair*), which consists of transistors Q_1 and Q_2. Differential pairs are used in many applications, and they are even implemented using vacuum tubes. The output of the differential pair is taken from the collector of Q_2 and applied to PNP common emitter amplifier Q_6. The output is taken from the emitter of Q_8 which is a class A, emitter follower stage.

Transistors Q_3, Q_7, and Q_9 act as constant current sources that bias up the various transistor amplifier stages. Transistors Q_4 and Q_5 form a high resistance load (an *active load*) for the differential pair, giving high gain to the circuit. In addition to providing additional gain, Q_6, Q_8, and R_5 also shift the no-signal, DC output to approximately zero volts. As mentioned previously, direct coupling eliminates the need for coupling capacitors. Correct operation of the circuit requires very close matching of transistor characteristics, which fortunately is easy to achieve with integrated circuit fabrication methods.

We can deduce the operation of the differential pair by recognizing the various transistor configurations and applying a chain of cause and effect reasoning. Imagine that source V_1 is active and V_2 is off, acting as a short to ground. The input signal drives the base of Q_1. The emitter is the output terminal which means that Q_1 is an emitter follower. The emitter of transistor Q_1 in turn drives the emitter of Q_2. The collector of Q_2 is its output, so Q_2 acts as a common base amplifier. Transistor Q_2 drives PNP common emitter amplifier Q_6 which finally drives emitter follower Q_8. Thus, we see that a positive-going input applied to this input terminal results in a negative-going output, so this is the inverting input terminal.

Now, imagine that source V_2 is active and V_1 is off, acting as a short to ground. The input signal drives the base of Q_2, while its collector serves as the output terminal. This means that Q_2 now acts as a common emitter amplifier for source V_2. The remaining transistors function as described before, providing additional gain and DC level shifting.

If we activate both V_1 and V_2 at the same time, it turns out that only the difference between the two input voltages will be amplified. This is where the term *differential pair* comes from. The differential pair forms a differential amplifier. The output voltage is given by the following equation, where A_d is the differential voltage gain and $(V_2 - V_1)$ is the differential input voltage, which is often denoted as V_{id}.

$$V_o = A_d(V_2 - V_1) \qquad (3.62)$$

Ideally, if we apply the same voltage to both inputs ($V_1 = V_2$), there should be no output response, because the difference is zero. In reality, there will be some change in the output voltage because of slight mismatches between transistors and other factors. Common-mode rejection ratio (CMRR) is the parameter that tells us how well the op amp rejects common-mode inputs. High CMRR is desirable, especially in high-precision measurement and control circuits.

The op amps discussed so far in this chapter are very commonly used in all areas of electronic circuit design. There are other types of op amps that are less commonly used but deserve some mention before we leave this topic.

Operational Transconductance Amplifiers

The *operational transconductance amplifier*, or *OTA* for short, is a special type of op amp that functions as a voltage-to-current converter. One of the more commonly used OTAs is the CA3086 (or LM3086). The schematic symbol, DIP package pin designations, and equivalent internal schematic for the CA3086 are shown in Fig. 3.40. Higher-performance OTAs such as the NE5517 and LM13700 are available, but the operating principles for these devices are the same as for the CA3080.

For the most part, the internal circuitry of the OTA is very similar to that of a standard op amp, but there are some important differences. The first significant difference is that the designer must determine an appropriate OTA input differential pair bias current, called I_{ABC}. This is done by adding an external resistor from pin 5 to ground, which sets up the current mirror composed of D_1 and Q_3.

The second major difference between the CA3080 and a standard op amp is that the output of the OTA behaves like a current source rather than a voltage source. Output transistor Q_6 functions as a current source, and Q_7 is a current sink. With no input signal, Q_6 and Q_7 sink and source equally, producing no net output current. The presence of an input signal causes an imbalance between Q_6 and Q_7, resulting in a net output current. The effective output resistance of the CA3080 is in the order of $10^7\ \Omega$, which definitely qualifies as a current source output.

Examples of the CA3080 used in the noninverting, inverting, and differential amplifier configurations are shown in Fig. 3.41. The following equations apply to all three circuits. Resistor R_{ABC} sets the amplifier bias current according to (3.63). The amplifier bias current should be limited to $I_{ABC} \leq 1$ mA.

$$I_{ABC} = \frac{V_{EE} - V_{BE}}{R_{ABC}} \tag{3.63}$$

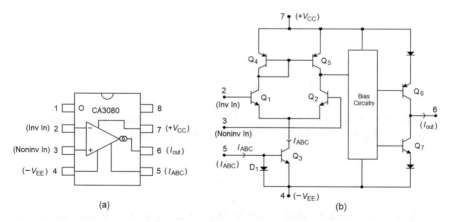

Fig. 3.40 CA3080 OTA. (**a**) DIP package pin designations. (**b**) Equivalent internal circuit

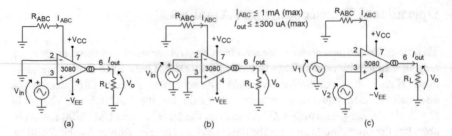

Fig. 3.41 OTA configurations. (**a**) Noninverting. (**b**) Inverting. (**c**) Differential

We usually assume $V_{BE} = 0.7$ V for silicon transistors. However, because the OTA generally operates at very low I_{ABC} (usually much less than 1 mA), perhaps a better approximation of base-emitter junction barrier potential in (3.63) would be $V_{BE} \cong 0.6$ V. The difference is really negligible for our purposes though, so we will continue to use $V_{BE} = 0.7$ V.

The input resistance of the amplifier is related to I_{ABC} through (3.64). In this equation, it's reasonable to assume that the transistors have $\beta = 100$ and $V_T = 26$ mV at normal operating temperatures.

$$R_{in} = \frac{4\beta V_T}{I_{ABC}} \tag{3.64}$$

Since the output of the CA3080 is a current and the input is a voltage (much like a JFET or MOSFET), the gain of the amplifier is of the form i_{out}/v_{in}, and the gain parameter is transconductance g_m. The transconductance of the CA3080 is given by

$$g_m = 19.2I_{ABC} \tag{3.65}$$

The output currents for each version of the OTA shown in Fig. 3.41 are given by the following equations:

$$\text{Noninverting} \quad I_{out} = g_m V_{in} \tag{3.66a}$$

$$\text{Inverting} \quad I_{out} = -g_m V_{in} \tag{3.66b}$$

$$\text{Differential} \quad I_{out} = g_m(V_2 - V_1) \tag{3.66c}$$

The output voltage for each circuit is given by direct application of Ohm's law where $V_o = I_{out}R_L$. Now, since $A_V = V_o/V_{in}$, we find that the voltage gain of the OTA is

$$A_V = g_m R_L \tag{3.67}$$

The CA3080 operates over a supply range of ± 4 V to ± 18 V and can drive its output to within about 1.5 V of either supply rail.

The output current is limited to a maximum of $I_{out(max)} = \pm300\,\mu A$. These limits must be kept in mind when a load is connected to the OTA.

An OTA Analysis Example

Let's analyze the circuit shown in Fig. 3.42 to see a practical application of the blizzard of equations that were just presented.

Applying Eqs. (3.63) through (3.67), the operating parameters for Fig. 3.42 are

$$I_{ABC} = \frac{V_{EE} - V_{BE}}{R_{ABC}} = \frac{15\,V - 0.7\,V}{56\,k\Omega} = 255\,\mu A$$

$$R_{in} = \frac{4\beta V_T}{I_{ABC}} = \frac{(4)(100)(26\,mV)}{255\,\mu A} = 40.8\,k\Omega$$

$$g_m = 19.2 I_{ABC} = 19.2 \times 255\,\mu A = 4.9\ mS$$

$$A_V = g_m R_L = 4.9\,mS \times 33\,k\Omega = 161.7$$

$$V_o = A_V V_{in} = 161.7 \times 50\,mV_{P-P} = 8.09\ V_{P-P}$$

$$I_{out} = g_m V_{in} = 4.9\,mS \times 50\,mV_{P-P} = 245\,\mu A_{P-P}$$

The calculated output current and output voltage are within maximum limits.

We could also determine the output voltage using the output current and Ohm's law, which gives us

Fig. 3.42 Example of a noninverting OTA circuit

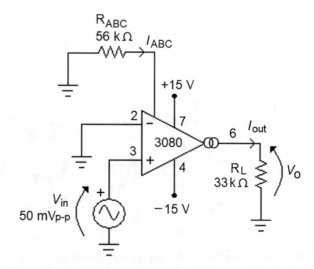

$$V_o = I_{out}R_L$$
$$= 245\ \mu A_{P-P} \times 33\ k\Omega$$
$$= 8.09\ V_{P-P}$$

In the preceding example, we did not take the frequency of the signal into account. The CA3080 has BW = 2 MHz and a very good slew rate of SR = 8 V/μs. These specs are more than sufficient to handle most audio frequency applications.

The OTA as a Voltage-Controlled Amplifier

The OTA is not nearly as convenient or easy to use as a standard op amp, and in most situations, the OTA would probably not be the device of choice. However, examination of the analysis equations shows that the gain of the OTA can be varied in direct proportion to the bias current I_{ABC}. This relationship allows the OTA to be used as a *voltage-controlled amplifier*, which is a function that is not easily achieved using any of the standard op amps covered in this chapter.

The circuit in Fig. 3.43a allows the gain of the OTA to be varied or modulated by a voltage source. Here, the modulating signal v_m causes a time-varying AC current i_{abc} to sum with the DC bias current I_{ABC}. The total instantaneous bias current is $i_{ABC} = I_{ABC} + i_{abc}$. This causes the gain of the OTA to vary directly with modulating voltage v_m. The resulting output signal is *amplitude modulated*.

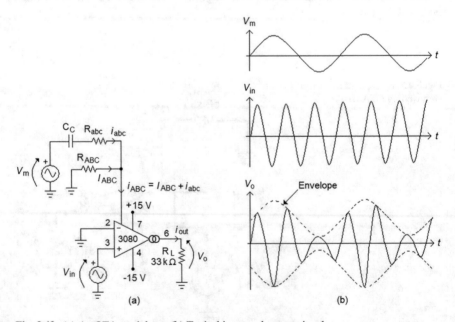

Fig. 3.43 (a) An OTA modulator. (b) Typical input and output signals

Representative input and output signals are shown in Fig. 3.43b. The peaks of the output signal vary with the modulating signal and form what is called the *envelope* of the output signal. Modulation is a nonlinear process that is used in a number of different guitar effects, as well as in radio and data communications applications. We will analyze these applications in greater detail in Chap. 5.

Current Difference Amplifiers

The final IC amplifier we will discuss is the *current difference amplifier* or *CDA* for short. These amplifiers are also sometimes called *Norton op amps* or *automotive op amps*. The most commonly encountered device is the LM3900 quad CDA. The pin designations for the LM3900 are shown in Fig. 3.44.

The equivalent internal circuitry for an LM3900 and connections for both inverting and noninverting configurations are shown in Fig. 3.45. The LM3900 is designed primarily to be used with a single-polarity power supply ranging from 4 to 32 V. The LM3900 has a respectable gain-bandwidth product of GBW = 2.5 MHz. The slew rate is asymmetrical with SR = 0.5 V/μs for positive-going outputs and SR = −20 V/μs on negative-going output swings. Output load current should be limited to about ±1 mA max. The LM3900 was used in a number of guitar effects circuits in the 1970s. Though it is now obsolete, the LM3900 is still readily available

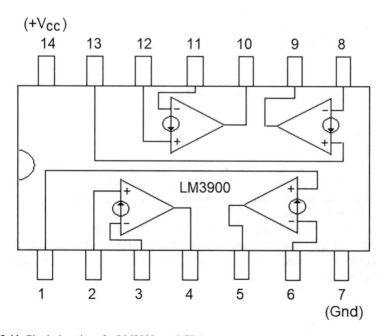

Fig. 3.44 Pin designations for LM3900 quad CDA

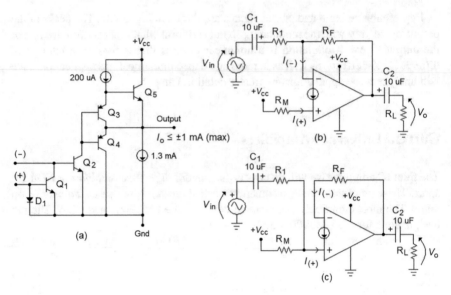

Fig. 3.45 (**a**) CDA equivalent internal circuit. (**b**) Inverting configuration. (**c**) Noninverting configuration

for about $0.50 USD from many sources, including Jameco Electronics (www.jameco.com).

In order to use the CDA, the designer must determine an appropriate value for R_M to set the bias current for input current mirror $D_1–Q_1$ at the noninverting input terminal. The mirror bias current $I_{(+)}$ is set using (3.68), which should look quite familiar by now.

$$I_{(+)} = \frac{V_{CC} - V_{BE}}{R_M} \qquad (3.68)$$

The mirror current should be limited to $I_{(+)} \leq 500\ \mu A$.

Once $I_{(+)}$ is established, the CDA output forces the current flowing into the inverting terminal to equal the current flowing into the noninverting terminal $I_{(-)} = I_{(+)}$. This is analogous to how a normal op amp with negative feedback forces the voltage at its inverting input to equal the noninverting input voltage (Op Amp Analysis Rule 2, $V_{(-)} = V_{(+)}$).

Resistor R_F is chosen to establish the desired quiescent output voltage V_{oQ}. Just as when we used a standard op amp with a single-polarity supply, the output of the CDA would normally be set for $V_{oQ} \cong V_{CC}/2$. Resistor R_F and V_{oQ} are related by the following equation:

$$V_{oQ} = I_{(+)}R_F + V_{BE} \tag{3.69}$$

For design purposes, we would solve (3.68) and (3.69) for R_M and R_F after choosing $I_{(+)}$ and V_{0Q}.

Examining the circuits in Fig. 3.45, we see that an input signal will cause variation of $I_{(-)}$ or $I_{(+)}$ for inverting and noninverting circuits, respectively. When the CDA forces the input currents to remain equal in response, an output voltage is produced.

Resistor R_1 is selected to set the voltage gain of the amplifier. The equation for the CDA voltage gain is the same for both inverting and noninverting configurations, except for a negative sign.

$$A_V = \frac{-R_F}{R_1} \quad \text{(inverting)} \tag{3.70a}$$

$$A_V = \frac{R_F}{R_1} \quad \text{(noninverting)} \tag{3.70b}$$

Lastly, the input resistance of the circuit is approximately equal to R_1 for both configurations.

$$R_{in} \cong R_1 \tag{3.71}$$

As stated previously, the LM3900 is now obsolete, but you will find these devices used in some vintage effects, so it's a useful device with which to become familiar.

Your Turn

We will finish this section with two short assignments for you. Refer to Fig. 3.45b. Assume that $V_{CC} = 9$ V, $R_M = 82$ kΩ, $R_F = 39$ kΩ, and $R_1 = 10$ kΩ. Verify that the main amplifier parameters work out to be

$$I_{(+)} = 101 \ \mu A$$
$$V_{oQ} = 4.65 \ V$$
$$A_V = -3.9$$
$$R_{in} = 10 \ k\Omega$$

Suppose we require $A_V \cong -10$, with $I_{(+)} \cong 150 \ \mu A$ and $V_{0Q} \cong 4.5$ V. Using standard 5% tolerance resistors, verify that appropriate values for R_M, R_F, and R_1 are

$$R_M = 56 \,\text{k}\Omega, \quad R_F = 27 \,\text{k}\Omega, \quad R_1 = 2.7 \,\text{k}\Omega$$

Miscellaneous Useful Circuits

Before we leave this chapter, we will take a look at a few circuits that are very useful and tie in perfectly with the material we have covered so far.

An Audio Test Oscillator

A guitar is not always the most convenient signal source, and if you don't happen to have an audio signal generator handy, the audio oscillator shown in Fig. 3.46 could be useful as a test signal source. You can use any of the op amps listed in the schematic, but the LM358 is recommended because it is designed specifically for use in single-polarity applications.

Op amp U1A and the three-stage RC network connected to its inverting input terminal make up a *phase-shift oscillator*. Because the circuit operates from a 9 V battery, U1A must be biased up to half the supply voltage by a voltage divider on the noninverting input.

When the circuit is initially powered up, wideband noise will be produced at the output of U1A. This noise is fed back through three low-pass, RC sections (R_1–R_4, C_1–C_3). Only one frequency will be phase-shifted by 180° through the three RC sections (–60° per section on average), which when applied to the inverting input results in positive feedback. The oscillation frequency is adjustable from about 200 Hz to 1 kHz.

The output of U1A will typically be around 7 to 8 $V_{P–P}$. In order to guarantee oscillation, U1A must have $A_{OL} > 29$. If the gain was just slightly greater than 29, the phase-shift oscillator would produce a very clean, low distortion sine wave output. Since U1A has $A_{OL} > 50,000$, oscillation is absolutely guaranteed.

While high gain guarantees oscillation, too much gain also causes distortion. In commercial signal generators, complex nonlinear feedback circuits are used to automatically adjust the gain of the op amp so that distortion is minimized. The approach we take in Fig. 3.46 is to simply use a second-order, low-pass filter to clean up the oscillator output signal. The corner frequency of the LP output filter works out to be about $f_C \cong 100$ Hz, which works quite well in this application, producing a reasonably clean sine wave output over all frequencies.

Unity gain op amp U1B buffers the output of the LP filter, preventing loading and distortion. The output of U1A has a DC offset of about +4.5 V, which serves as the bias voltage at the noninverting input of U1B. Output coupling cap C_7 blocks the DC output of U1B while passing the oscillator signal.

With R_9 adjusted for max amplitude, the output voltage varies from about 0.5 $V_{P–P}$ at 1 kHz to 1 $V_{P–P}$ at 200 Hz. This should be high enough for testing the circuits in this book, but if you'd like to try your hand at a bit of circuit design, you could modify buffer U1B for higher gain, say $A_V = 3$ or 4, to obtain higher output voltages.

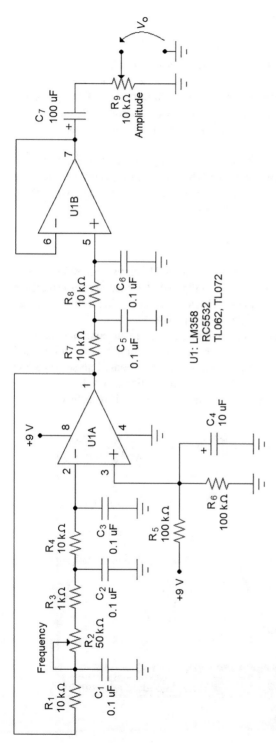

Fig. 3.46 Audio test oscillator

A Closer Look at the Oscillator Output Signal

The output waveform of the buffered and filtered oscillator is shown in the scope display of Fig. 3.47. The oscillator was adjusted for $V_0 \cong 500$ mV$_{P-P}$, at $f \cong 500$ Hz. There is very noticeable distortion of the sine wave output signal. I listened to the signal using an external amplifier/speaker, and the distortion was definitely audible (especially if you know what a good, clean sine wave sounds like), but not too objectionable. It's clean enough for general circuit testing purposes, but definitely not a high-purity sine wave.

The screen capture in Fig. 3.48 shows the frequency spectrum of the oscillator waveform in Fig. 3.47. In case you would like to look into this topic more deeply, the digital oscilloscope mathematically computes the signal spectrum using a numerical technique called the *fast Fourier transform* or *FFT*.

dBV$_{rms}$

Recall that we looked at the waveform spectra for a BJT class A amplifier back in Fig. 3.16. The spectrum looks a lot different in Fig. 3.48 because the vertical axis is scaled in dBV$_{rms}$, which means that the vertical axis is actually logarithmic, similar to the horizontal frequency axis of a filter response plot.

Log scaling allows both large- and small-amplitude signal components to be viewed simultaneously. High-amplitude components are compressed, while small signals are expanded. The price we pay for this scaling is increased noise and a less intuitive view of the harmonic amplitudes. Most newer digital oscilloscopes have FFT capability and allow the choice of either linear or log vertical scaling.

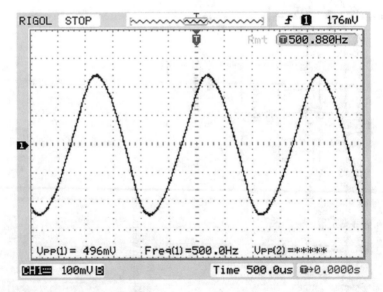

Fig. 3.47 Oscillator output waveform

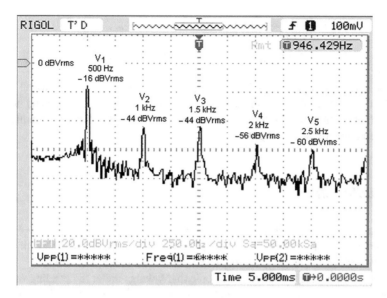

Fig. 3.48 Spectrum of the oscillator output signal

Although it may seem somewhat intimidating, all that dBV_{rms} scaling means is that the 0 dB reference level is 1 V_{rms}. Assuming that we know the rms value of some voltage V_X, we would calculate dBV_{rms} using

$$dBV_{rms} = 20 \log V_{X(rms)} \tag{3.72}$$

If you know the peak amplitude of the voltage, dBV_{rms} is calculated by factoring the conversion constant $\sqrt{2}$ into the formula as follows:

$$dBV_{rms} = 20 \log \left(\frac{V_{X(pk)}}{\sqrt{2}} \right) \tag{3.73}$$

In this formula, we are using the *common* or *base-10 log function*. Voltages that are greater than 1 V_{rms} have positive dBV_{rms} values. Voltages with amplitude less than 1 V_{rms} have negative dBV_{rms} values.

If we wish to calculate the peak amplitude $V_{X(pk)}$ of a signal component expressed in dBV_{rms}, we rearrange (3.73) to produce

$$V_{X(pk)} = \sqrt{2} \log^{-1} \left(\frac{dBV_{rms}}{20} \right) \tag{3.74}$$

where $\log^{-1}$ is the base-10 *antilog* or *inverse log* function. Let's apply (3.74) to calculate the peak amplitudes of the fundamental V_1 (500 Hz) and second harmonic V_2 (1 kHz) components shown in the signal spectrum. Scope vertical sensitivity is set to 20 dBV_{rms}/division.

$$V_1 = \sqrt{2} \ \log^{-1}(-16/20) \qquad V_2 = \sqrt{2} \ \log^{-1}(-44/20)$$
$$= 1.414 \times 0.1585 \qquad\qquad = 1.414 \times 0.00631$$
$$= 244 \ \text{mV}_{\text{pk}} \qquad\qquad\quad = 8.92 \ \text{mV}_{\text{pk}}$$

The harmonics of this signal (V_2 through V_5) are very small relative to the 500 Hz fundamental (V_1). These components would be very hard to see using a linear voltage scale that would also allow us to view the 500 Hz fundamental at a reasonable height on the display.

The Rail Splitter

A rail splitter is a circuit that is used to convert a single-polarity power supply into a bipolar supply by creating a virtual ground halfway between the positive and negative rails of the single-polarity supply, hence the name "rail splitter."

In some ways, the rail splitter is a topic that belongs under the heading of power supplies and voltage regulators, which we covered back in Chap. 1. However, a rail splitter is basically a noninverting DC amplifier with unity gain. A simple rail splitter is shown in Fig. 3.49.

The operation of the rail splitter is simple. Equal-value resistors R_1 and R_2 form a voltage divider that splits the input supply voltage in half. The gain of the op amp is unity ($A_V = 1$), so its output terminal is held exactly halfway between the upper and lower supply rails. The op amp forces this virtual ground to exist regardless of loading conditions. Resistor R_F is usually chosen to be about equal to the equivalent resistance at the noninverting input (50 kΩ) to minimize output offset voltage.

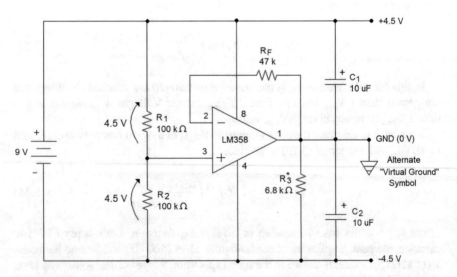

Fig. 3.49 Simple op amp rail splitter

Resistor R_3 allows a small output current to flow, which stabilizes the op amp under no-load conditions. This resistor is optional, and any value from 4.7 to 10 kΩ would be ok. Capacitors C_1 and C_2 serve to supply current during fast transient changes in load current.

Usually, the input voltage source is a battery, which is a floating source, so ground may be designated arbitrarily. Here, we are defining the output of the op amp as our ground reference. The top rail is +4.5 V with respect to virtual ground and the bottom rail is −4.5 V. Because the virtual ground is not necessarily at the same potential as earth ground, or chassis ground in a larger system, it is a good idea to use a different virtual ground symbol, so there is no confusion.

The op amp determines the maximum current source/sink capability of the ground reference. In this case, using an LM358, the maximum current is limited to about ±50 mA, which is sufficient for most low-power applications.

Special rail splitter ICs are available as well, such as the TLE2426 from Texas Instruments. The TLE2426 is a rail splitter that is available in an 8-pin DIP as well as a 3-pin TO-92 package, like the 2N3904 transistor. The TL2426 can sink or source up to 20 mA with an input up to 40 V.

A rail splitter can also be used to convert a typical single-polarity, two-prong, wall-wart-type DC supply into a bipolar supply. As long as the wall-wart is isolated from earth ground by an internal power transformer, the DC output terminals are floating and may be treated like a battery. Rail splitters are useful because some circuit designs can be drastically simplified if a bipolar power supply is available. The use of a rail splitter can eliminate the need for coupling capacitors, allowing a circuit to process DC voltages. This is an important consideration in the design of certain effects circuits.

High-Power Rail Splitter

A high-power rail splitter is shown in Fig. 3.50. Transistors Q_1 and Q_2 buffer the output of the op amp which increases the current and power handling capacity of the

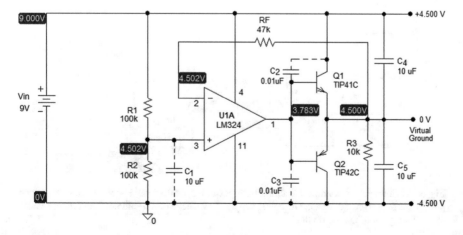

Fig. 3.50 High-power rail splitter circuit

circuit to $\pm I_{C(max)}$ and $P_{D(max)}$ for the transistors. The TIP41C and TIP42C are complementary transistors in TO-220 packages, rated for $I_{C(max)} = 6A$ and P_D $_{(max)} = 65$ W (with a *very* large heat sink). Without a heat sink, a realistic maximum power dissipation limit would be 1 or 2 W, which is still far greater than using the op amp alone.

The circuit of Fig. 3.50 was simulated in PSpice using an LM324, which is just a quad LM358. The voltages highlighted in gray are those produced by the simulator using the bottom supply rail as ground. Virtual ground referenced values are shown on the output terminals.

The addition of output transistors to the basic rail splitter has a tendency to cause the circuit to oscillate. To help prevent this oscillation from occurring, capacitors may be added as shown in 3.50. Capacitor C_1 provides a low-impedance AC path to ground for the noninverting input. Capacitors C_2 and C_3 cause the gain of the transistors to roll off at low frequency, helping to suppress oscillation. Resistor R_3 also helps stabilize the circuit by providing a light load that turns on Q_1 slightly. We will see a nearly identical version of the high-power rail splitter in Chap. 4 when power amplifiers are discussed, so we will hold off a more detailed discussion of the operation of the output transistors until then.

Use with Pedal Boards and Daisy Chain Power Cords

A word of caution is in order. Pedal boards are a nice way to organize and carry effects units and stomp boxes. Most pedal boards come with a supply module that can power many typical effects units via a daisy chain-type power cord. These cords can also be purchased separately, or you may have even built one yourself.

Connecting effects via this daisy chain method is generally not a problem, except for certain configurations of rail splitter-powered effects units. The diagram of Fig. 3.51 illustrates a situation that may result. The voltages enclosed in parentheses

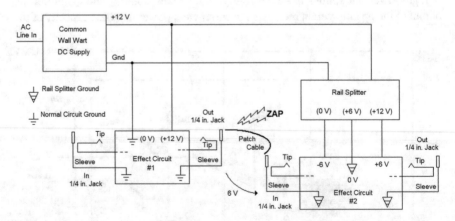

Fig. 3.51 Short circuit condition resulting from powering standard single-polarity and rail splitter-powered effects from a common DC power supply. Jack sleeve of rail splitter effect is connected to rail splitter ground. Do *not* do this!

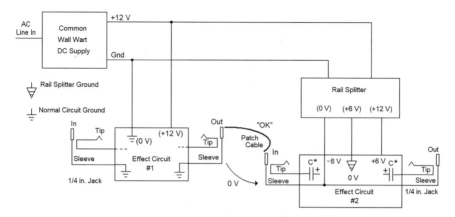

Fig. 3.52 Rail splitter in/out jack sleeve connection to solve grounding problem. Be sure to capacitively couple input and output of Effect Circuit #2

are relative to normal circuit ground as defined by the single-polarity effect circuit. Note that the virtual ground created by the rail splitter is at a potential of 6 V with respect to the normal, single-polarity ground. Connecting these two effects will short the rail splitter, preventing the effect circuit from operating properly and possibly destroying the splitter op amp. Since the sleeve terminal of the rail splitter unit output jack is connected to virtual ground like the input jack sleeve, the same problem would occur if the order of the effects units were reversed.

One possible solution to this grounding problem is to connect the sleeve terminals of the effects unit powered by the rail splitter to the negative supply rail. This is shown in Fig. 3.52. In this system, there is no potential difference between the sleeve terminals of any of the input/output jacks of the various effects units in the signal processing chain.

Because the sleeves of the input/output jacks are now connected to the negative rail of the splitter, there will be large DC offset voltages at the input and output terminals unless coupling capacitors are used. This shouldn't be a problem; simply include coupling capacitors in your design. Unless your circuit is powered by a 9 V battery, your design strategy must be well thought out to prevent possible problems when rail splitters are used. If you are not sure, it is best to avoid daisy chain power with a rail splitter-powered effect.

Charge Pumps

Charge pumps (often also called *switched-capacitor voltage converters*) provide an alternative method of deriving a bipolar power supply from a single-polarity source such as a 9 V battery. A charge pump increases and/or inverts a voltage by charging capacitors and then reconnecting the charged capacitors in series to sum to the

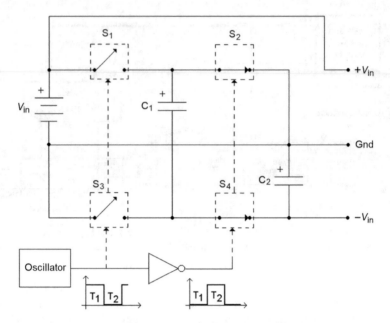

Fig. 3.53 Charge pump equivalent circuit

desired output voltage. The diagram in Fig. 3.53 shows a typical charge pump circuit. The switches are implemented with MOSFETs, which are opened and closed electronically by two antiphase clock signals. For the following description of operation, assume that $C_1 = C_2$, and the switches have negligible on-state resistance.

During time interval T_1, switches S_1 and S_3 are closed and capacitor C_1 charges to V_{in}. Switches S_2 and S_4 are open during this time interval. The clock switches states during T_2 which opens S_1 and S_3 and closes S_2 and S_4. Now, C_1 dumps part of its charge into C_2. This process repeats every clock cycle, and since the clock frequency is about 30 kHz, C_2 charges up to V_{in} very quickly. The positive terminal of capacitor C_2 is connected to ground, which is the negative terminal of V_{in} forming a series circuit. The negative terminal of C_2 forms the negative output terminal.

As long as current drain is not too high, a stable bipolar output is maintained. The larger capacitors C_1 and C_2 are, the higher the output current that can be supplied. However, capacitor charging current pulses increase in direct proportion to capacitance, which requires higher-power MOSFET switches. Commercially available charge pump ICs list maximum allowable capacitor values in their data sheets.

There are a number of commercially available switched-capacitor voltage converters, with common examples being the LTC1044, the ICL7660, and the MAX1044. The MAX1044 is a charge pump converter that is available in an 8-pin DIP, suitable for use in low-power effects circuit applications. A schematic showing the MAX1044 configured to produce a ±9 V power supply is shown in Fig. 3.54.

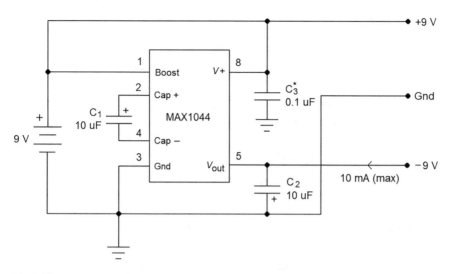

Fig. 3.54 MAX1044 configured to produce a ±9 V bipolar power supply

At a maximum negative supply rail current of 10 mA, the negative output voltage will change to about –8.5 V. The oscillator frequency of the MAX1044 is approximately 30 kHz with the Boost pin connected to the positive supply voltage as shown. Capacitor C_3 is a bypass capacitor that helps to reduce clock noise at the output of the converter. The value of C_3 is not critical and may range from 0.1 to 10 µF.

Charge Pump vs. Rail Splitter

There are two main advantages of the charge pump over the rail splitter. First, we don't have to worry about grounding problems when daisy-chaining effects circuits from a common DC supply. This is because the negative terminal of the input supply (which is normally ground) is also ground for the output of the charge pump. The second advantage is that we are effectively doubling the input DC supply voltage (without using a transformer!), so we have a lot of voltage headroom to work with. The rail splitter divides the input supply voltage between positive and negative output rails which limits output voltage headroom for the devices it powers.

The rail splitter has two advantages over the charge pump supply. First, the typical op amp can supply greater load current than a typical charge pump like the MAX1044. The LM358, for example, can supply up to ±50 mA. The second advantage of the rail splitter is that it is a pure DC circuit. There is no clock that can potentially introduce noise into the signal chain. Having used the MAX1044 in many designs, I recommend it over the rail splitter whenever possible. It's a great chip.

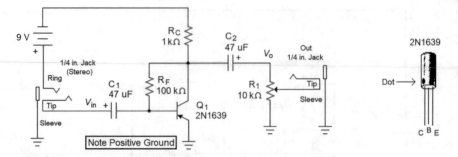

Fig. 3.55 Simple class A amplifier using a 2N1639 PNP germanium transistor

Class A, Collector Feedback, Germanium Transistor Amplifier

We will finish this chapter with the examination of the class A amplifier shown in Fig. 3.55. As shown, this circuit could be used as a preamplifier or overdrive. This circuit uses a biasing technique called *collector feedback* biasing. The transistor used here, the 2N1639, is a PNP germanium device. I chose the 2N1639 simply because I had a few dozen of them on hand, but just about any general purpose, PNP germanium transistor could be used successfully in this circuit.

The first thing to note about this amplifier is that the circuit has a positive ground. This was a standard practice on some vintage effects that used PNP germanium transistors. The positive ground isn't a problem if the circuit is powered from a battery or its own external DC supply. However, daisy-chaining with other circuits from a common DC supply will cause shorting and should be avoided. This problem can be resolved by moving the circuit ground to the negative supply rail. This modification is shown in the next application.

The 2N1639 is fairly typical of vintage PNP transistors. Measurements of five randomly selected 2N1639s showed a beta variation from 77 to 147, so $\beta = 100$ is a reasonable average value, which we will use in subsequent calculations.

Collector feedback biasing is an especially simple technique that can be used with any BJT, germanium or silicon. The following analysis is based on a germanium transistor making the usual approximation $V_{BE(Ge)} = 0.3$ V[3]. If a silicon transistor was used, we would simply substitute the appropriate barrier potential $V_{BE(Si)} = 0.7$ V into (3.75).

---

[3]To be technically correct, V_{CC}, V_{BE}, and V_{CE} for PNP transistors are negative values. To prevent confusion, we will use positive values for these voltages and in the analysis equations.

DC Analysis

The collector current of the transistor is found as follows:

$$I_{CQ} = \frac{V_{CC} - V_{BE}}{R_C + \dfrac{R_F}{\beta}}$$

$$= \frac{9\,V - 0.3\,V}{1\,k\Omega + \dfrac{100\,k\Omega}{100}} \qquad (3.75)$$

$$= 4.35\,mA$$

We find V_{CEQ} as usual, using Ohm's law and Kirchhoff's voltage law.

$$\begin{aligned} V_{CEQ} &= V_{CC} - I_{CQ}R_C \\ &= 9V - (4.35\,mA \times 1\,k\Omega) \\ &\cong 4.7V \end{aligned}$$

Notice that $V_{CEQ} = 4.7$ V is almost exactly half of the supply voltage ($V_{CC}/2 = 4.5$ V). This means that the Q-point of the amplifier is almost perfectly centered on the DC load line. One of the nice things about the collector feedback biasing arrangement is that it's easy to design. In order to obtain a centered Q-point, we simply use the following design equation:

$$R_F = \beta R_C \quad (\text{for centered Q-point}) \qquad (3.76)$$

The dynamic emitter resistance is found as usual.

$$\begin{aligned} r_e &= 26\,mV/I_{CQ} \\ &= 26\,mV/4.35\,mA \\ &= 6\,\Omega \end{aligned}$$

AC Analysis

The equivalent collector AC load resistance is now found to be

$$\begin{aligned} R'_C &\cong R_C \parallel R_1 \\ &= 1\,k\Omega \parallel 10\,k\Omega \\ &= 909\,\Omega \end{aligned}$$

Because this is a common emitter amplifier, the voltage gain is inverting. There is no external emitter resistance, so the voltage gain is given by

$$A_V \cong \frac{-R'_C}{r_e}$$
$$= -909/6$$
$$= -152$$

For several reasons, the input resistance of this amplifier is quite low. To understand this, we first determine the effective value of R_F as seen looking into the input of the amplifier. This is

$$R'_F = \frac{R_F}{1 + |A_V|}$$
$$= \frac{100,000}{1 + 152} \qquad (3.77)$$
$$= 654 \ \Omega$$

This effective reduction in the value of R_F seen looking into the input of the amp is a result of a phenomenon called the *Miller effect*. The approximate overall input resistance of the amplifier may now be calculated using

$$R_{in} \cong \beta r_e \ \| \ R'_F$$
$$= (100 \times 6\,\Omega) \ \| \ 654\,\Omega \qquad (3.78)$$
$$= 313\,\Omega$$

The output clipping points are calculated using (3.26) and (3.27).

$$V_{o(\max)} = I_{CQ}R'_C$$
$$= 4.35 \text{ mA} \times 909 \ \Omega$$
$$= 4 \text{ V}$$

$$V_{o(\min)} = -V_{CEQ}$$
$$= -4.7 \text{ V}$$

Experimental Results

I constructed this circuit and tested it using three different 2N1639s. The results are tabulated below.

Parameter	Theory	Experimental measurements		
		Q_1	Q_2	Q_3
I_{CQ}	4.35 mA	4.4 mA	5.2 mA	4.5 mA
V_{CEQ}	4.7 V	4.6 V	3.8 V	4.5 V
A_V	−152	−122	−134	−130
$V_{o(max)}$	4 V	5 V	5 V	5 V
$V_{o(min)}$	−4.7 V	−3 V	−3 V	−3 V

Experimental measurements

Parameter	Theory	Q_1	Q_2	Q_3
$V_{o(P-P)}$	8 V_{P-P}	6 V_{P-P}	6 V_{P-P}	6 V_{P-P}

The measured Q-point voltage and current values are reasonably close to the theoretical values. Transistor-to-transistor beta variation accounts for most of the differences. The measured voltage gain is consistently lower than predicted, but still nice and high, making this a useful amplifier.

The maximum positive output voltage swing is higher than predicted, while the negative swing is smaller. This is due to the presence of the collector feedback resistor which was not accounted for in the equations. As a practical matter, we are really just after ballpark estimates for these signal swing limits, and the predicted values are close enough for our purposes.

When driven to its output voltage limits, this amplifier produces the familiar nonlinear squashing/stretching distortion that we observed for previous BJT amplifiers. A scope capture showing this is shown in Fig. 3.56. Compare this waveform with that of Fig. 3.16. Notice that the distortion of this waveform is inverted; the squashing/stretching characteristic is flipped over. This occurs because we have reversed polarities by using a PNP transistor with a positive ground.

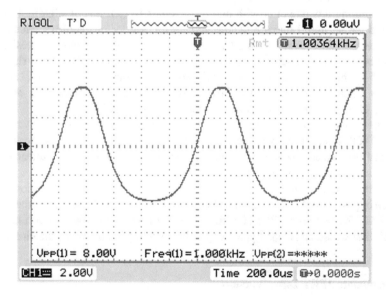

Fig. 3.56 Output of the PNP germanium, collector feedback, CE amplifier at max output

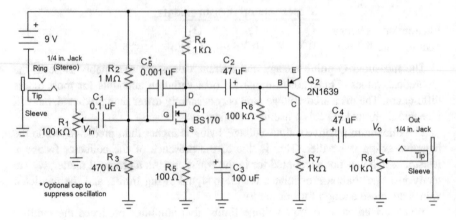

Fig. 3.57 Two-stage MOSFET/PNP germanium BJT amplifier

Practical Use of the Amplifier

The input resistance of this amplifier (313 Ω) is really too low for stand-alone use with a guitar. The main reason for this is because the loading caused by the PNP amp would severely reduce the output level and high frequency response of the guitar pickups, resulting in a very muddy sound.

Combining this amplifier with one of the other BJT, JFET, or MOSFET stages covered earlier could make a unique and useful booster or overdrive circuit. For example, we can use the MOSFET amplifier of Fig. 3.30 as a high input resistance input stage along with our PNP germanium amp as the output stage. This is shown in Fig. 3.57. The PNP section of the circuit has been redrawn upside down to accommodate a negative ground. This allows daisy-chaining power with other negative ground effects but does not change the operation of the PNP transistor.

Overall, the input resistance is $R_{in} \cong 100$ kΩ, and the voltage gain is a very high $A_V \cong 1000$. Because the voltage gain is so high, the amplifier can be driven into clipping very easily. This necessitates the use of an input-level adjustment pot. If you're into massive amounts of gain, this one is for you.

A simple modification that could be made to the circuit is the elimination of the source bypass capacitor C_3. This would reduce the overall voltage gain of the circuit to $A_V \cong 250$, which is still pretty high. Another possibility would be to include a switch that would allow the bypass cap to be switched in and out of the circuit.

An Alternate JFET Input Version

An alternate version of the amplifier using a JFET in the first stage is shown in Fig. 3.58. In this circuit, the input-level adjustment pot is replaced with a gain adjustment pot in series with the JFET source bypass capacitor. When the gain pot is set for minimum resistance, the overall gain of the amp is $A_V \cong 180$. When set for

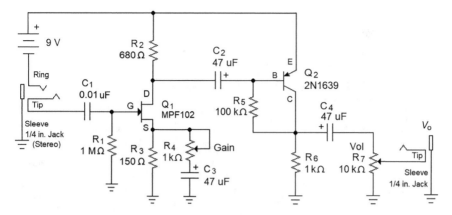

Fig. 3.58 Two-stage JFET/PNP germanium BJT amplifier

maximum resistance, overall gain drops to $A_V \cong 60$. As you can probably tell, I am a fan of the JFET.

Mixing Magnetic and Piezo Pickups

A few years ago, I was approached to design a preamp/mixer that allows simultaneous use of magnetic and piezoelectric pickups in an acoustic/electric guitar. The idea was to allow the player to continuously adjust the relative mixture of the two pickups, as well as the overall volume of the output. The resulting circuit is shown in Fig. 3.59. The circuit was designed to operate from a single 9 V battery, using the output jack as a power switch.

In this circuit, each pickup has its own preamp. Recall that piezoelectric pickups produce a smaller signal than most magnetic pickups, typically around 50 mV vs. 200 mV or more. Because of this difference in signal amplitude, the magnetic pickup preamp has a gain of $A_{V(mag)} = 3.2$, while the piezo preamp has $A_{V(piezo)} = 5.7$. These gain values worked well with the pickups used in the guitar project but may require adjustment for different pickup characteristics.

The preamp outputs are combined via 20 kΩ pot, R_{11}. As the wiper is moved toward a given preamp, that output signal is shunted progressively more to ground. Resistors R_9 and R_{10} prevent the pot from directly shorting the preamp outputs, reducing maximum current drain. At maximum rotation to either extreme, the output will consist of either the magnetic or piezo pickup signal. When the wiper of R_{11} is centered, the magnetic and piezo preamp outputs are mixed equally.

This circuit was used by a group involved in an NSF grant that uses guitar building as a vehicle for STEM education (www.guitarbuilding.org). You should check out their site!

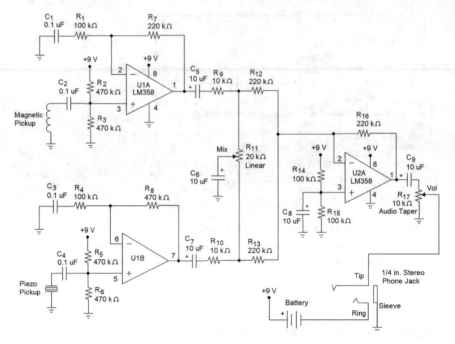

Fig. 3.59 Magnetic/piezo pickup preamp/mixer for acoustic guitar

Final Comments

Whew. This is a good time to stop and let everything sink in, take a deep breath, and maybe go back to playing the guitar for a while to unwind. I have to admit that if you are not already familiar with electronics either as a hobby, by education, or by vocation, this was probably a very intense and abstract chapter. Some very powerful (and hopefully interesting) ideas have been introduced here. You will see these concepts applied over and over again as we examine the operation of power amplifiers (transistor and vacuum tube) and effects circuits.

Don't feel discouraged if there were sections that you didn't understand. It may be partially my fault for not giving clear enough explanations, but when you come right down to it, a lot of this material comes pretty close to being rocket science. When possible, I would rather err on the side of presenting a little too much technical analysis rather than leave you wondering how something works.

This chapter was a good preparation for the upcoming topics that we will study, and nearly all of the circuits we looked at here will appear in one form or another in other applications later in the book.

Summary of Equations

Gain-Defining Equations

$$A_V = \frac{v_o}{v_{in}} \tag{3.1}$$

$$A_i = \frac{i_o}{i_{in}} \tag{3.2}$$

$$A_P = \frac{p_o}{p_{in}} \tag{3.3}$$

Voltage, Current, and Power Gain in Decibels

$$A_{V(dB)} = 20 \log \left(\frac{v_o}{v_{in}}\right) \quad \text{or} \quad A_{V(dB)} = 20 \log A_V \tag{3.4}$$

$$A_{i(dB)} = 20 \log \left(\frac{i_o}{i_{in}}\right) \quad \text{or} \quad A_{i(dB)} = 20 \log A_i \tag{3.5}$$

$$A_{P(dB)} = 10 \log \left(\frac{p_o}{p_{in}}\right) \quad \text{or} \quad A_{i(dB)} = 10 \log A_P \tag{3.6}$$

Input Voltage with Source Loading

$$v_{in} = v_s \left(\frac{R_{in}}{R_{in} + R_S}\right) \tag{3.7}$$

Amplifier Output Voltage with Loading

$$v_o = A_V v_{in} \left(\frac{R_L}{R_L + R_o}\right) \tag{3.8}$$

Amplifier Bandwidth

$$BW = f_H - f_L \tag{3.9}$$

Slew Rate (Rate of Change) of a Linear Function

$$SR \cong \frac{\Delta V_o}{\Delta t} \, (V/s) \tag{3.10}$$

Amplifier Slew Rate (Maximum Possible Rate of Change of v_o)

$$SR = \frac{dv_o}{dt}\Big|_{max} \quad (V/s) \tag{3.11}$$

Equivalent Value: Two Resistors in Parallel

$$R_{eq} = \frac{1}{\frac{1}{R_1} + \frac{1}{R_2}} \tag{3.12a}$$

or

$$R_{eq} = \frac{R_1 R_2}{R_1 + R_2} \tag{3.12b}$$

Effective External Base Resistance to Ground: Voltage Divider Bias

$$R'_B = R_1 \parallel R_2 \tag{3.13}$$

Biasing Voltage Applied to Base: Voltage Divider Bias

$$V_{Th} = V_{CC}\left(\frac{R_2}{R_1 + R_2}\right) \tag{3.14}$$

Q-Point Collector Current and V_{CE}: Voltage Divider Bias

$$I_{CQ} = \frac{V_{Th} - V_{BE}}{\frac{R'_B}{\beta} + R_E} \tag{3.15}$$

$$I_{CQ} \cong \frac{V_{Th} - V_{BE}}{R_E} \tag{3.19}$$

$$V_{CEQ} = V_{CC} - I_{CQ}(R_C + R_E) \tag{3.16}$$

Emitter Dynamic Resistance (Internal to BJT)

$$r_e = \frac{26 \text{ mV}}{I_{CQ}} \tag{3.17}$$

Stable BJT Q-Point Criteria

$$\beta R_E \gg R'_B \tag{3.18}$$

Typical Beta for General-Purpose BJT

$$\beta \cong 100 \tag{3.20}$$

Forward Voltage Drop Across Silicon PN Junction

$$V_{BE} \cong 0.7\,\text{V} \tag{3.21}$$

AC External Collector Resistance

$$R'_C = R_C \parallel R_L \tag{3.22}$$

Common Emitter Input Resistance

$$R_{in} = R'_B \parallel \beta\left(r_e + R'_E\right) \tag{3.23}$$

Common Emitter Output Resistance

$$R_o = R_C \tag{3.24}$$

Common Emitter Voltage Gain

$$A_V = \frac{-R'_C}{r_e + R'_E} \tag{3.25}$$

Single-Stage, Class A, BJT Positive Output Clipping Voltage

$$V_{o(\max)} = I_{CQ}R'_C \tag{3.26}$$

Single-Stage, Class A, BJT Negative Output Clipping Voltage

$$V_{o(\min)} = -V_{CEQ} \tag{3.27}$$

Corner Frequency of Input HP Network

$$f_{C(in)} \cong \frac{1}{2\pi R_{in} C_1} \tag{3.28}$$

Corner Frequency of Output HP Network

$$f_{C(out)} \cong \frac{1}{2\pi R'_C C_2} \tag{3.29}$$

Corner Frequency of Emitter Bypass HP Network

$$f_{C(E)} \cong \frac{1}{2\pi r_e C_3} \tag{3.30}$$

AC Equivalent External Emitter Resistance (Partial Bypass)

$$R'_E = R_E \parallel R_X \tag{3.31}$$

JFET Self-Bias Source Resistor (for $I_{DQ} \cong I_{DSS}/2$)

$$R_S \cong 1/g_{m0} \tag{3.32}$$

JFET Transconductance ($I_{DQ} = I_{DSS}/2$)

$$g_m \cong 0.75 g_{m0} \tag{3.33}$$

Voltage Gain of Source-Bypassed CS Amplifier

$$A_V = -g_m R'_D \tag{3.34}$$

AC Equivalent External Drain Resistance

$$R'_D = R_D \parallel R_L \tag{3.35}$$

JFET Input Resistance

$$R_{in} = R_G \tag{3.36}$$

JFET Output Resistance

$$R_o = R_D \tag{3.37}$$

JFET Q-Point Drain-to-Source Voltage

$$V_{DSQ} = V_{DD} - I_D(R_D + R_S) \tag{3.38}$$

Corner Frequency of JFET Input HP Network

$$f_{C(in)} = \frac{1}{2\pi R_G C_S} \tag{3.39}$$

Corner Frequency of JFET Output HP Network

$$f_{C(out)} = \frac{1}{2\pi R'_D C_2} \tag{3.40}$$

Corner Frequency of Source Bypass HP Network

$$f_{C(S)} = \frac{1}{(2\pi)\left(\frac{1}{g_m} \parallel R_S\right) C_3} \tag{3.41}$$

Overall Cascaded Voltage Gain

$$A_V = A_{v(1)} A_{v(2)} \tag{3.42}$$

Definition of Transconductance

$$g_m = \left. \frac{dI_D}{dV_{GS}} \right|_{V_{DS}=\text{const.}} \tag{3.43}$$

Approximate Transconductance Calculation

$$g_m \cong \frac{\Delta I}{\Delta V} \tag{3.44}$$

Device Transconductance Equations

$$\text{BJT}: \ I_C = I_S e^{V_{BE}/\eta V_T} \tag{3.45}$$

$$\text{JFET}: \ I_D = I_{DSS}\left(1 - \frac{V_{GS}}{-V_P}\right)^2 \tag{3.46}$$

$$\text{MOSFET}: \ I_D = k(V_{GS} - V_T)^2 \tag{3.47}$$

Quadratic Formula for MOSFET Drain Current

$$I_{DQ} = \frac{-B - \sqrt{B^2 - 4AC}}{2A} \tag{3.48}$$

$$A = R_S^2, \ B = -\left(2R_S(V_G - V_T) + \frac{1}{k}\right), \ C = (V_G - V_T)^2$$

MOSFET Transconductance

$$g_m = 2k(V_{GS} - V_T) \tag{3.49}$$

Noninverting Op Amp Equations

$$A_V = \frac{R_F + R_1}{R_1} \quad \text{or} \quad A_V = 1 + \frac{R_F}{R_1} \tag{3.50}$$

$$\beta = \frac{R_1}{R_1 + R_F} \tag{3.52}$$

$$R_{in(F)} = R_{in}(1 + \beta A_{OL}) \tag{3.53}$$

$$R_{o(F)} = \frac{R_o}{1 + \beta A_{OL}} \tag{3.54}$$

$$\text{BW} = \text{GBW}\left(\frac{R_1}{R_1 + R_F}\right) \tag{3.55}$$

Inverting Op Amp Equations

$$A_V = \frac{-R_F}{R_1} \tag{3.51}$$

$$\beta = \frac{R_1}{R_F} \tag{3.56}$$

$$R_{in\ (F)} = R_1 \tag{3.57}$$

$$R_{o(F)} = \frac{R_o}{1 + \beta A_{OL}} \tag{3.58}$$

$$BW = GBW \left(\frac{R_1}{R_1 + R_F} \right) \tag{3.59}$$

Power Bandwidth (SR in Volts/Second)

$$BW_P = \frac{SR}{2\pi V_{o(max)}} \tag{3.60}$$

Input Resistance of Fig. 3.36

$$R_{in} \cong R_C \tag{3.61}$$

Differential Amplifier Output Voltage Equation

$$V_o = A_d(V_2 - V_1) \tag{3.62}$$

CA3080 OTA Equations

$$I_{ABC} = \frac{V_{EE} - V_{BE}}{R_{ABC}} \quad \text{(current mirror)} \tag{3.63}$$

$$R_{in} = \frac{4\beta V_T}{I_{ABC}} \tag{3.64}$$

$$g_m = 19.2 I_{ABC} \tag{3.65}$$

$$\text{Noninverting} \quad I_{out} = g_m V_{in} \tag{3.66a}$$

$$\text{Inverting} \quad I_{out} = -g_m V_{in} \tag{3.66b}$$

$$\text{Differential} \quad I_{out} = g_m(V_2 - V_1) \tag{3.66c}$$

$$A_V = g_m R_L \tag{3.67}$$

LM3900 Current Difference Amplifier Equations

$$I_{(+)} = \frac{V_{CC} - V_{BE}}{R_M} \quad \text{(current mirror)} \tag{3.68}$$

$$V_{oQ} = I_{(+)}R_F + V_{BE} \tag{3.69}$$

$$A_V = \frac{-R_F}{R_1} \quad \text{(inverting)} \tag{3.70a}$$

$$A_V = \frac{R_F}{R_1} \quad \text{(noninverting)} \tag{3.70b}$$

$$R_{in} = R_1 \tag{3.71}$$

Decibel Voltage Gain Conversions

$$dBV_{rms} = 20 \log V_{X(rms)} \tag{3.72}$$

$$dBV_{rms} = 20 \log \left(\frac{V_{X(pk)}}{\sqrt{2}} \right) \tag{3.73}$$

$$V_{X(pk)} = \sqrt{2} \log^{-1} \left(\frac{dBV_{rms}}{20} \right) \tag{3.74}$$

Collector Feedback Biasing

$$I_{CQ} = \frac{V_{CC} - V_{BE}}{R_C + \frac{R_F}{\beta}} \tag{3.75}$$

$$R_F = \beta R_C \tag{3.76}$$

$$R'_F = \frac{R_F}{1 + |A_V|} \tag{3.77}$$

$$R_{in} \cong \beta r_e \parallel R'_F \tag{3.78}$$

Chapter 4
Solid-State Power Amplifiers

Introduction

In this chapter, we will examine the basic operation of some common power amplifier circuits. Recall that the capacitively coupled, class A amplifiers covered in the last chapter have a maximum theoretical efficiency of 25%. This means that if you were to design a perfect class A amplifier that delivered 10 W to a load, the amplifier itself would dissipate 30 W. Nearly all of this power would be dissipated by the output transistor(s), which would require a large heat sink. Recall also that most of the discrete transistor amplifiers examined in Chap. 3 were common emitter (CE) amplifiers. The output resistance of the CE tends to be high—typically thousands to tens of thousands of ohms. This makes the CE amplifier unsuited for driving a low-resistance load, like a loudspeaker, which will typically have a resistance of 8 Ω or less.

Transistor power amplifiers are almost always designed using the emitter follower (or source follower for MOSFETs) configuration. This is done because the emitter follower can easily be designed to have very low output resistance, which is good for driving high-current, low-resistance loads. Additionally, class B and AB biasing are usually used in order to obtain higher efficiency.

The Basic Push-Pull Stage

Most transistor power amplifiers use a push-pull output topology. The circuit shown in Fig. 4.1 has a very long name; it is a *class B, push-pull, complementary-symmetry, emitter follower*. Most people just call it a push-pull amplifier. Here's how the amplifier works. Assuming that the circuit has been powered up and has reached equilibrium, coupling capacitors C_1 and C_2 will have charged to half of the supply voltage. Under no-signal conditions, the base and emitter terminals of both

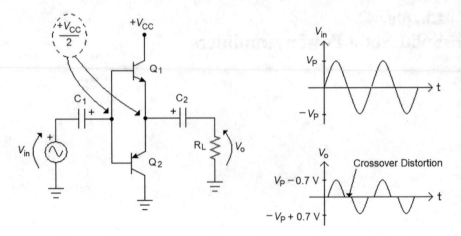

Fig. 4.1 Simple push-pull amplifier stage and input/output waveforms

transistors are sitting at $V_{CC}/2$ volts. Since the B-E junctions of Q_1 and Q_2 both have $V_{BE} = 0$ V, the transistors are in cutoff, which is class B operation.

When v_{in} goes positive, the signal is coupled to the bases of the transistors through C_1. Once the input voltage reaches about 0.7 V, the B-E junction of NPN transistor Q_1 will become forward biased, turning on the transistor. The B-E junction of PNP transistor Q_2 is reverse biased, and it remains in cutoff, continuing to act as an open circuit.

With Q_1 turned on, capacitor C_2 couples the output signal to the load. Since Q_1 is an emitter follower, the voltage gain is always less than one, and the output signal will be somewhat smaller in amplitude than the original input signal. Also, because v_{in} has to overcome the barrier potential of Q_1, there is an additional loss of about 0.7 V at the output.

When the input signal goes negative by about 0.7 V, PNP transistor Q_2 is turned on, and Q_1 returns to cutoff. During this negative half-cycle, since Q_1 is in cutoff, the power supply is isolated from the output, and the output coupling capacitor now supplies stored energy to the load. The PNP transistor is also acting as an emitter follower, so the voltage gain is less than one, and because class B biasing is used, 0.7 V is lost on negative output voltage peaks as well.

Because the input signal must overcome the B-E junction barrier potentials, there is a flat spot where the output signal would normally be crossing zero volts. This is called *crossover distortion*, which is shown on the output waveform of Fig. 4.1. Crossover distortion is most severe at low signal levels and, in my opinion, really sounds terrible.

Class AB: Eliminating Crossover Distortion

Because of its high crossover distortion, the circuit in Fig. 4.1 is not generally suitable for use in audio applications. In order to reduce or even completely eliminate crossover distortion, we must bias the transistors on slightly. This is class AB biasing. A commonly used class AB, push-pull amplifier stage is shown in Fig. 4.2.

Assuming that the base currents of Q_1 and Q_2 are negligible, the circuit works as follows. Current I_1 flows down through the biasing network. The voltage drops across D_1 and D_2 are large enough to bias Q_1 and Q_2 on just slightly. As before, when v_{in} goes positive, transistor Q_1 turns on harder, and Q_2 is driven into cutoff. On negative signal swings, Q_2 turns on harder and Q_1 turns off. The forward voltage drops across the diodes cancel out the transistor B-E barrier potentials, eliminating crossover distortion.

Resistors R_1 and R_2 are chosen to set I_1 to a few milliamps or so. Ideally, we would like to use perfectly matched pairs for D_1-Q_1 and D_2-Q_2. This would allow us to use the simple current mirror equation (see Fig. 3.40, Eq. (3.63), or Fig. 3.35, Eq. (3.68), to review this) to set $I_{CQ} \cong I_1$.

Close matching is not practical using discrete devices, so we will take the brute-force approach of using relatively large resistor values for R_1 and R_2 (around 10 kΩ) to keep idling current I_{CQ} down to a few mA which is sufficient to eliminate crossover distortion.

The voltage gain of the circuit is

$$A_V = \frac{R_L}{R_L + r'_e} \cong 1 \tag{4.1}$$

where r'_e is the *average* internal resistance of the emitter. For typical power transistors, r'_e is usually around 0.1 Ω or less. For low-power transistors like the 2N3904 or 2N2222, r'_e is usually about 5 Ω. In practice, the voltage gain of the class AB, push-pull, emitter follower is often approximated as being unity ($A_V \cong 1$) as indicated in (4.1).

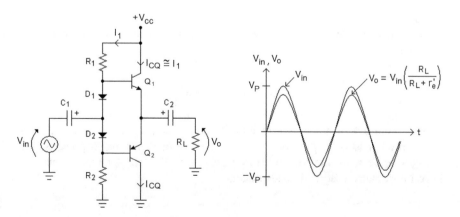

Fig. 4.2 Class AB, push-pull amplifier with input/output waveforms

The output resistance of the push-pull amplifier is quite low and can be approximated by the equation

$$R_o \cong r'_e \tag{4.2}$$

Disregarding the effect of biasing components R_1, R_2, D_1, and D_2, the input resistance of the push-pull stage is approximately

$$R_{in} \cong \beta(R_L + r'_e) \tag{4.3}$$

Output Power Determination

A power amplifier must be able to deliver high current to a low-impedance load; therefore, power transistors are used to implement the push-pull pair. The 2N3055 and MJ2955 (see Table 3.1) are inexpensive complementary devices that are very commonly used in push-pull designs. Assuming the transistors act like perfect short circuits when saturated, the maximum load current for the output stages shown in Figs. 4.1 and 4.2 is given by

$$I_{L(max)} \cong \frac{V_{CC}}{2R_L} \tag{4.4}$$

Ideally, the peak power that can be delivered to a load is given by

$$P_{o(pk)} \cong \frac{V_{CC}^2}{4R_L} \tag{4.5}$$

We usually prefer to express amplifier output power in rms (root-mean-squared) form. RMS power is the equivalent power that would be delivered by a DC source to cause the same amount of heating in a load.

For a sinusoidal signal, the rms power is simply half of the peak power, so all we need to do is divide (4.5) by two, and we have the maximum rms output power equation

$$P_{o(rms)} = \frac{P_{o(pk)}}{2} \cong \frac{V_{CC}^2}{8R_L} \tag{4.6}$$

Maximum output power is a function of power supply voltage and load resistance. In order to get high output power levels, we must use higher power supply voltages or decrease the load resistance. In practical terms, 4 ohms is about the lowest impedance used with loudspeakers.

Also keep in mind that these equations assume that the peak-to-peak output of the amplifier can reach the supply rails. In practice, the maximum output voltage swing (*output compliance*) may fall several volts short of each supply rail. If you are using supply voltages of 25 V or more, the error is not too significant. However, if you are only using say a 10 V supply, the power output calculation is a bit optimistic.

Bipolar Power Supply Operation

Recall from Chap. 3 that amplifier coupling capacitors form high-pass filters. Because they are used to drive low-resistance loads, power amplifiers that operate from a single-polarity power supply require large output coupling capacitors. For example, if we are driving an 8 Ω speaker and we wish response to extend down to 50 Hz, the required coupling capacitor works out to be about 400 μF (using $C = 1/2\pi f_C R$).

We can eliminate the need for coupling capacitors by using a bipolar power supply to power the amplifier. This is shown in Fig. 4.3. Because the circuit is symmetrical, the upper and lower halves of the circuit should split the power supply voltage evenly, resulting in the input and output terminals being at 0 V with respect to ground. The power supply voltages will typically range from about ±12 V to ±50 V or more.

The voltage gain, input resistance, and output resistance equations for this amplifier are the same as those for the single-polarity power supply version. However, the output voltage, current, and power equations must be modified slightly.

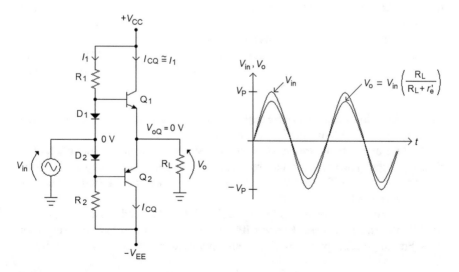

Fig. 4.3 Push-pull stage with bipolar power supply

$$I_{L(max)} \cong \frac{V_{CC}}{R_L} \tag{4.7}$$

$$P_{o(pk)} \cong \frac{V_{CC}^2}{R_L} \tag{4.8}$$

$$P_{o(rms)} \cong \frac{V_{CC}^2}{2R_L} \tag{4.9}$$

These are just the single-polarity supply equations, scaled appropriately because the bipolar power supply gives us twice as much voltage to work with.

Power Transistors

The output stage transistors of a power amplifier are likely to be subjected to voltages, currents, and power dissipation levels that are near their design limits. The transistors used in this application are classified as power transistors. In general, any transistor that is in a TO-3, TO-220, or similar package (see Fig. 3.12) is considered to be a power transistor. Of course, if you are in doubt, simply look up the data sheet, which will indicate the intended purpose of the device.

In order to handle high current and high power dissipation, power transistors have much greater silicon die area than general-purpose devices. For example, according to the Central Semiconductor Corp., Process CP235 Data Sheet, the 2N3055 power transistor has a silicon die (or chip) area of about 0.0112 in^2, while the die area for the 2N3904 is about 0.000221 in^2 (Process CP192V Data Sheet). Figure 4.4 illustrates the relative size of these transistor chips.

The 2N3904 is an epitaxial planar transistor, while the 2N3055 is a glass-passivated mesa transistor. These terms describe the physical geometry and construction of the transistor chips. The base and emitter regions of the chips are *interdigitated* with contacts on the top of the chip. The collector is on the bottom side of the chip, which is electrically connected to the case. This information is certainly not critical to us, but it is interesting.

Because power transistors have such large surface area, they also have proportionately higher junction capacitances. This is one of the reasons that power transistors generally have slower switching speeds and lower cutoff frequencies than smaller transistors.

Power transistors are typically implemented as BJTs and enhancement-mode MOSFETs. There are special power JFETs available, but these are not very common. For all practical purposes, JFETs are only used in low-power applications.

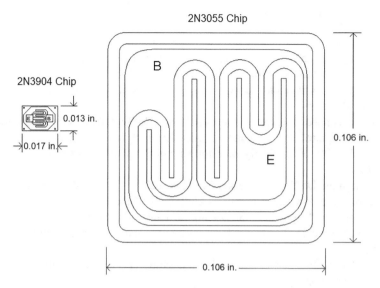

Fig. 4.4 Relative size of 2N3904 and 2N3055 silicon dice (chips)

Although this is an oversimplification, power BJTs are approximately just scaled-up versions of typical low-power transistors. This is not true of power MOSFETs. Power MOSFETs are typically fabricated from hundreds to several thousand small MOSFET cells, fabricated on the same silicon chip, operating in parallel. This technique results in lower on-state resistance than a single large area device.

Another class of transistor that you may have heard of is the insulated gate bipolar transistor, or IGBT. IGBTs are power transistors that are a hybrid of the BJT and the enhancement-mode MOSFET. IGBTs are available in both N- and P-channel versions. Although it is possible to design IGBT audio amplifiers, they are not well suited for linear operation.

Composite Transistors

We can connect two or more transistors together to increase current or power handling capability or to increase gain. The resulting compound device is normally treated as a single transistor. We will look at two such devices here: the Darlington and Sziklai transistors.

Darlington Transistors

The Darlington transistor is a composite BJT, either NPN or PNP, made from two transistors connected as shown in Fig. 4.5. Note that the PNP Darlington has been drawn with its emitter on the top. This is the way it will normally be found in schematic diagrams.

You can connect discrete transistors to form a Darlington, but they are commonly available integrated into a standard three-terminal transistor package (2N2646, TIP102, TIP106, etc.). Nearly all commercially available Darlingtons are power transistors.

Although the Darlington is made from two BJTs, it is treated as a single device. The output-side transistor Q_B carries almost all of the collector current and will dissipate most of the power in the Darlington, so it will typically be a large geometry power transistor chip. The input-side transistor Q_A may be a smaller chip, similar to the 2N3904.

The effective B-E barrier potential of a Darlington is approximately twice that of a normal transistor.

$$V_{BE(D)} \cong 1.4 \text{ V}$$

The total effective beta of the Darlington transistor is approximately

$$\beta_D \cong \beta_A \beta_B$$

A Darlington transistor constructed from typical BJTs, each with $\beta = 100$, would have total beta $\beta_D = 10,000$. Referring back to Table 3.2, we find that for the TIP102 and TIP106 Darlingtons, beta may vary from 1000 to 20,000. The extremely high beta of the Darlington results in very high input resistance and low base bias current. Both are very desirable in most amplifier applications.

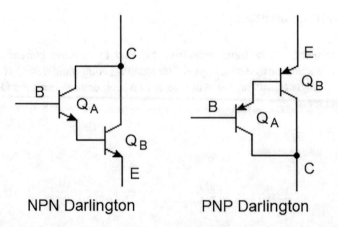

NPN Darlington   PNP Darlington

Fig. 4.5 Complementary NPN and PNP Darlington transistors

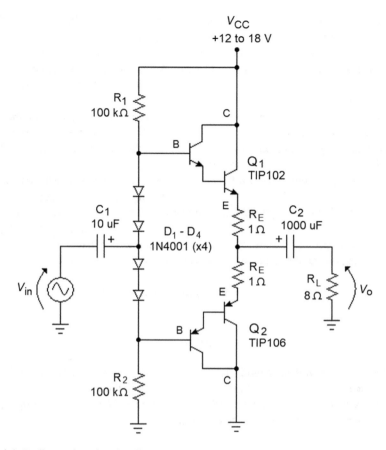

Fig. 4.6 Darlington-based push-pull power output stage

A push-pull stage using complementary Darlington transistors is shown in Fig. 4.6. Transistor Q_1 is an NPN Darlington and Q_2 is PNP. Because the Darlingtons have $V_{BE(D)} \cong 1.4$ V, a total of four diodes are required in the biasing circuit to eliminate crossover distortion.

As you probably expect by now, resistors R_1 and R_2 are equal in value, setting the idling current of the circuit to a value in the low-milliamp range. Because the Darlingtons have very high beta, R_1 and R_2 may range from around 50 kΩ to the megohm range. This helps to keep the input resistance of the stage very high. Typical input resistance with an 8 Ω load would be around 25–50 kΩ.

While Darlington transistors have the nice features of high beta and high input resistance, they also have some less than desirable characteristics. Darlingtons have slow switching speed and very poor high-frequency characteristics—much worse than standard power BJTs. This is a major concern in high-fidelity audio and high-speed switching applications, but is not a problem in most of the applications we will be studying.

Darlingtons also have much higher temperature sensitivity than normal BJTs. Low-value emitter resistors, called *swamping resistors*, are included in Fig. 4.6 to increase the temperature stability of the circuit. Emitter swamping resistors can also be used in standard push-pull stages like Fig. 4.3 to increase thermal stability and increase input resistance, at the expense of decreased voltage gain and output compliance.

Thermally coupling the biasing diodes to the Darlingtons, by mounting them on the same heat sink, also improves thermal stability. This way, as the Darlingtons heat up, the biasing diodes heat up as well. This reduces the bias voltage, which tends to cancel the effect of the temperature increase.

In general, push-pull emitter followers are very susceptible to parasitic oscillation (see Fig. 3.38), but this is even more likely in Darlington push-pull stages. Because of this, circuit layout and power supply bypassing is even more critical than usual.

Finally, although we are primarily concerned with power amps in this chapter, you can use a Darlington transistor in a class A amplifier like that of Fig. 3.14. Using a Darlington, the input resistance and voltage gain equations are modified as follows:

$$R_{in} = R'_B \parallel \beta_D(2r_e + R'_E)$$

$$A_V = -R_C/(2r_e + R'_E)$$

The base bias voltage would also have to be increased to compensate for the higher barrier potential of the Darlington. The values of the voltage divider resistors are usually scaled up by a factor of 10 or more because the high value of β_D makes the base terminal a very high resistance point.

Sziklai Transistors

NPN and PNP equivalent *Sziklai* pairs are shown in Fig. 4.7. Here's how the NPN Sziklai works. A positive voltage applied to the base terminal forward biases the B-E junction of NPN input transistor Q_A. In turn, the collector of the NPN pulls current from the base of the PNP output transistor Q_B, forward biasing its B-E junction. The emitter of the PNP output transistor behaves like the collector of a normal transistor. Effectively, the small NPN input transistor forces the PNP power transistor to behave like an NPN power transistor.

The PNP Sziklai works in a similar manner. Note that the PNP Sziklai transistor is drawn with its emitter on top, which is standard drafting convention.

Unlike Darlingtons, Sziklai transistors have the same B-E barrier potential (0.7 V) as a single transistor. The effective beta of the Sziklai transistor β_S is approximately

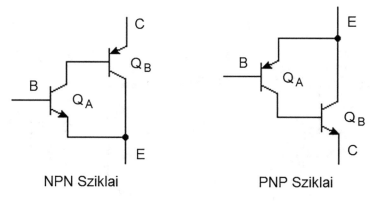

NPN Sziklai   **PNP Sziklai**

Fig. 4.7 Complementary Sziklai composite transistors

$$\beta_S \cong \beta_A \beta_B$$

As with the Darlington, typical beta for the Sziklai transistor may range from 1000 to 20,000. Generally, I prefer to use a conservative estimate of about 1000 for overall Sziklai or Darlington beta. Sziklai transistors are not as sensitive to temperature as Darlingtons, but they are still much more thermally sensitive than normal BJTs. Sziklai transistors also have slightly worse high-frequency performance than equivalent Darlington devices.

A Complete Power Amplifier

A complete Sziklai power amplifier is shown in Fig. 4.8. I designed this amp many years ago, and it has been built successfully dozens of times by students in my electronics classes. This circuit works extremely well with bipolar power supply voltages from ± 10 to ± 18 V; however, it will even operate with supplies as low as ± 5 V with reduced power output.

Both op amps are set up for gain

$$
\begin{aligned}
A_{V(1)} = A_{V(2)} &= \frac{R_F + R_1}{R_1} \\
&= \frac{100 \text{ k}\Omega + 10 \text{ k}\Omega}{10 \text{ k}\Omega} \\
&= 11
\end{aligned}
$$

The output stage has approximately unity gain $A_{V(3)} \cong 1$, so the overall gain of the amplifier is

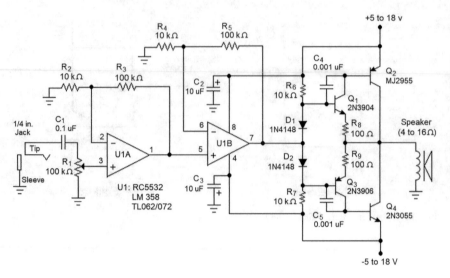

Fig. 4.8 Practical guitar amplifier

$$A_V = A_{V(1)}A_{V(2)}A_{V(3)}$$
$$= 11 \times 11 \times 1$$
$$= 121$$

With this gain, the amplifier to be driven to full output by an input of $V_{in} = 41$ mV$_{pk}$ for ± 5 V supplies, and $V_{in} = 150$ mV$_{pk}$ for ± 18 V supplies. The overall gain can be increased by reducing the value of R_2 or R_4 or increasing R_3 or R_5.

The input resistance of the amplifier is $R_{in} = R_1 = 100$ kΩ. Capacitor C_1 is not required but was included to prevent the amplifier from responding should a DC voltage inadvertently be applied to the input and to provide some protection for the circuit. Capacitors C_2 and C_3 are *decoupling capacitors*. These capacitors help keep the op amp stable and should be mounted physically close to the op amp package.

A modified class AB push-pull output stage is used in this circuit. Transistor pairs Q_1-Q_2 and Q_3-Q_4 are Sziklai pairs, which are used because of their extremely high beta ($\beta_S \geq 1000$ typically). This is important because the typical op amp may not have sufficient current sink/source capability to drive a simple emitter follower, push-pull output stage with a low-resistance load. Sziklai pair Q_1-Q_2 behaves like a single NPN transistor, while Q_3-Q_4 is equivalent to a PNP transistor.

Resistors R_8 and R_9 are swamping resistors. As mentioned previously, these resistors swamp out mismatched barrier potentials between the diodes and transistors Q_1 and Q_3. It is very tedious and time-consuming to match diodes with transistors using a curve tracer. And as a practical matter, most hobbyists don't have access to a curve tracer anyway. With this in mind, I chose values for resistor pair R_{11}-R_{12} large enough to keep the idling current of the output stage sufficiently low for just about any randomly chosen 1N4148s and 2N3904/3906s.

It is worth mentioning again that power output stages are notoriously susceptible to high-frequency oscillation, especially when constructed using protoboards or point-to-point wiring. In order to help ensure that oscillation does not occur, capacitors C_4 and C_5 reduce the high-frequency gain of transistors Q_1 and Q_3.

I chose to use Sziklai transistors rather than Darlingtons in this circuit primarily because only two diodes are necessary to obtain class AB biasing. Darlingtons are also somewhat more temperature sensitive. Darlington transistors also have slightly better frequency response than equivalent Sziklai pairs. This would make a Darlington-based amplifier more prone to parasitic oscillation than the Sziklai amp.

As a final point, it is not necessary to use a regulated supply to power the amplifier. There are several reasons that an unregulated supply is acceptable. First, at low-power levels where ripple voltage would be most noticeable as 120 Hz hum on the output, the amplifier presents a very light load on the supply. This means that very little ripple voltage will be present on the supply rails. In addition, op amps and push-pull stages are relatively insensitive to ripple voltage to start with. In high-performance amplifiers, the low-level amplification stages, which are more sensitive to ripple and other noise, may be powered from separate regulated supplies.

Output Stage Analysis

It is worth taking the time to do a more complete analysis of the output stage of the amplifier. We will make the following assumptions:

1. The amplifier is driving an 8 Ω load.
2. The amplifier output can swing from rail to rail.
3. The output transistors (Q_2 and Q_4) are mounted on a heat sink.
4. The amplifier operates at 50% efficiency at maximum output.

The maximum theoretical efficiency of a class AB stage is approximately 78%, so the assumption of 50% efficiency is reasonable in a practical, real-world situation. This also simplifies the analysis a bit because when $\eta = 50\%$, the power transistors will dissipate the same power as the load (split evenly between Q_2 and Q_4). Output power and current and transistor power dissipation levels for various supply voltages are presented in Table 4.1.

These are pretty impressive output power numbers for such a simple circuit, but the output transistors are doing all of the heavy lifting here, so their specifications are critical. Referring back to Table 3.1, we find that the 2N3055 and MJ3955 transistors have maximum ratings of $I_{C(max)} = 15$ A, $BV_{CEO} = 60$ V, and $P_{D(max)} = 115$ W.

Table 4.1 Max amplifier power and current values, $R_L = 8\ \Omega$, $\eta = 50\%$	Supply voltage		
	± 5 V	± 12 V	± 18 V
$P_{o(rms)}$	1.5 W$_{rms}$	9 W$_{rms}$	20.25 W$_{rms}$
P_{DQ2}, P_{DQ4}	750 mW$_{rms}$	4.5 W$_{rms}$	10.125 W$_{rms}$
$I_{L(max)}$	525 mA	1.5 A	2.25 A

These voltage and current ratings are more than sufficient for this application at all supply voltages. The power dissipation ratings of the transistors seem high enough as well, but it's a good idea to verify that this is actually true.

Transistor Thermal Analysis

When a transistor dissipates power, its junction temperature will increase. The amount of temperature increase depends on how efficiently heat can be removed from the transistor. The flow of heat is analogous to the flow of current, and similarly, resistance to the flow of heat is called *thermal resistance*, designated by theta θ. The thermal resistance between the actual C-B junction of the transistor and the TO-3 case in which it is mounted is denoted as θ_{JC}.

Power transistors are usually mounted on a heat sink of some kind to help keep them from overheating. A typical heat sink designed for use with a TO-3 package is shown in Fig. 4.9.

If you want to do a detailed thermal analysis, you will have to look up the specifications for the heat sink you are using. The heat sink in Fig. 4.9a (Digikey Part No. 345-1052-ND) has a thermal resistance from its surface to ambient air of $\theta_{SA} = 2.4\,°\text{C/W}$. This means the heat sink will increase in temperature by 2.4 °C for every watt of power it absorbs from a transistor.

The metal case of the TO-3 transistor package is electrically connected to the collector, and the typical heat sink is made of aluminum. To prevent possible shorting of the power supply and other shock hazards, the power transistors should be insulated from the heat sinks. Usually a thin mica or silicone washer is used as an insulator. The mica washer is coated with thermally conductive silicone grease to reduce thermal resistance. Thermal grease is not required for silicone washers. The mounting arrangement for a transistor, using a mica washer, is shown in Fig. 4.10a.

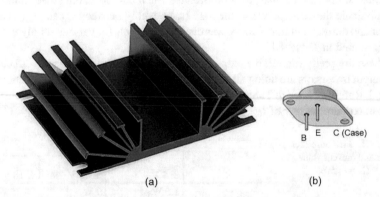

B   E   C (Case)

(a)   (b)

Fig. 4.9. (a) Typical TO-3 heat sink, $\theta_{SA} = 2.4°\text{C/W}$. (b) TO-3 transistor package

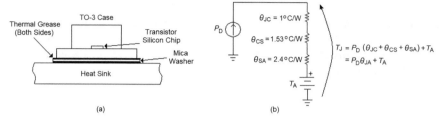

Fig. 4.10 (a) Mounting a TO-3 case on a heat sink. (b) Thermal circuit equivalent

Table 4.2 Transistor junction temperatures: $T_A = 25\ °C$ (77 °F)	Supply voltage		
	± 5 V	± 12 V	± 18 V
$P_{o(rms)}$	1.5 W_{rms}	9 W_{rms}	20.25 W_{rms}
P_{DQ2}, P_{DQ4}	740 mW_{rms}	4.5 W_{rms}	10.125 W_{rms}
T_{J2}, T_{J4}	29 °C	47 °C	75 °C
$P_{D(derated)}$	112 W	100 W	82.2 W

In a thermal circuit, power is equivalent to current, temperature is equivalent to voltage, and thermal resistance is equivalent to electrical resistance. The thermal circuit of the transistor, mounted on a heat sink, is shown in Fig. 4.10b. The thermal resistances shown, $\theta_{JC} = 1\ °C/W$ and $\theta_{CS} = 1.53\ °C/W$, are typical for transistors in TO-3 packages. T_A is ambient or room temperature.

The power dissipated by the transistor, shown as current source P_D, flows through several thermal resistances to ambient air, which is represented by a battery in this model. The ground symbol represents 0 °C, which is a convenient reference temperature. Junction temperature is found using the equivalent of Ohm's law and Kirchhoff's laws, treating the thermal resistances just as we would series resistors in an electrical circuit. The junction temperature equation is

$$T_J = P_D(\theta_{JC} + \theta_{CS} + \theta_{SA}) + T_A \tag{4.10}$$

Using this equation, the junction temperatures of the output transistors were calculated for various power supply voltages and listed in Table 4.2.

The maximum power dissipation rating of the transistor must be reduced or *derated* as junction temperature increases. Using the 2N3055 and MJ2955 for this example, maximum power dissipation must be decreased by 0.657 W/°C above 25 °C. This value is subtracted from the maximum power rating of the transistor (115 W in this case) to find the *derated power dissipation* limit. As an equation, this is written as

$$P_{D(\text{derated})} = P_{D(\max)} - 0.657(T_J - 25°C) \tag{4.11}$$

Applying the derating constant with the junction temperatures calculated in Table 4.2, we find that the power transistors will operate safely at full power. As long as P_{DQ2}, $P_{DQ4} < P_{D(derated)}$ and the ambient temperature does not get too high, the transistors should not overheat. If you are going to the trouble to actually calculate junction temperatures, to be conservative, it's a good idea to keep the transistor power dissipation less than 50% of the derated value. As a general rule, a heat sink can never be too big.

Parallel-Connected Power Transistors

There are times when either the maximum power dissipation or maximum current ratings for a transistor are too low for a given application. In these cases, power transistors can be connected in parallel to increase power and current handling capability. When transistors are connected in parallel, the current and power ratings increase directly with the number of transistors used. Connecting transistors in parallel does not increase the maximum voltage ratings for the transistors.

Thermal Runaway

We can't simply connect transistors in parallel because even slight differences in B-E junction barrier potential, leakage current, and beta, from one device to the next, will cause the transistors to share collector current unevenly. This is called *current hogging*. The transistor that carries more current will get hotter than the others, which will cause it to carry even more current, which will cause it to heat up more, and so on. This self-reinforcing current/temperature cycle is called *thermal runaway*. Thermal runaway can continue until the hotter transistor carries almost all of the current, overheats, and finally burns up. If there are several more transistors in parallel, the next transistor carrying the most current will go into thermal runaway and so on.

Bipolar transistors are especially susceptible to thermal runaway, primarily because junction leakage current increases exponentially with temperature. This effectively reduces the barrier potential of the B-E junction. Beta also increases with temperature, which reinforces the thermal runaway cycle. MOSFETs are much less susceptible to thermal runaway, making them the preferred power transistor used in applications like car audio, where high ambient temperatures are common.

Push-Pull Stage with Parallel Transistors

A push-pull output stage, using parallel-connected power transistors in Sziklai pairs, is shown in Fig. 4.11. Assuming pair Q_4-Q_6 are 2N3055s, they are equivalent to a

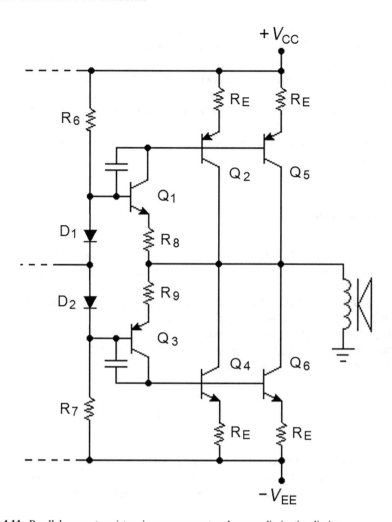

Fig. 4.11 Parallel power transistors increase current and power dissipation limits

single transistor with $I_{C(max)} = 30$ A and $P_{D(max)} = 230$ W. Transistor pair Q_2-Q_5 would most likely consist of MJ2955s with identical equivalent ratings.

Swamping resistors R_E are placed in series with the emitters of the transistors to compensate for mismatch (β, V_{BE}, I_S, and r'_e) between devices, helping prevent current hogging and thermal runaway. Typical values for the emitter resistors range from 1 Ω down to 0.1 Ω. Using higher values for R_E increases stability of the circuit but reduces gain and output voltage compliance and wastes power. In general, we try to use the lowest value swamping resistors we can get away with.

The emitter resistors can dissipate quite a bit of power, so wire-wound, ceramic resistors rated from 1 to 10 W are common. Higher power output will require higher power ratings of the emitter resistors. It is also a common practice to use noninductive wire-wound resistors as well to help prevent parasitic oscillation.

There are many variables you can experiment with in terms of controlling the idling current of the output transistors in this circuit. For example, the values of R_8 and R_9 could be increased to perhaps as high as 1 kΩ, or resistors R_6 and R_7 could be increased in value. Another alternative would be to use diodes such as 1N4001s in place of 1N4148s. 1N4001 diodes have a lower forward voltage drop at a given current level than 1N4148s and are even less likely to allow output idling current to be excessive.

Adding a Tone Control

Any of the tone control circuits discussed in Chap. 2 could be used in this amplifier. Here, the improved single-pot tone control of Fig. 2.24 will be used. The circuit is shown in Fig. 4.12.

The addition of a passive tone control circuit will result in an overall reduction in the gain of the amplifier. Originally, the overall gain of the amplifier was $A_V = 121$. Adding the tone control with an insertion loss of about 12 dB is equivalent to multiplying the overall gain by $A_{V(tone)} = 0.25$ (see Table 2.2). This reduces the overall gain of the original circuit to

$$A'_V = A_{V(\text{tone})}A_V$$
$$= 0.25 \times 121$$
$$= 30.25$$

The sensitivity of the amplifier is now too low to be driven to full output by a typical guitar pickup. To compensate for the loss caused by the tone control circuit, the gain of the original amplifier must be increased by a factor of

$$1/A_{V(\text{tone})} = 1/0.25$$
$$= 4$$

We can make up the gain in several different ways. Here, we will increase the gain of the first stage by a factor of 4. Multiplying R_2 by the fractional gain of the tone control will accomplish this goal.

$$A_{V(\text{tone})}R_2 = 0.25 \times 10\,\text{k}\Omega$$
$$= 2.5\,\text{k}\Omega$$

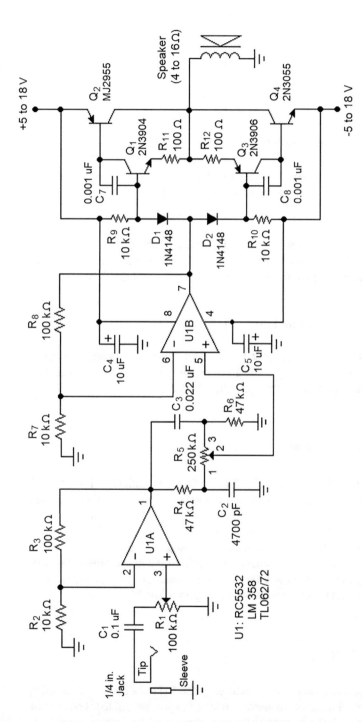

Fig. 4.12 Complete amplifier with tone control

Using the next lower standard value, $R_2 = 2.2$ kΩ, gives us

$$A_{V(1)} = \frac{R_3 + R_2}{R_2}$$
$$= \frac{100k + 2.2k}{2.2k}$$
$$= 46.5$$

This increases the overall gain to

$$A_V = A_{V(1)}A_{V(\text{tone})}A_{V(2)}A_{V(3)}$$
$$= 46.5 \times 0.25 \times 11 \times 1$$
$$= 128$$

This is a bit greater than the original gain, which is acceptable.

If we had used the Baxandall tone control circuit of Fig. 2.26, which has a loss of 23 dB (0.0708), then the gain of the original amplifier would have to be increased by a factor of

$$1/A_V(\text{ tone }) = 1/0.0708 \cong 14$$

In this case, it would be preferable to split the required additional gain between U1A and U1B (set $A_{V(1)} = A_{V(2)} = 41.2$), or possibly even leaving U1A and U1B alone, and inserting another op amp gain stage with $A_V = 14$.

Amplifier Stability Issues

Amplifier instability often manifests itself in the form of oscillation. The phase-shift oscillator presented in the previous chapter is a good example of a circuit for which we went to great lengths in order to ensure oscillation. However, any circuit that contains an amplifier has the potential to oscillate. Refer back to Fig. 3.38 for an example of one form of parasitic oscillation that might be observed in an unstable circuit. This sort of oscillation can cause RF interference and can cause components to overheat. Amplifier stability is a very complex topic, and we can only touch the surface of some typical causes and cures for amplifier instability here.

Ground Reference

Ideally, any component that is connected to ground should have zero impedance between itself and all other components connected to ground. All leads connected to ground should be at the same potential, that is, zero volts. Resistance in the ground path causes unintended voltage drops to be created, which can lead to instability.

When components are connected to ground via paths of varying resistance, *ground loops* can be formed. When current flows through a ground loop, the components connected at the ends of the loop will be at different potentials. This happens at DC and at signal frequencies as well. The end results can range from excessive 60/120 Hz hum to circuit oscillation.

Star Grounding

A technique that is sometimes used to help eliminate ground loop problems is *star grounding*. In star grounding, one point is used as the common ground, and all connections to ground are made to this point. Star grounding is not always convenient, but it is a very effective grounding method. A star ground point in an amplifier chassis is shown in Fig. 4.13.

Motorboating

Oscillation may occur at any frequency from sub-audio well into the megahertz range. Oscillation can cause severe output distortion and overheating of output power transistors.

A type of oscillation called *motorboating* occurs at frequencies that range from around 1 to10 Hz. The term motorboating is used because the amplifier will produce

Fig. 4.13 Multiple lines connected to a common star ground point

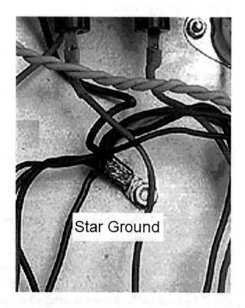

Star Ground

an output that literally sounds like an idling motor boat. Motorboating is usually caused by excessive resistance in the power supply ground returns of an amplifier. This is usually the result of poor solder joints or mechanical connections to ground. If you have an amplifier that is motorboating, grounding problems are the first things to look for. Motorboating and high-frequency oscillation are common when circuits are built on protoboards, especially when long interconnecting wires are used, and contact resistance causes ground loops to be formed.

Decoupling Capacitors

The internal resistance of an ideal voltage source is zero ohms. One of the assumptions that we make when performing an AC signal analysis on an amplifier is that the power supplies are ideal voltage sources, and so the supply rails are zero impedance paths to ground. Various factors such as the length, resistance, and stray inductance of wires or PCB traces leading to the power supply can cause motorboating.

In order to minimize the effects of non-ideal supply rail behavior, decoupling capacitors are often placed close to various gain stages in an amplifier. These were used in several circuits that we examined previously, including the amplifiers of Figs. 4.8 and 4.12, and with the voltage regulators used way back in Chap. 1. The decoupling capacitors help ensure that power supply rails behave as signal ground points.

The Zobel Network

Loudspeakers are complex electromechanical systems in themselves, but they appear primarily as inductive loads on the amplifier. If a power amplifier is going to oscillate, it will usually oscillate at tens or hundreds of kilohertz. The speaker, behaving primarily as an inductance, will have very high impedance at these frequencies. In many amplifier designs, especially those that do not employ global negative feedback, the gain of the amplifier will increase in direct proportion to load impedance. Should high-frequency oscillation occur, these factors can cause the gain of the amplifier to increase drastically, producing an inaudible high-amplitude output signal that can damage to the amplifier.

A series RC network connected from the amp output to ground, called a *Zobel network*, helps to keep the load impedance being driven by the amp from increasing to excessively high values at high frequencies. The basic idea behind the Zobel network is to cause the impedance being driven by the amplifier to level off at approximately R_Z ohms at very high frequencies. This prevents the amplifier gain from increasing, as well as damping high-frequency oscillation. A Zobel network suitable for use with the power amplifiers covered in this chapter is shown in Fig. 4.14 Resistor R_Z should normally be rated at 1 watt or more for this circuit

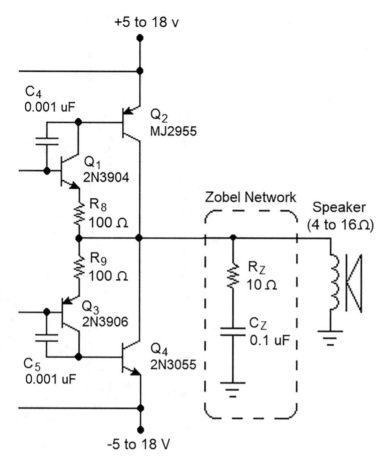

Fig. 4.14 Adding a Zobel network for increased amplifier stability

It is worth repeating that because the voltage gain of many amplifier topologies increases in proportion to load impedance, a power amplifier should never be operated, even for short periods of time, without a load. An open-circuit load has infinite impedance, which can cause gain to increase to astronomical values, which can damage the amplifier. This is especially costly in the case of vacuum tube amplifiers where the output tubes, output transformers, or both can be destroyed.

MOSFET Output Stages

MOSFETs (metal-oxide-semiconductor FETs) can be used in place of bipolar power transistors in amplifier circuits. A simple push-pull, source follower arrangement is shown in Fig. 4.15. The MOSFETs here are functionally equivalent to complementary BJTs.

Fig. 4.15 Push-pull, source follower, power MOSFET output stage

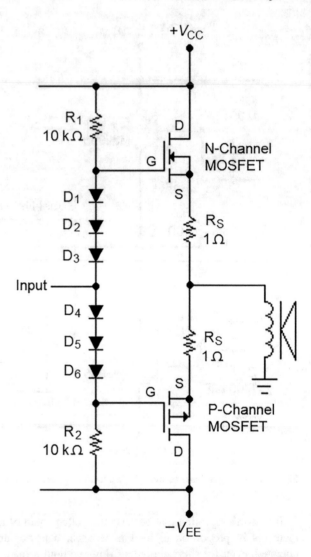

Recall that MOSFETs have several advantages over BJTs. The main advantage is that MOSFETs have extremely high input resistance, which makes them easy to drive. MOSFETs are also much less susceptible to thermal runaway than BJTs, and they tend to share current more equally when connected in parallel.

A MOSFET push-pull stage requires a higher biasing voltage than the BJT version in order to eliminate crossover distortion. The typical MOSFET will not begin to turn on until the gate-to-source voltage exceeds the threshold voltage: $V_{GS} > V_T$. For most MOSFETs, $V_T \geq 2$ V. This is why so many diodes are used in Fig. 4.11. Sometimes Zener diodes or LEDs are used to bias up MOSFET output stages, rather than long strings of standard silicon diodes.

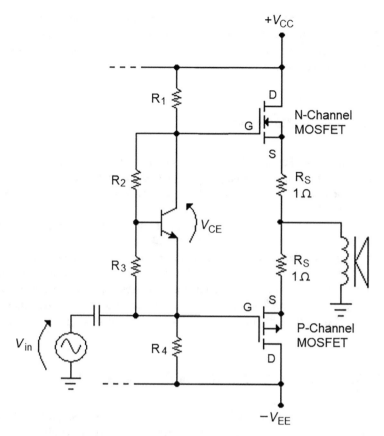

Fig. 4.16 V_{BE} multiplier biasing of a MOSFET push-pull stage

The V_{BE} Multiplier

The need for long strings of biasing diodes in Darlington and MOSFET push-pull stages may also be avoided through the use of a circuit called a V_{BE} *multiplier*. A biasing circuit using a V_{BE} multiplier is shown in Fig. 4.16.

The value of V_{CE}, which is the bias voltage for the output MOSFETs, is given by

$$V_{CE} = V_{BE}\left(1 + \frac{R_2}{R_3}\right) \tag{4.12}$$

The name V_{BE} multiplier comes from the fact that the output voltage V_{CE} is a multiple of the forward drop across the transistor B-E junction. In addition to replacing a long string of diodes, the V_{BE} multiplier has the advantage of allowing for an easy adjustment of output idling current by using a potentiometer for either R_2 or R_3.

You may have heard of a technique called *servo biasing*, which is an interesting topic that you may wish to investigate. Servo biasing is used in more sophisticated BJT, MOSFET, and vacuum tube power amp designs, but it is not very commonly used in guitar amplifiers.

The Rail Splitter Revisited

Hopefully you recall that the rail splitter is a circuit that is used to convert a single-polarity power supply into a bipolar supply. The output current sink/source capability of the basic rail splitter was increased by adding a push-pull output stage, as was shown in Fig. 3.50.

A simplified rail splitter using a push-pull output is shown in Fig. 4.17. When load conditions require the ground terminal to sink current, the op amp output drives negative, turning on PNP transistor Q_2. This holds the inverting input terminal at the same voltage as the noninverting input. When load conditions require the ground terminal to source current, the op amp output drives positive, turning on NPN transistor Q_1. Load currents of several amps can easily be handled by this circuit with proper choice of output transistors and adequate heat sinking.

Enclosing Base-Emitter Junction in the Feedback Loop

The similarities between the rail splitter of Fig. 4.17 and the push-pull audio amplifiers we covered in this chapter are obvious. There are some differences that

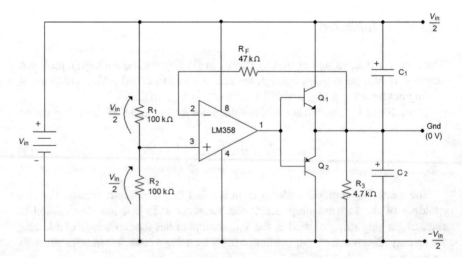

Fig. 4.17 High-current rail splitter

are very significant though. Rail splitters do not use a biasing diode/resistor arrangement to cancel the B-E barrier potential of transistors Q_1 and Q_2. Instead, the B-E junctions are enclosed inside the feedback loop of the op amp, which reduces the effective barrier potential to

$$V'_{BE} \cong \frac{V_{BE}}{A_{OL}} \qquad (4.13)$$

Since the open loop gain is so large ($A_{OL} \geq$ 100,000), the effective barrier potential V'_{BE} is practically zero volts.

Resistor R_3 should be included if the rail splitter is to be operated with no external load to draw current from the output. This resistor presents a light load on the rail splitter, which in turn causes the op amp to turn on Q_1 just slightly. If there was no load at all on the circuit, transistors Q_1 and Q_2 are in cutoff, acting like open circuits. This effectively results in an open feedback loop around the op amp, causing the voltage gain to increase to A_{OL} (around 100,000 or so). This could cause the circuit to oscillate. The load presented by R_3 helps prevent this but has negligible effect on normal operation.

Changes in the input voltage will have a direct effect on the output of the rail splitter. For example, if V_{in} should decrease by 10%, the output rails will also decrease by 10%. A stable, well-filtered input voltage is necessary to obtain stable, clean output voltage rails. Also, changes in the voltage applied to the noninverting input terminal will also be reflected directly to the output of the circuit. We can take advantage of this to use the circuit as an amplifier.

Finally, all of the precautions regarding virtual ground, earth ground, and chassis ground discussed in Chap. 3 are even more important for this circuit because of the potential for very high short circuit currents should incorrect ground connections be made.

Converting the Rail Splitter to an Amplifier

It is possible to use the rail splitter as the basis for an audio amplifier. The circuit of Fig. 4.18 illustrates this application. The original output filter capacitors C_1 and C_2 must be removed; otherwise, they will act as low-impedance paths to the signal rails, which will short the output signal to ground. Resistor R_3 should be kept in the circuit to help prevent oscillation. This is basically a capacitively coupled, class B amplifier. This circuit is class B because there is no bias applied to the push-pull transistors.

The input signal is coupled to the noninverting input of the op amp and superimposed on the DC bias voltage produced by the voltage divider. Essentially, we have now converted the rail splitter into a circuit that is equivalent to a single-supply op amp like that of Fig. 3.36.

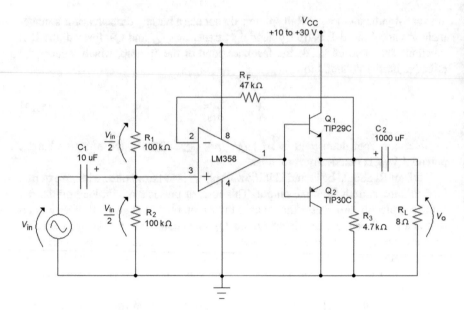

Fig. 4.18 Rail splitter modified for use as a single-polarity operated power amplifier

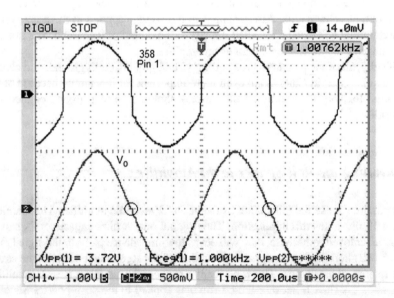

Fig. 4.19 Waveforms for the amplifier of Fig. 4.18

I breadboarded this amplifier using an LM358 op amp with TIP29C and TIP30C transistors. The power supply was +15 V, and the load was a 10 Ω resistor. With $V_{in} = 2.0$ V_{P-P}, 1 kHz, waveforms from the output of the op amp (pin 1) and the output voltage across R_L are shown in the scope capture of Fig. 4.19.

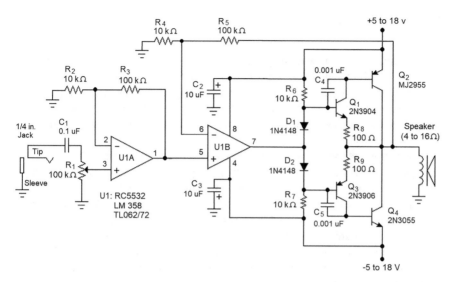

Fig. 4.20 Amplifier with output transistors in op amp feedback loop, employing class AB biasing

Slew Rate-Induced Crossover Distortion

The voltage gain of the amplifier was almost exactly unity as expected. There were some very slight hints of crossover distortion visible on the output waveform, which are circled on the lower scope trace. Although small in amplitude, you would be able to hear this distortion.

The problem here is not that the gain of the op amp isn't high enough to eliminate the barrier potential of the B-E junctions. Rather, the LM358 cannot slew its output terminal fast enough to blast through the barrier potential dead-band without us seeing (or hearing) it with this output signal. The upper trace shows how the op amp output pin slews as fast as it can to overcome the output transistor barrier potentials at zero-crossing on the nearly vertical portion of the waveform.

The crossover distortion becomes more severe at higher output amplitudes and higher signal frequencies. The brute-force solution to this problem is to use a faster op amp such as a TL071, which has SR = 13 V/μs. Keep in mind that this kind of circuit is very susceptible to oscillation, and a faster op amp will make this even more likely. A more effective way to eliminate this slew distortion and reduce the possibility of instability and oscillation would be to add the usual resistor/diode biasing network, effectively making this a class AB biased circuit. This is shown in Fig. 4.20.

Final Comments

Power amplifiers can be extremely complex circuits, and they are usually much more difficult to analyze and design than small-signal amplifiers. We have only looked at a very small sample of the many different power amp topologies in use. In particular, we have not studied class A power amplifiers because they are rarely implemented using transistors. This is not the case in vacuum tube amplifier designs, which we will examine later.

Aside from the usual circuit analysis/design stuff presented, there are two important general observations to take from this chapter. First, power amp output stages tend to be unstable and can be very tricky to work with. If you have breadboarded any of the circuits presented in this chapter, it is very likely that you have encountered parasitic oscillation and perhaps some transistor overheating. Circuit layout, even using PC board construction techniques, is critical. The second thing to keep in mind is that adequate heat sinking is very important in power amp design. It's always a good idea to use a larger heat sink than you think you might need. Basically, your heat sink can never be too big.

Summary of Equations

Emitter Follower, Push-Pull Amplifier Equations

$$A_V = \frac{R_L}{R_L + r'_e} \cong 1 \tag{4.1}$$

$$R_o \cong r'_e \tag{4.2}$$

$$R_{in} \cong \beta(R_L + r'_e) \tag{4.3}$$

Push-Pull Output Current and Power Equations (Single-Polarity Supply)

$$I_{L(max)} \cong \frac{V_{CC}}{2R_L} \tag{4.4}$$

$$P_{o(pk)} \cong \frac{V_{CC}^2}{4R_L} \tag{4.5}$$

$$P_{o(rms)} = \frac{V_{CC}^2}{8R_L} = \frac{P_{o(pk)}}{2} \qquad (4.6)$$

Push-Pull Output Current and Power Equations (Bipolar Supply)

$$I_{L(max)} \cong \frac{V_{CC}}{R_L} \qquad (4.7)$$

$$P_{o(pk)} \cong \frac{V_{CC}^2}{R_L} \qquad (4.8)$$

$$P_{o(rms)} \cong \frac{V_{CC}^2}{2R_L} \qquad (4.9)$$

Transistor Junction Temperature Equation

$$T_J = P_D(\theta_{JC} + \theta_{CS} + \theta_{SA}) + T_A \qquad (4.10)$$

Transistor Power Derating Factor for 2N3055/MJ2955

$$P_{D(derated)} = P_{D(max)} - 0.657(T_J - 25°C) \qquad (4.11)$$

V_{BE} Multiplier Equation

$$V_{CE} = V_{BE}\left(1 + \frac{R_2}{R_3}\right) \qquad (4.12)$$

Effective Barrier Potential of Diode in Op Amp Feedback Loop

$$V'_{BE} \cong \frac{V_{BE}}{A_{OL}} \qquad (4.13)$$

Chapter 5
Effects Circuits

Introduction

There are many different types of guitar effects circuits, and I like to break them down into three main categories. First are overdrive and distortion circuits. These circuits are usually fairly simple, and their basic operation is relatively easy to understand. The second group includes more advanced effects such as phase shifters, wah-wah pedals, vocoders, compressors, reverbs, and so on. Many of these circuits can be quite complex. The third category is miscellaneous circuits that could be considered effects including equalizers and noise gates.

The one thing that all of these effects have in common is that they alter the guitar signal in some way. Before we look at the circuits, we will cover a little more background on signals, spectral analysis, and nonlinearity. If you have diligently studied the previous chapters, much of this introductory material will be review. Some relatively advanced concepts must be used to understand the deeper, nitty-gritty operation of many effects circuits. I tried to present a balanced approach using qualitative descriptions whenever possible and, though sometimes it may not seem so, keeping advanced math to a minimum.

Signals and Spectra

Signals can be examined in either the time domain or the frequency domain, and we have seen examples of both representations in previous chapters. Most often, the signals that we are interested in are voltages. In the time domain, we are looking at a voltage whose amplitude is a function of time $v(t)$. The oscilloscope is used to view signals in the time domain.

In the frequency domain, we are looking at voltage as a function of frequency $V(f)$. This representation tells us the spectral or frequency content of a signal.

D. J. Dailey, *Electronics for Guitarists*,
https://doi.org/10.1007/978-3-031-10758-0_5

A spectrum analyzer, or more commonly these days, a digital oscilloscope with FFT (fast Fourier transform) capability, is used to view signals in the frequency domain. Most people don't have trouble understanding and visualizing typical time-domain signals, but the frequency domain is a more abstract and unfamiliar concept.

Time, Period, Frequency, and Pitch

As a musician, you are probably familiar with all of these terms, but we will define them more precisely now. Frequency can be defined as the number of cycles a *periodic signal* completes per second. Frequency is given the units Hertz (Hz).

$$f = \text{Cycles/second (Hertz, Hz)} \qquad (5.1)$$

The term *pitch* is the musical equivalent of frequency, but instead of using the units Hertz, pitches are given names such as A2, C#, Dm, etc. Some frequencies that you may be familiar with are listed in Table 5.1. These are the frequencies and pitches for standard guitar tuning. Standard string numbering starts with the high E string. We usually use 1 kHz as the standard test frequency when working with audio equipment.

The signals shown in Fig. 5.1 are examples of periodic signals. The period of a signal is represented with the letter *T*.

The period of a signal is the time it takes for that signal to complete one cycle. In this section, we are assuming that the signals we are looking at are periodic. That is, they repeat the same basic shape over and over again.

The waveform in Fig. 5.1a is a sine wave. The waveform in Fig. 5.1b is a cosine wave. There is a phase difference of 90° between the sine and cosine waves. We could say that the cosine leads the sine by 90° or that the sine lags the cosine by 90°. It depends on which waveform you want to use as your reference. In this book, we will use the sine wave as the reference. In general, if we don't know or care about the phase of the signal, we refer to these as *sinusoidal* waveforms, or just sinusoids.

The waveform in Fig. 5.1c is also a sine wave, but it has greater amplitude than the other waveforms, and its period is shorter. Period and frequency are reciprocals of each other. As equations, these relationships, which are probably very familiar to you at this point, are

Table 5.1 Standard guitar string tuning frequencies and pitches

String	1	2	3	4	5	6
Frequency (Hz)	329.63	246.94	196.00	146.83	110.00	82.41
Pitch	E4	B3	G3	D3	A2	E2

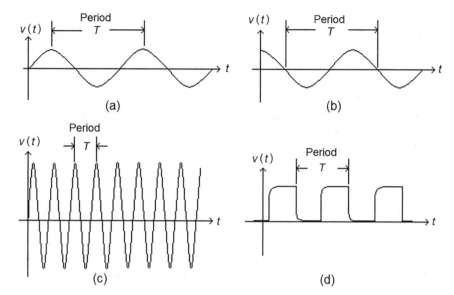

Fig. 5.1 Examples of periodic signals

$$f = \frac{1}{T} \tag{5.2}$$

and

$$T = \frac{1}{f} \tag{5.3}$$

The waveform of Fig. 5.1c has the highest frequency of the waveforms shown here, because it has the shortest period.

The waveform in Fig. 5.1d is a square wave with some distortion of the corners. This waveform also has a DC offset which causes it to be shifted or translated upward so that the amplitude never goes below zero volts. Recall that we used coupling capacitors to block these DC levels in amplifier circuits.

Sinusoids in the Time Domain

In general, a sinusoidal voltage is given by the following unfriendly looking trigonometric equation:

$$v(t) = V_P \sin\left(2\pi ft + \Phi\right) \tag{5.4}$$

In this equation, $v(t)$ is the voltage at any instant in time, V_P is the peak amplitude, f is the frequency in Hz, t is time in seconds, and Φ is the phase of the signal. The constant 2π is used because fundamentally, the math is based on angular units expressed in *radians*. We are all familiar with angular measurement in degrees, and we know that there are 360° in a circle. Using radian measure, there are 2π radians in a circle: $360° = 2\pi$ rad.

The constant 2π converts frequency in Hertz into frequency expressed in radians per second, which is designated by lowercase Greek omega ω. In the form of an equation, this is written as

$$\omega = 2\pi f \qquad (5.5)$$

Although frequency expressed in rad/s is less familiar than cycles/s (i.e., Hz), it is used extensively in mathematics, physics, and electronics. For example, you may recall that inductive reactance is found using the equation $|X_L| = 2\pi f L$, where the radian frequency is disguised in the formula. Because it is more familiar, we will express frequency in Hertz most of the time throughout this book.

Waveform Shape, Symmetry, and Harmonic Relationships

The sine function is an example of what we call an *odd function*. The graph of an odd function has symmetry about the origin. Mathematically, an odd function obeys the relationship

$$v(t) = -v(-t).$$

The cosine waveform is an *even function*. Even functions have symmetry about the vertical axis. This means the left and right sides of the graph are mirror-image reflections of one another. The mathematical expression of even symmetry is

$$v(t) = v(-t).$$

Examples of even and odd functions are shown in Fig. 5.2a, b. Some functions that have neither even nor odd symmetry are shown in Fig. 5.2c. Incidentally, a function that is neither even nor odd can be broken down into even and odd parts using

$$v_{\text{even}}(t) = \frac{v(t) + v(-t)}{2} \qquad v_{\text{odd}}(t) = \frac{v(t) - v(-t)}{2}$$

The frequency-domain representation (which we will simply call a spectrum from here on) of any signal that is not precisely sinusoidal in shape will consist of more than one (and possibly infinitely many) sinusoidal frequency components.

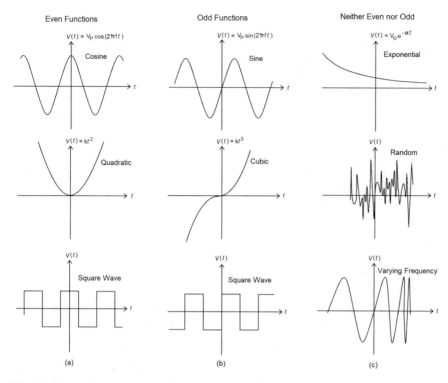

Fig. 5.2 Examples of functions with even, odd, and neither even nor odd symmetry

Square waves, like those at the bottom of Fig. 5.2, consist of odd harmonics of the fundamental square wave frequency. For example, if the period of the square wave is $T = 1$ ms, the fundamental frequency is 1 kHz, and the spectrum consists of only odd harmonics (including the fundamental): 1 kHz, 3 kHz, 5 kHz, 7 kHz, and so on. Ideally, these harmonics go on forever, but they also get smaller, approaching zero amplitude as we go higher in frequency.

If the square wave is an even function, as in the bottom waveform of Fig. 5.2a, the harmonics will all be cosine waves. If the square wave was inverted, the spectrum would consist of inverted (negative) cosines as well.

If the square wave is an odd function, like in Fig. 5.2b, the harmonics in the signal spectrum will all be sine waves, because the sine is an odd function. If the square wave was shifted left or right so that the symmetry was neither even nor odd, its spectrum would still consist of discrete odd harmonics, but they would be phase-shifted sines or cosines.

This is a good place to summarize the discussion so far: the symmetry of a periodic signal tells us phase information. That is, symmetry tells us if the waveform spectrum consists of sines, cosines, or something in between. The actual shape of the waveform, i.e., sinusoidal, square, exponential, periodic, nonperiodic, etc., tells us information about the harmonic content. That is, the basic shape of the waveform tells us how the signal energy is distributed across the frequency spectrum.

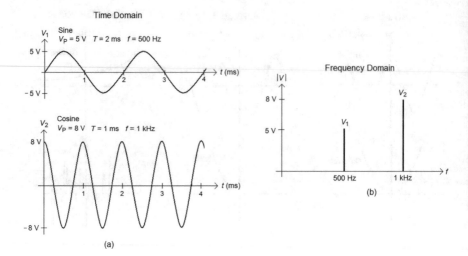

Fig. 5.3 Sine and cosine waves. (**a**) Time domain. (**b**) Frequency domain

Of the functions shown in Fig. 5.2, the periodic waveforms, and perhaps the swept sinusoidal, and random noise waveforms are most likely to be present in audio signals.

The quadratic, cubic, and exponential functions aren't good candidates for audio signals, but could be useful in representing control signals or transfer functions. The function $v = kt^2$ is an even function, while $v = kt^3$ is odd. For *power functions* like these, you can tell by the exponents as well as their graphs whether the function is even or odd. These are not periodic functions; they don't repeat a basic shape at regular intervals. The spectrum of a nonperiodic function such as a single pulse consists of a continuum of sines and/or cosines, rather than discrete frequency components.

This is some pretty abstract territory we are exploring, so let's take a look at the frequency spectra of a few sinusoidal waveforms to get a better handle on things. Sinusoidal waveforms are especially nice to work with because they have the simplest frequency-domain representations. The time-domain waveforms to be used are shown in Fig. 5.3a. The sine wave has $V_P = 5$ V, $T = 2$ ms, and $f = 500$ Hz. The cosine wave has $V_P = 8$ V, $T = 1$ ms, and $f = 1$ kHz.

The frequency spectrum for an ideal sinusoid is simply a spike of zero width, called an *impulse*. In Fig. 5.3, the height of the impulse equals the peak amplitude of the sinusoid. Sometimes the rms voltage is used, as we saw back in Fig. 3.48. The impulse is located at the frequency of the sinusoid on the frequency axis. The spectrum for these particular sinusoids is shown in Fig. 5.3b.

Notice that the spectrum plot of Fig. 5.3b does not tell us anything about the relative phase of the input signals; it only tells us amplitude and frequency information. A separate phase plot can be generated to provide this information.

When we are looking at audio signals, it's usually not as important that we know whether a signal is composed of sines or cosines. Rather, it is usually more important for us to know about the frequencies of the components that make up the signal.

Transfer Function Symmetry and Harmonic Distortion

The concept of the transfer function was introduced briefly in Chap. 2, when we talked about potentiometers, and also in Chaps. 3 and 4, when amplifier operation was analyzed. Recall that a transfer function is the relationship (equation or graph) that describes the output/input behavior of a circuit. The graph of a device or circuit transfer function can tell us a lot about the harmonic content to expect in an output signal produced by that circuit.

For an ideal amplifier, voltage gain remains constant regardless of how hard the amplifier is driven. In reality, all amplifiers exhibit some nonlinear behavior that causes voltage gain to vary as the output voltage changes. One place where this is obvious is when an amplifier clips. Once clipping occurs, the output will not increase further no matter how much the input voltage is increased.

The use of negative feedback will cause an amplifier to behave more linearly, so in applications where linear operation is important (high-fidelity amplifiers, electronic measuring and test equipment, medical instrumentation, etc.), lots of negative feedback is used. Op amps come close to having perfectly linear transfer functions right up until clipping because they use massive amounts of negative feedback in most applications.

The transfer function of a typical op amp with $A_V = 5$ is shown in Fig. 5.4. The slope of the transfer function graph is equal to the voltage gain of the amplifier. In this case, clipping occurs at $V_o = \pm13$ V, which would be typical for an LM741 or TL071 op amp using a ±15 V power supply. By the way, a function that is

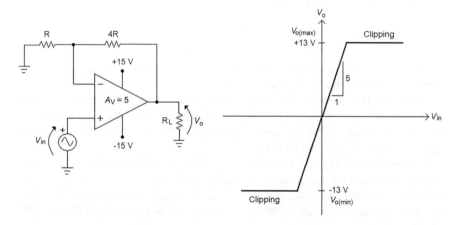

Fig. 5.4 Op amp with $A_V = 5$ and the graph of its transfer function

constructed by connecting different lines and curves end to end, like the transfer function in Fig. 5.4, is called a *piecewise continuous function*.

As long as the amplifier stays on the sloped portion of its transfer characteristic, the output waveform will be an exact replica of the input, except that it is scaled in amplitude by the voltage gain. The output spectrum of a linear amplifier will not contain any frequency components that were not in the original input signal spectrum.

When clipping occurs, the top and bottom of the output signal will be flattened symmetrically, distorting the shape of the signal. This is a form of nonlinearity which, because it gives the transfer function graph odd symmetry, will cause odd harmonics to be created in the output.

The important concept to understand here is that you can tell what kind of harmonics will be created in the output signal by noting the symmetry of the nonlinear transfer function graph. The relationships between transfer function symmetry and harmonic content are summarized as follows:

> **Odd Nonlinear Transfer Function Symmetry**
> If the transfer function graph has odd symmetry (it is an odd function), the output will contain only odd harmonics.
> **Even Nonlinear Transfer Function Symmetry**
> If the transfer function graph has even symmetry (it is an even function), the output will contain only even harmonics.
> **Neither Even Nor Odd Nonlinear Transfer Function Symmetry**
> If the transfer function graph does not have either even or odd symmetry, then in general the output signal will contain both even and odd harmonics.

The creation of new output frequency components that are integer multiples of the input frequency is called *harmonic distortion*. Harmonics are integer multiples of the fundamental frequency.

An Odd Symmetry Example

You may remember from the previous chapter that the class B, push-pull amplifier will produce an output signal that exhibits crossover distortion. Crossover distortion occurs because the input signal must overcome the barrier potential of the base-emitter junctions of the push-pull transistors.

An approximate graph of the transfer function for a push-pull stage, using silicon transistors, is shown in Fig. 5.5. The transfer function for the push-pull amplifier is an odd function; it exhibits symmetry about the origin. Because of this, the output signal should only contain odd harmonics.

The amplifier of Fig. 5.5 was simulated with $V_{in} = 2$ V_{pk} and $f = 1$ kHz using PSpice. The resulting output waveform and its spectrum are shown in Fig. 5.6. As expected, the time-domain waveform shows crossover distortion, and the frequency spectrum contains only odd harmonics of the 1 kHz fundamental (3 kHz, 5 kHz, 7 kHz, etc.).

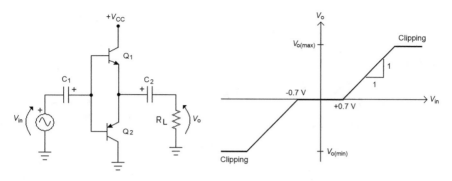

Fig. 5.5 Class B, push-pull stage and a graph of its transfer function

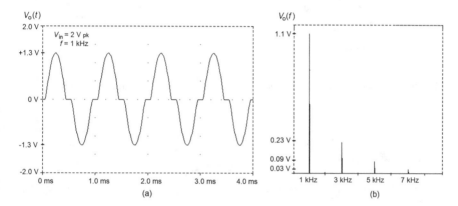

Fig. 5.6 (**a**) Waveform with crossover distortion. (**b**) Frequency spectrum of waveform

If we were to simulate driving the amplifier into clipping, the output spectrum would still contain only odd harmonics, but the relative amplitudes of the higher harmonics would increase.

An Even Symmetry Example

So far we have only seen circuits that have odd transfer functions. One of the few circuits that has even transfer function symmetry is the full-wave rectifier. A full-wave rectifier and its transfer function are shown in Fig. 5.7a. The flat line at the origin of the graph is caused by the barrier potential of the diodes in the bridge.

If we could eliminate barrier potential, an ideal full-wave rectifier would be formed. An ideal full-wave rectifier performs the *absolute value function*, which is shown in block diagram form in Fig. 5.7b. We will see some very cool uses for the absolute value circuit later in this chapter.

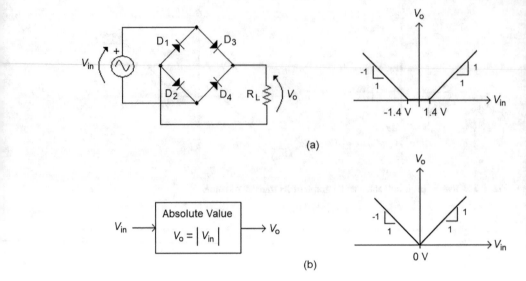

Fig. 5.7 Even symmetry transfer functions. (**a**) FW rectifier. (**b**) Absolute value circuit

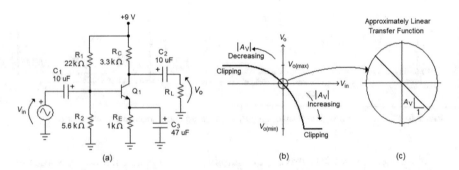

Fig. 5.8 (a) Class A amplifier. (b) Transfer function. (c) Small-signal, linear area of operation

An Example of Neither Even Nor Odd Symmetry

The transfer functions of the BJT, JFET, and MOSFET class A amplifiers that were covered back in Chap. 3 have neither even nor odd symmetry. A BJT, class A amplifier circuit and its transfer function are shown in Fig. 5.8. For small signals (V_{in} within the circled area around the origin), the transfer function is approximately linear, and the amplifier produces negligible distortion.

Imagine that we are applying a 1 kHz sinusoidal signal to the amplifier. As the amplitude of the input signal varies, the gain of the amplifier will change as well. When v_{in} goes positive (v_o goes negative), the voltage gain magnitude increases. This occurs because the slope of the transfer function graph gets steeper when we move to the right from the origin.

When v_{in} goes negative (v_o goes positive), the gain magnitude decreases because the transfer function slope becomes more horizontal as we move to the left. The larger the signal, the further we move from the origin, causing increased nonlinearity.

The behavior of the transistor is exponential e^x, but as stated earlier for small signals, the gain is approximately linear. As the output signal gets larger, the nonlinearity starts to become significant. At first, the voltage gain becomes approximately proportional to v_{in}^2. Since v_{in}^2 is an even function, the second harmonic (2 kHz) pops up first in the output spectrum.

As the amplitude of the output signal increases, and the nonlinearity increases, the cubic term becomes larger, and the gain is now proportional to both v_{in}^2 and v_{in}^3. So we see that both even and odd harmonics are created due to exponential behavior of the transistor, which has neither even nor odd symmetry. Increasing the amplitude even further will cause v_{in}^4 and higher terms to be created. Until clipping rears its head.

Once clipping occurs, initially both even and odd harmonics will increase. Driving the amplifier even further into clipping will eventually cause a greater increase in odd harmonics because the transfer function begins to be dominated by output clipping, which has odd symmetry. These characteristics were visible in the scope photos of Fig. 3.16, which are repeated in Fig. 5.9 for convenience.

Intermodulation Distortion

Up until now, we have only looked at distortion using a single sinusoidal input signal. Musical instrument signals are complex combinations of sinusoids that have very specific frequency relationships to one another. We have seen that even picking a single string results in a complex waveform, and not a simple sinusoid.

When we drive a linear amplifier with several frequencies at one time, the output contains exactly the same frequency components as the original input, but the amplitudes of the frequency components are increased (it is an amplifier, after all). A tone control circuit may be used to alter the relative amplitudes and phase relationships of the frequencies, but it will not alter the frequencies themselves. A nonlinear amplifier will produce additional output frequency components that are the sums and differences of the input frequencies. These new frequencies are generally not integer multiples of the input frequencies and so are not harmonically related.

The exact nature of the nonlinearity will determine the output frequency content. The math behind this stuff gets ugly very fast, so a relatively simple example will demonstrate formally how intermodulation distortion occurs. Let's assume that we have an input signal consisting of the following two pure sinusoidal tones:

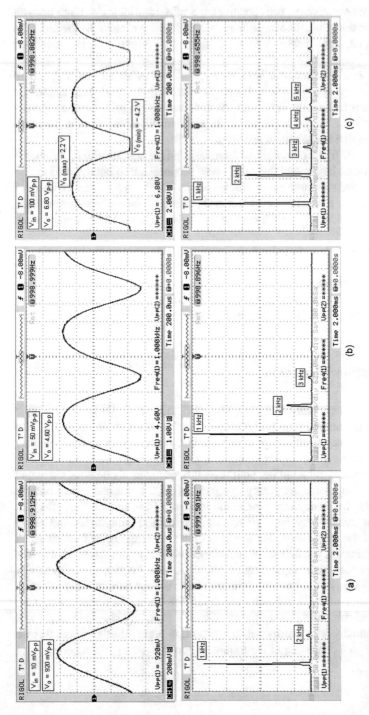

Fig. 5.9 Nonlinearity of amplifier gain distorts v_o and creates harmonic distortion

$$V_1 : 110 \text{ Hz(A2)}, V_P = 3\text{V}$$
$$V_2 : 165 \text{ Hz(E3)}, V_P = 1\text{V}$$

This is a nice sounding pitch combination, and the formal mathematical description of this input signal is written as

$$v_{in} = 3 \sin (2\pi 110t) + 1 \sin (2\pi 165t)$$

We will apply this signal to a hypothetical amplifier that has a linear gain of five and a quadratic gain term with a coefficient of two. As an equation, the output voltage is given by

$$v_o = 5v_{in} + 2v_{in}^2$$

The output voltage v'_o due to the linear gain term $5v_{in}$ is simply

$$v'_o = 5[3 \sin (2\pi 110t) + 1 \sin (2\pi 165t)]\text{V}$$
$$= 15 \sin (2\pi 110t) + 5 \sin (2\pi 165t)\text{V}$$

The output signal v''_o that is produced by the quadratic gain term is quite a bit messier to work with. Basically, we need to calculate the following product:

$$v''_o = 2[(3 \sin (2\pi 110t) + 1 \sin (2\pi 165t)) \times (3 \sin (2\pi 110t) + 1 \sin (2\pi 165t))]$$

The right side of the equation is expanded using the good-old FOIL method and then reduced to a usable form with the aid of the following trigonometric identity for the product of two sine functions:

$$A \sin (\omega_1 t)B \sin (\omega_2 t) = \frac{AB}{2}[\cos (\omega_1 - \omega_2)t - \cos (\omega_1 + \omega_2)t] \qquad (5.6)$$

Without showing the detailed steps of the calculation, we obtain the following output voltage equation from the nonlinear (quadratic) term:

$$v''_o = 10 - 9\cos (2\pi 220t) + 6\cos (2\pi 55t) - 6\cos (2\pi 275t) - 1\cos (2\pi 330t)\text{V}$$

The net output voltage is the sum of the linear and nonlinear responses, which is written here with the terms ordered in ascending frequency.

$$v_o = v'_o + v''_o$$
$$= 10 + 6\cos (2\pi 55t) + 15 \sin (2\pi 110t) + 5 \sin (2\pi 165t) - 9\cos (2\pi 220t)$$
$$-6\cos (2\pi 275t) - 1\cos (2\pi 330t)\text{V}$$

Working from left to right, the first term 10 is a DC voltage that would be blocked by a coupling capacitor. The 55 Hz term is the difference between the two input frequencies, which also happens to be note A1. The 110 and 165 Hz (A2 and E3) terms are the output components we would obtain from an undistorted, linear amplifier. The 220 Hz (A3) term is one octave above A2. The 275 Hz term is just slightly below C#4 (277.2 Hz), and the final 330 Hz component is precisely note E4. So, in terms of musical pitch, here's what we have at the output:

A1,A2,E3,A3,C#4 (almost), and E4

This happens to form a very nice, melodic chord. If you happen to have a piano handy, you could listen to it.

Time-domain waveforms for signals like these are hard to visualize, but they can easily be plotted using a computer program such as MATLAB, or PSpice, or even using a graphics calculator such as the trusty old TI-86. Here, PSpice was used to simulate the amplifier. With no distortion, we obtain the waveforms shown in Fig. 5.10a.

When the quadratic distortion term is included, we obtain the output voltage plot shown in Fig. 5.10b. This distortion would actually sound pretty nice to most people, but combining more frequency components and driving the amplifier into

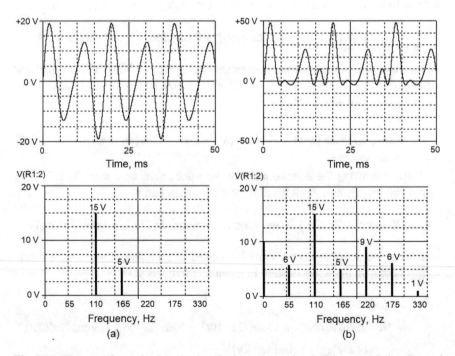

Fig. 5.10 Time- and frequency-domain plots for (a) undistorted and (b) intermodulation distorted output waveforms

cubic (kv_{in}^3) and higher distortion will create a much more complex output spectrum that will almost certainly contain dissonant frequency components. Whether this distortion would sound good or bad, appropriate or inappropriate, is a matter of opinion. If you are feeling very energetic, you might want to see what happens if a cubic gain term such as $v'''_o = 0.5v_{in}^3$ is included in the response.

Influence of Amplifier Design on Distortion

There are many factors that affect amplifier distortion characteristics. Some of these factors are device characteristics (tubes vs. BJTs vs. FETs), biasing and Q-point location (class A vs. class B vs. class AB), basic circuit topology (single-ended, push-pull, diff amp, etc.), and the use of local and/or global feedback. Less obvious factors such as non-ideal power supply operation can also affect amplifier overdrive characteristics.

Effects of Negative Feedback

The overall "sound" of an amplifier as well as its distortion characteristics often depends more on how negative feedback is used in the amp than on the types of devices used in the design. Recall that negative feedback makes the amplifier behave more linearly, which gives the amplifier transfer function nearly perfect odd symmetry, as was shown back in Fig. 5.4.

Generally, amplifier designs that utilize a lot of negative feedback will be very clean, exhibiting very low distortion when not overdriven. This approximates the characteristics of the hypothetical ideal amplifier which is quite desirable in high-fidelity audio applications but will normally result in a flat, sterile sounding guitar amplifier. It is worth restating that amplifiers that use large amounts of negative feedback also tend to exhibit hard, symmetrical clipping which results in harsh sounding overdrive characteristics consisting almost completely of odd-order harmonics.

Solid-state amplifiers have a reputation for producing harsh sounding distortion, which makes sense when you consider the design of the typical solid-state power amplifier. In order to achieve maximum power output and decent efficiency and keep cost low, solid-state amplifiers are often built using combinations of op amps and/or discrete transistors that drive a class AB output stage. Op amps normally require the application of huge amounts of negative feedback because they have such high open loop gain.

Many important parameters of discrete transistors such as beta and leakage current vary greatly from one device to another. Many of these parameters are strongly temperature dependent as well. Because of this, significant levels of negative feedback are typically employed throughout most solid-state amplifier designs which swamps out the effects of parameter variation.

Tube-based guitar amplifier designs generally employ very little, if any, negative feedback. Because of this, the nonlinearities of the tubes used in the amp are not suppressed. Even when not being overdriven, the tube amplifier designed with little or no negative feedback will exhibit more distortion than the typical solid-state amplifier.

Single-Ended vs. Push-Pull

The common emitter and common source, class A amplifiers of Chap. 3 are also sometimes called *single-ended amplifiers*. A single-ended amplifier usually has a single output device (BJT, FET, or tube) driving the load. In audio applications, single-ended amplifiers are always class A. As stated before, solid-state power amplifiers are almost never class A designs, but are usually class B or AB, push-pull designs.

The push-pull output stage has more symmetrical overdrive characteristics than the single-ended output stage. The resulting transfer characteristic is dominated by odd symmetry and so will produce primarily odd-order harmonic distortion. The typical transistor-based class AB push-pull stage will exhibit symmetrical *hard clipping* when overdriven. Hard clipping means that there is a sharp discontinuity where the peaks of the output signal flatten out. The combination of massive amounts of negative feedback, odd transfer function symmetry, and hard clipping causes the output to contain high-amplitude odd harmonics, which sound harsh and give the amplifier the infamous "transistor" sound when overdriven.

As we will soon see in Chap. 7, the output stage of a vacuum tube-based guitar amp will either be a single-ended class A design or a push-pull configuration (possibly class A as well). Like the class A transistor amplifier transfer function of Fig. 5.8b, the vacuum tube-based single-ended output stage transfer function has neither even nor odd symmetry and so will tend to produce both even and odd harmonics, with the second harmonic dominating initially. This is generally perceived as a nice, warm-sounding distortion.

Compared to solid-state push-pull stages, vacuum tube push-pull tube stages tend to exhibit *soft clipping*. This means that the peaks of the output signal flatten out in a smoother, more gradual way. Soft clipping is due partly to the inherent characteristics of tubes and because of the minimal use of negative feedback used in the design of tube guitar amps. So, although the tube-based push-pull stage will typically have mainly odd transfer function symmetry, soft clipping and unsuppressed tube nonlinearities cause odd harmonics to grow more slowly.

Effects of Device Transfer Characteristics on Distortion

We now need to revisit the BJT and FET transfer characteristic equations which we first looked at back in Chap. 3. And since it will be extremely important to us in the next chapter, we will also introduce the transfer characteristic equation for the vacuum tube triode as well.

BJTs

Bipolar transistors have an exponential transfer characteristic given by

$$I_C \cong I_S e^{V_{BE}/\eta V_T} \tag{5.7}$$

The various parameters are defined as follows. Lowercase italic variables (i_C, v_{BE}, etc.) represent the variation, or AC component of a given parameter, caused by the input signal.

I_C Total collector current, $I_{CQ} + i_C$
I_S Saturation or leakage current of the transistor
V_{BE} Total voltage drop from base to emitter, $V_{BE(on)} + v_{BE}$
V_T Thermal equivalent voltage ($V_T \cong 26$ mV at typical operating temperatures)
η Emission coefficient. $1 \leq \eta \leq 2$. Important in logarithmic and exponential circuits

As we noted in Chap. 3, the important thing to notice in (5.7) is that the presence of an input signal causes the exponent of the equation to vary. The collector current is exponentially related to the input voltage.

FETs

The drain current equations for the JFET and the MOSFET are square-law functions. That is, in each case, the drain current is proportional to the square of the input signal voltage. The JFET drain current is given by

$$I_D = I_{DSS} \left(1 - \frac{V_{GS}}{-V_P} \right)^2 \tag{5.8}$$

where the various parameters are defined as

I_D Total drain current, $I_{DQ} + i_D$
I_{DSS} Maximum drain current, $V_{GS} = 0$ V, $V_{DS} \geq V_P$
V_{GS} Total drop from gate to source, $V_{GSQ} + v_{GS}$
V_P Pinch-off voltage

The enhancement-mode MOSFET drain current is given by

$$I_D = k(V_{GS} -- V_T)^2 \tag{5.9}$$

The parameter definitions are

I_D Total drain current, $I_{DQ} + i_D$
k Device proportionality constant
V_{GS} Total voltage drop from gate to source, $V_{GSQ} + v_{GS}$
V_T Threshold voltage

Triodes

The schematic symbol for a vacuum tube *triode* is shown in Fig. 5.11. The name triode is derived from the fact that there are three terminals, the *cathode*, *grid*, and *plate* (or *anode*). The grid controls the flow of current from the cathode to the plate, much the same way that the gate of the JFET controls drain current.

You may recall from Chap. 1 that vacuum tube diodes behave according to the Child-Langmuir law. There is a similar relationship for the triode. The plate current of a triode is given by

$$I_P = k(V_P + \mu V_G)^{3/2} \tag{5.10}$$

I_P Total plate current, $I_{PQ} + i_P$
k Perveance. Equivalent to proportionality constant k for the MOSFET
V_P Plate voltage[1]

Fig. 5.11 Schematic symbol for the vacuum tube triode

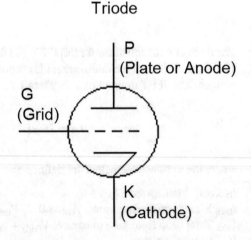

Triode

P
(Plate or Anode)

G
(Grid)

K
(Cathode)

---

[1] Technically, these should be V_{PK} (plate-to-cathode voltage) and V_{GK} (grid-to-cathode voltage), but it is a common practice in tube amp literature to omit the reference to the cathode in this notation.

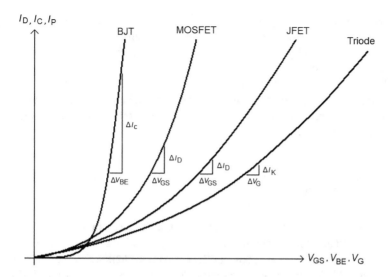

Fig. 5.12 Transconductance curve comparison for BJT, JFET, MOSFET, and triode

μ Amplification factor. Maximum possible voltage amplification
V_G Total grid voltage $V_{GQ} + v_G$ [7]

The significant characteristic to note here is that the grid voltage controls the plate current according to the 3/2 power law.

A comparison of BJT, JFET, MOSFET, and triode transconductance curves is shown in Fig. 5.12. Keep in mind that these curves are just for comparison purposes to emphasize the similarities and differences between the transfer characteristics of these devices. For example, depending on the triode selected, the actual transconductance curve might initially be much steeper than that of the typical JFET, but it will still be a 3/2 law curve.

The important thing to note here is that the BJT curve has a very radical change in slope compared to either the FETs or the triode. The triode curve is gentler than either the exponential function of the BJT or the square law (parabolic) curves of the FETs. So we see that the triode is the most linear device, followed by the FETs and then the BJT which is the most nonlinear.

All things being equal, the more linear the device behaves, the less severe both harmonic and intermodulation distortion will be.

Remember, the nonlinearity of a given device is not as significant in either solid-state or tube amplifiers that use lots of negative feedback. Negative feedback swamps out the nonlinearity, except when clipping occurs. If you want flexibility in tweaking the character of distortion produced by your amplifier, it is best not to use much, if any, overall negative feedback.

There is much debate on the relative sound qualities of amplifiers built using tubes, BJTs, and FETs. Although there are many good sources of information on this subject, there are also many opinions based on poor to nonexistent understanding of basic circuit operation and design principles. Hopefully, I have provided you with enough technical background to sort reasonable and technically plausible information from nonsense.

When you come right down to it, guitar amplifier sound quality is totally subjective, and what sounds good to one person might sound horrible to another. In general, however, most guitarists, myself included, prefer the sound of tube amps over solid-state amps.

Effect Bypassing

Before we dive into effects circuit analysis and design, let's first discuss methods of effect bypassing, which is simply switching a given effect in and out of the signal path. There are two basic methods of effect bypass switching. The simplest method is to place a SPDT switch at the output of the effect, as shown in Fig. 5.13a. The advantage of this simple bypassing approach is that a very inexpensive SPDT switch is used. A significant disadvantage is that the effect is always connected to the signal source, which may be a guitar or another effects circuit. This may cause excessive source loading and amplitude variation when effects are switched in and out of the signal chain.

The switching arrangement of Fig. 5.13b is called *true bypass*. Here, a DPDT switch connects or disconnects the input and output of the effect circuit from the signal chain. This has the advantage of eliminating loading effects on the guitar as effects are switched in and out of the signal chain.

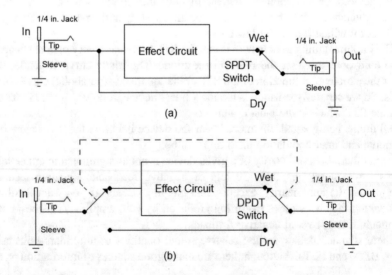

Fig. 5.13 (a) Simple effect bypassing. (b) True bypassing

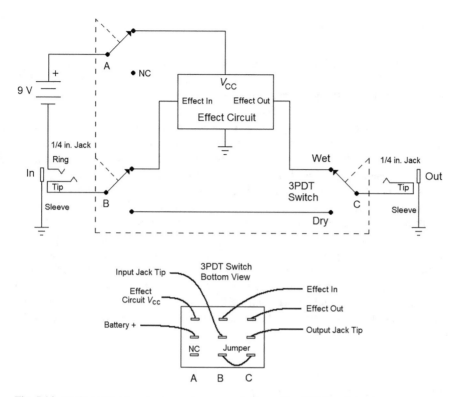

Fig. 5.14 Implementing true bypass and power switching with a 3PDT switch

Many effects circuits have power switching implemented via the input jack. It is also nice to be able to leave a long string of effects boxes connected with patch cords and have power controlled via the effect bypass switch. A typical switching arrangement that implements true bypass and power switching is shown in Fig. 5.14.

True bypass with power switching requires a three-pole, double-throw (3PDT), push-on, push-off switch. The lower part of Fig. 5.14 shows the wiring connections to the terminals on the back of the switch. These switches are available specifically for stomp-box applications for 4 or 5 dollars. This arrangement requires the input jack to be present and the effect to be switched in via the 3PDT switch for the battery to power up the circuit.

In order to reduce complexity, true bypass/power switching will only be shown occasionally on the effects schematics presented in this chapter. Keep in mind, however, that you can always add this feature to effects circuits that you build.

Overdrive Circuits

An overdrive circuit is simply a preamplifier that boosts the guitar signal to a high enough level that the input stage of the main amp is overdriven. The overdrive circuit itself does not necessarily create any distortion. Rather, the distortion is generated in the front-end stage of the main amplifier.

The sound of the distortion produced by overdrive will vary from one amplifier to another. Amplifiers that use op amps at the front end will most likely produce the typical harsh, odd harmonic distortion. If the amplifier has a class A transistor input stage, the distortion produced will usually be somewhat less harsh sounding. If the front end is a typical class A tube amplifier, the overdrive will tend to produce a warm, bluesy sounding distortion. Personally, most of the time, I prefer the more subtle sound of overdrive to that of other types of distortion.

If the overdrive circuit gain is high enough, and if the circuit has an output level control, it may be possible to produce distortion in the overdrive circuit itself, without overdriving the main amp. Technically, this is now a distortion effect. It is also possible to distort both the overdrive circuit and the main amp. I'm not sure what this would be called, but you can see that sometimes it is difficult to distinguish one type of effect from another. In this section, we will concentrate strictly on the overdrive aspect of circuit operation.

Single-Stage Transistor Overdrives

The cool thing about designing with discrete transistors is that you can devise so many unique overdrive circuit configurations. There are hundreds of different combinations of silicon, germanium, NPN and PNP BJTs, JFETs, and MOSFETs that can be used. Each circuit will tend to have its own unique character. This section presents a few of these possibilities.

Detailed Q-point and gain calculations will not be presented for the circuits shown here, but you should be able to apply the analysis techniques and equations presented back in Chap. 3 to verify the stated values. Of course, building and testing the circuits is a great way to verify operation. These low-power, battery-operated circuits are quite safe and fun to experiment with.

Simple BJT Overdrive

A very simple overdrive circuit is shown in Fig. 5.15. Depending on the winding resistance of the guitar pickups, this circuit provides a maximum effective gain of about 50 or so, which should be more than sufficient to overdrive any typical amplifier. The gain of the transistor may be adjusted with the 500 Ω potentiometer.

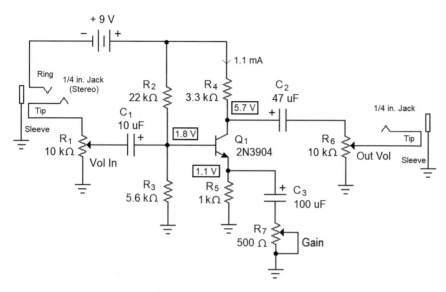

Fig. 5.15 Simple single-stage overdrive circuit. Potentiometer R_7 adjusts gain

A potential problem with the circuit of Fig. 5.15 is that the input impedance of the amplifier is quite low ($R_{in} \cong 1.2$ kΩ for this circuit). This places a heavy load on the guitar pickup, which tends to reduce output and reduces high frequency response. The low input resistance of this circuit can also cause severe level adjustment problems when multiple effects are cascaded. However, this simple circuit works reasonably well as an overdrive and, because of its relative simplicity, makes a good starter project.

Sziklai Overdrive Circuit

A few other possible single-transistor overdrive circuits that have higher input resistance are now presented. The circuit in Fig. 5.16 uses a Sziklai transistor, which in this case results in an amplifier with a minimum input resistance of about $R_{in} \cong 50$ kΩ and voltage gain that can be varied approximately over the range 5 to 100. Approximate DC voltages, with respect to ground, are shown at various points in the circuit. This circuit is preferable to the single BJT overdrive of Fig. 5.15.

Darlington Overdrive Circuit

A Darlington version of the overdrive circuit is shown in Fig. 5.17. This circuit is very similar to the Sziklai circuit. The minimum input resistance is a bit higher at $R_{in} \cong 80$ kΩ, but the voltage gain is lower, approximately adjustable over the range of 5–45.

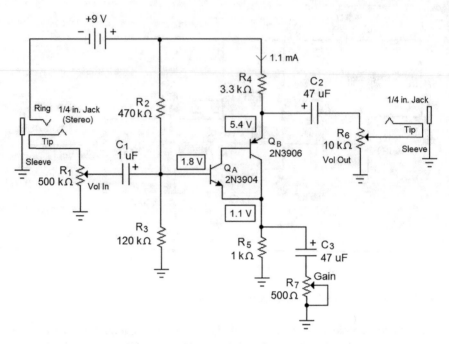

Fig. 5.16 Single Sziklai transistor overdrive with variable gain

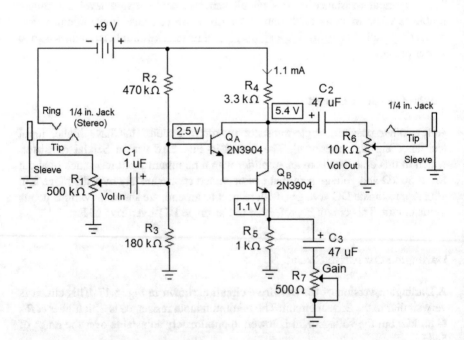

Fig. 5.17 Single Darlington transistor overdrive with variable gain

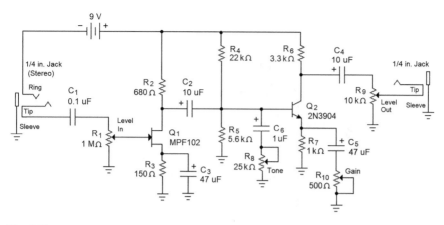

Fig. 5.18 Two-stage, JFET/BJT variable gain overdrive circuit

Multiple-Stage Overdrive Circuits

Using multiple stages allows us to combine different amplifier configurations to obtain more unique response characteristics, as well as very high gain if desired. High gain also allows the use of various tone control circuits without adversely affecting overdrive capability.

JFET/BJT Overdrive

The two-stage circuit of Fig. 5.18 is adapted from the piezo pickup preamplifier of Chap. 3. Because of the JFET front end, this overdrive circuit has high input resistance, $R_{in} = 1$ MΩ. Overall voltage gain is quite high, ranging from about 30 with gain set to minimum to over 300.

Because the voltage gain of this circuit is so high, it is practical to use more sophisticated tone control circuits in place of the simple treble-cut circuit shown. For example, the Vox-type tone control has a loss of 18 dB, which is a factor of 0.126. For the circuit of Fig. 5.18, this reduces the maximum voltage gain from 300 down to $A_{V(max)} = 300 \times 0.126 = 37.8$. This is quite a significant loss in gain, but still more than adequate for an overdrive.

MOSFET/PNP Germanium Overdrive

The two-stage MOSFET/PNP germanium BJT amplifier we examined in Chap. 3 also can serve as an overdrive circuit. The input resistance is fairly high at $R_{in} = 100$ kΩ, so loading of other effects is not a concern. The overall voltage gain of the original circuit was about 1000 which is really too high for our purposes. Fortunately, adding a Baxandall tone control circuit introduces enough loss to bring the gain down to manageable levels. The circuit is shown in Fig. 5.19.

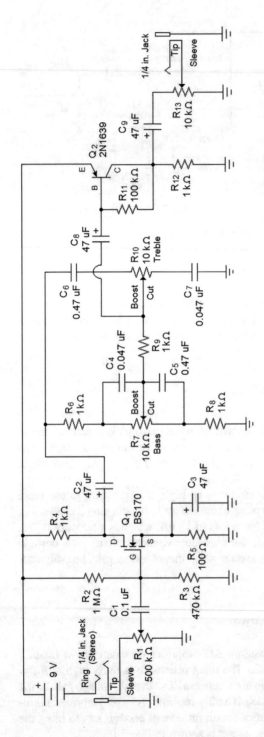

Fig. 5.19 Two-stage overdrive with Baxandall tone control

The component values for the tone control section in Fig. 5.19 are vastly different than those originally suggested for Fig. 2.28. The original component values were chosen based upon typical input resistances found in vacuum tube amplifiers, which are generally around 100 kΩ or greater. Here, we are dealing with the very low input resistance of the PNP transistor stage, $R_{in} \cong 313\,\Omega$, so the tone circuit was modified to match this low resistance level. Accounting for the loss of the tone circuit (around 30 dB), the overall gain is about 30.

An Op Amp Overdrive Circuit

We've been concentrating on overdrive circuits using discrete transistors, but op amps can be used just as well. The circuit shown in Fig. 5.20 will operate with any of the dual op amps listed in the schematic and others too. If you decide to build this circuit, you can experiment with different op amps to see if there are any perceivable differences in sound quality from one device to another. The gain of the circuit is A_V = 11 with component values shown, but this is adjustable by changing either R_2 or R_F. Personally, I prefer using discrete transistors over op amps in both overdrive and distortion circuits, but that is just my personal preference.

You may wish to try devising your own overdrive circuit using a combination of op amps and discrete transistors. The design possibilities are nearly endless, and you might come up with a very cool, nice sounding circuit.

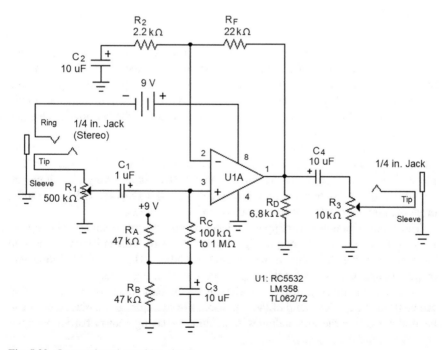

Fig. 5.20 Op amp-based overdrive circuit

Distortion Circuits

Distortion circuits are not necessarily designed to overdrive the main amplifier but rather distort the guitar signal prior to being applied to the amp. Of course, simultaneous overdrive and distortion is possible, but we will be concentrating specifically on distortion circuits in this section.

Fuzz vs. Distortion

There is often some debate about the differences between fuzz and distortion. Any circuit that alters the harmonic content of a signal is by definition distorting that signal, so fuzz is simply a form of distortion. The difference between classical fuzz and distortion is really subjective, and there is no unambiguous, quantitative way to differentiate between them.

Fuzz boxes were some of the earliest guitar distortion effects used, and they tended to be very simple BJT circuits that introduced distortion which sounded much different than overdriven tube amps of the day. The introduction of more sophisticated distortion circuits and digital signal processing (DSP)-based effects has made the distinction between fuzz and distortion somewhat irrelevant.

Diode Clippers

There are two very simple ways of introducing distortion into the guitar signal. One way is to overdrive one of the amplifiers somewhere in the signal chain. Another relatively simple way to introduce distortion is to clip the output of an amplifier with diodes as shown in Fig. 5.21. Don't be distracted by the op amp portion of the circuit. The section we are interested in here is the circled part of the schematic. This is a classic wave shaping circuit called a *diode clipper*, or *limiter*, in some applications.

The operation of the clipper is very simple; assuming silicon diodes are used, when the output of the op amp reaches about +0.7 V, diode D_1 will conduct, preventing further increase of the output. On negative output swings, D_2 limits the output to −0.7 V. Almost any diodes may be used, with 1N4148, 1N914, 1N400x, or germanium diodes (1N34 or similar) being very common.

The simple silicon diode clipper produces relatively hard, symmetrical clipping. As you might expect, the output signal will consist of the fundamental and odd harmonics, so the distortion is fairly harsh sounding. The transfer function and output voltage waveforms for this clipper are shown in Fig. 5.22a.

There are a number of different variations on the basic diode clipper circuit that can be used to alter the sound of the distortion. For example, germanium diodes can be used in place of the silicon diodes. In addition to having a lower barrier potential

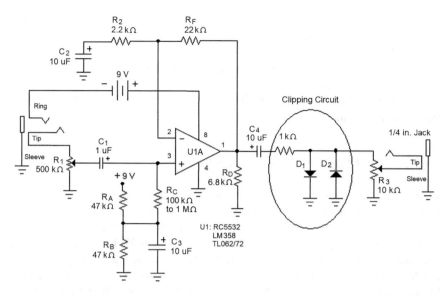

Fig. 5.21 Diode clipper distortion circuit

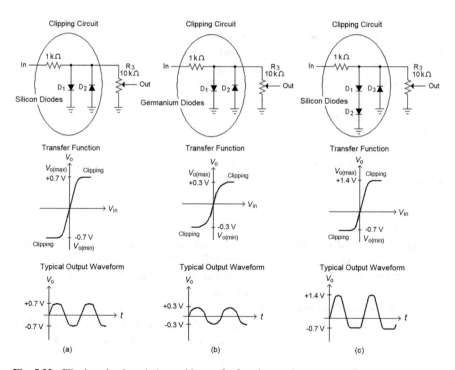

Fig. 5.22 Clipping circuit variations with transfer functions and output waveforms

than silicon, germanium diodes have a smoother, more gradual transconductance curve, so they clip somewhat more softly than silicon diodes. This is shown in Fig. 5.22b.

Asymmetrical positive and negative clipping levels can be obtained by placing different numbers of diodes in series, as shown in Fig. 5.22c. This causes both even and odd harmonics to be created initially, but once the circuit is heavily into clipping, odd harmonics dominate the output spectrum.

Asymmetrical Clipper with Power Indicator

It isn't necessary to use an op amp in the design of a clipping circuit. Discrete transistor amplifiers can be used just as well. The circuit of Fig. 5.23 is a two-stage amplifier using our favorite old standbys, the JFET input stage with a BJT common emitter stage driving a diode clipper. The 1N34 shown as D_1 in the schematic is a germanium diode, and D_2, the 1N4148, is silicon, but just about any silicon or germanium diodes you have on hand would work. Though it is subjective, I prefer the sound of the distortion produced by the asymmetrical clipper to that of the symmetrical version.

With the clipper diodes oriented as shown and the distortion pot wiper set to the bottom position, the output signal will begin to clip at about 0.3 V on positive swings and at about -0.7 V for the negative-going output swing. The 2N3904 positive output swing is limited to about 2 V_{pk}, while negative output swing is about -4 V_{pk} with the 1 kΩ distortion pot wiper set to the upper end. If you would like to experiment a bit, you could try different diodes or reversing the relative orientation of the diodes as shown in the schematic.

This circuit includes the usual bypass/power switching and something new, a LED power indicator. The LED current is equal to I_{DQ} for the JFET, which will be about 3 mA, so the LED will be fairly bright. It is best to stick with a red LED in this circuit, which will have a forward voltage drop of about 1.6 V at this current level. In order to compensate for the voltage drop across the LED, the JFET drain resistor was decreased slightly from previous examples. By the way, the LED will not increase battery current drain, because it is in series with the drain of the JFET.

The JFET stage is used primarily to provide high input resistance so that loading is negligible. Because of the reduced drain resistor value and the rather low input resistance of the BJT stage, the gain of the JFET is about unity. The gain of the BJT stage is about $A_V \cong 25$ so that, overall, the circuit has good sensitivity and could be driven effectively by nearly any magnetic or piezo pickup.

Overall, I really like the sound of this distortion circuit. Optional 0.1 μF capacitor C_6 provides a small amount of filtering and smooths out the clipping slightly. This capacitor can be eliminated or switched in and out of the circuit with an SPST switch.

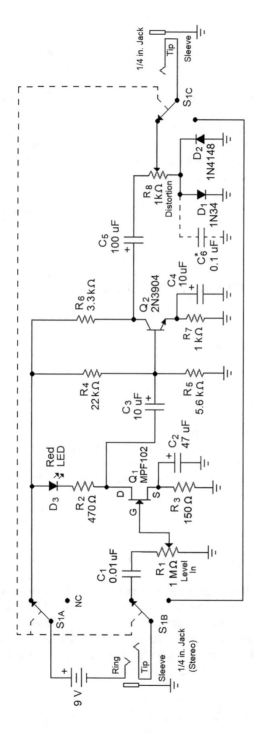

Fig. 5.23 Discrete BJT/diode clipper distortion circuit, with bypass and power indicator

Adjustable Op Amp Distortion Circuit

The circuit in Fig. 5.24 is a variation on the diode clipper that I designed many years ago. The circuit was simulated using PSpice, producing the output waveforms shown on the right side of the illustration. The circuit was also tested in the lab and found to produce very distinctively different sounding distortion at various settings of potentiometer R_6.

When a sinusoidal input signal is applied, with R_6 set to about 2.5 kΩ, the output distortion is very symmetrical as shown in the center waveform display. At either extreme of rotation, the output waveform clips asymmetrically, which causes even harmonics to be created, significantly changing the sound of the output signal.

If you feel like experimenting with this circuit, swapping in a germanium diode (1N34 or similar) or perhaps an LED for D_1 or D_2 could be interesting. The actual clipping level obtained with a given LED will be related to the LED color. Recall from discussions in Chap. 1 that in general, the shorter the wavelength of the light (the bluer the light), the higher the LED barrier potential V_F will be. For example, a typical red LED may have $V_F \cong 1.8$ V, while a similar green LED may have $V_F \cong$ 2.2 V, and a blue LED could have $V_F \cong 3$ V or more.

Logarithmic Amplifiers

It is also possible to cause clipping by placing diodes in the feedback loop of the amplifier as shown in Fig. 5.25. This circuit is a form of *logarithmic amplifier* or simply a log amp. The output of the op amp is proportional to the natural logarithm of the input voltage. For the following discussion, keep in mind that when the input signal goes positive, D_1 is forward biased, while D_2 is reverse biased. That is, D_1 conducts, while D_2 acts like an open circuit. Conversely, a negative-going input causes forward biasing of D_2 while reverse biasing D_1. The diode that is reverse biased may be ignored.

The gain of the op amp is dominated by the forward-biased diode, and we can ignore the reverse-biased diode. However, we don't treat the forward-biased diode as a short-circuit or even a simple linear resistance.

Here's a qualitative way of thinking about the operation of the log amp. At low forward bias current levels, diodes have high dynamic resistance r_D; therefore, in this circuit, a small input voltage (positive or negative) will experience very high voltage gain. As the input signal increases, the output of the op amp increases, and the forward-biased diode carries more current, causing its forward resistance to drop, reducing the voltage gain of the circuit. In a nutshell, the log amp has very high gain for small input voltages and low gain for large signals. The voltage gain decreases exponentially as the input voltage increases.

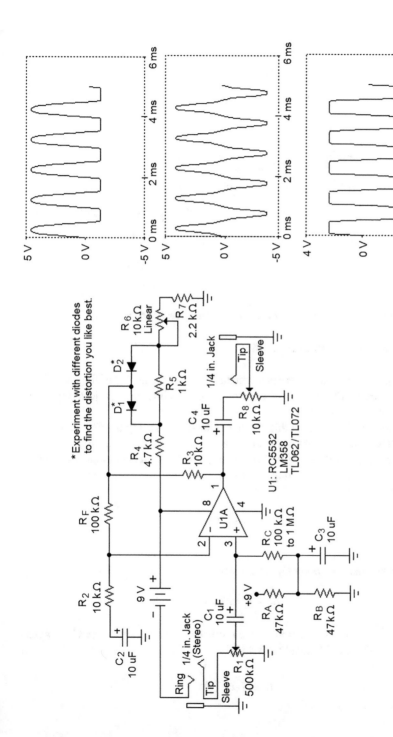

Fig. 5.24 Op amp-based distortion circuit

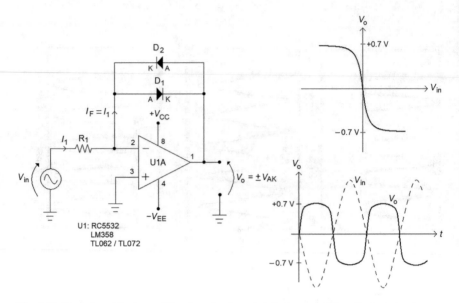

Fig. 5.25 Basic logarithmic amplifier, transfer characteristic curve, and input/output signals

Log Amp Output Equation Derivation

For those that are interested, here is the math behind the behavior of the logarithmic amplifier. If you don't want to suffer through the algebra, you can just skip ahead without incurring any penalties.

The logarithmic behavior of this circuit can be derived using the Golden Rules of op amp analysis and the transconductance equation for a forward-biased diode. Assume that V_{in} is positive. D_2 will be reverse biased, so it acts as an open circuit and may be ignored. The inverting input terminal of the op amp is a virtual ground, so the current into R_1 is

$$I_1 = V_{in}/R_1$$

No current enters the inverting input; the op amp forces this current to flow through the feedback loop via D_1, so we have

$$I_F = I_1 = V_{in}/R_1$$

The transconductance equation for the diode (sometimes called *Shockley's equation*, which is virtually identical to (3.45)) is

$$I_F \cong I_s e^{V_{AK}/\eta V_T} \tag{5.11}$$

In order to maintain virtual ground at the inverting input, the op amp output voltage must be equal to

$$V_o = -V_{AK}$$

Finally, a useful equation for V_o is found by solving (5.11) for V_{AK} and back substituting, which gives us

$$V_o = \eta V_T \ln \left(\frac{V_{in}}{R_1 I_S} \right) \tag{5.12}$$

The main thing to notice in (5.12) is that the output voltage is proportional to the natural logarithm of the input voltage. The operation of the circuit is exactly the same for negative-going input voltages, except that diode D_2 will be conducting, and D_1 acts as an open circuit.

The analysis of the log amp shows a very interesting phenomenon where the transfer function of the device enclosed within the feedback loop of the op amp will be transformed into its inverse function at the output of the op amp. The logarithm and exponential functions are inverses of one another. In this circuit, the exponential behavior of the forward-biased diode is converted to logarithmic behavior at the output of the op amp. If we were to place a device with quadratic (x^2) transconductance in the feedback loop, the output voltage would be proportional to the square root, $V_o = k\sqrt{x}$. Pretty cool!

Log Amp Distortion Circuit

I had an idea for a distortion circuit based on a log amp, combined with a summing amplifier. The circuit is shown in Fig. 5.26. In this circuit, the output of the log amp is summed with the original input signal. Resistor R_2 limits the maximum gain of the log amp to $A_V = -10$. The input resistance to this circuit is $R_{in} = R_1 \parallel R_4 = 5$ kΩ.

At low input voltage signal levels ($V_{in} \leq 50$ mV$_{pk}$), the output of the summing amp is relatively undistorted. At these small-signal levels, the diodes are nearly open circuits, and the log amp behaves like an inverting amp with $A_V = -10$. Voltage V_2 has negligible effect on the output at this point.

As the instantaneous value of the input voltage goes above about 50 mV$_{pk}$, the gain of the log amp begins to drop off. At this point, V_1 and V_2 begin to cancel, flattening the output voltage peaks.

As the input voltage increases beyond about 200 mV, the output of the log amp essentially stops increasing. V_2 keeps increasing and begins to swamp out the log voltage V_1. At this point, the summer output voltage will switch polarity, inverting the peaks of the output signal. This is shown in the output voltage waveforms of Fig. 5.27.

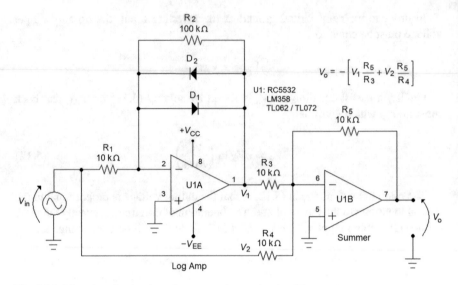

Fig. 5.26 Distortion circuit using a log amp and summing amplifier

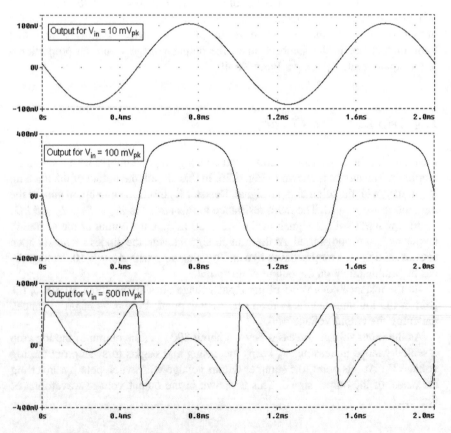

Fig. 5.27 Output waveforms from log amp distortion circuit

I included this circuit because the output waveforms are interesting, and the log amp is an important and useful circuit in its own right, but I don't like the sound of the distortion it produces; it's way too intense for my taste. Adding a tone control or even a simple adjustable treble-cut filter would probably help reduce some of the harshness of the distortion. Also, mixing different diodes in the circuit will produce asymmetry that will affect the sound of the distortion.

In the second edition of this book, I suggested that readers might want to experiment with the circuit to improve the sound of the effect. Several readers responded and sent me their designs, which was great. I also received some interesting designs and feedback at the Facebook page for the book as well (just search for Electronics for Guitarists).

Phase Shifters

One of the coolest sounding effect circuits is the *phase shifter* or *phaser*. The phaser effect is actually a form of a circuit called a *swept comb filter*. A comb filter has a frequency response with multiple peaks and dips, appearing much like the teeth of a comb, which is how the name was derived. Mixing the original instrument signal with a phase-shifted version produces a comb response. Sweeping the comb response back and forth produces the familiar swooshing phase shifter sound.

The basic block diagram for a phase shifter is shown in Fig. 5.28. The block labeled LFO is a variable low-frequency oscillator. The frequency of this oscillator determines how fast the phaser sweeps back and forth. The relative amount of the phase-shifted *wet* signal that is summed with the original *dry* signal determines the depth of the phaser effect.

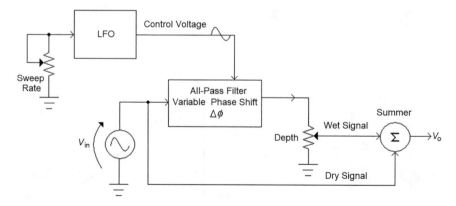

Fig. 5.28 Simplified block diagram for a phase shifter or phaser

The All-Pass Filter

The heart of the phase shifter is an interesting circuit called an *all-pass filter*. The schematic diagrams for two variations of the all-pass filter and their corresponding phase response plots are shown in Fig. 5.29. All filters have a phase response that varies with frequency, but the interesting thing about the all-pass filter is that its amplitude response is flat. The all-pass filter does not attenuate different frequencies like HP, LP, or BP filters do. The output phase equation for Fig. 5.29a is

$$\varphi = 180^0 - 2\,\tan^{-1}\!\left(\frac{f_{in}}{f_o}\right) \tag{5.13}$$

For the circuit of Fig. 5.29b, the output phase equation is

$$\varphi = -2\,\tan^{-1}\!\left(\frac{f_{in}}{f_o}\right) \tag{5.14}$$

The notation $\tan^{-1}$ is the *inverse tangent* or *arctangent* function. The critical frequency f_o for an all-pass filter is defined as the frequency at which the output phase is halfway between the min and max values. For both filters, this frequency is given by

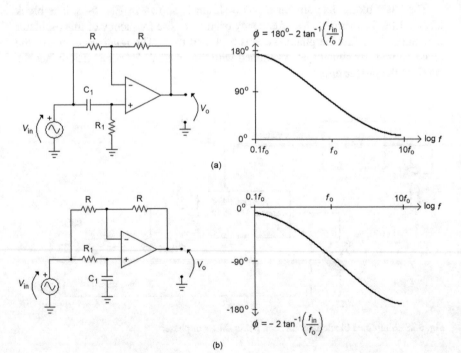

(a)

(b)

Fig. 5.29 All-pass filters. (**a**) Positive (leading) phase shift. (**b**) Negative (lagging) phase shift

$$f_o = \frac{1}{2\pi R_1 C_1} \tag{5.15}$$

The values of the gain-setting resistors, both labeled R in Fig. 5.29, are not critical as long as they are set equal to each other. The more closely these resistors are matched, the flatter the amplitude response will be. Note that R_1, which sets the critical frequency of the filter, can have any value; it is not related to the other two resistors.

As we saw in Fig. 5.28, in order to create a classic phaser, we need to sweep the response of the all-pass filter with an oscillator. If you examine (5.15), the only way we can change the critical frequency of the all-pass filter is by varying either R_1 or C_1. We need to be able to vary one of these parameters electronically, so that the phase shifter constantly sweeps f_o up and down. Voltage-controlled capacitors called varactor diodes are available, but these have very low capacitance (typically less than 100 pF). Varactor diodes also operate at very low signal levels, with peak voltages typically around 10 mV or so. At audio frequencies, the value of capacitor C_1 will typically need to be around 0.1 μF or so, which is impractical to implement with a varactor, and we are usually working with signals of several volts in amplitude which also rules out the varactor.

Optocouplers

Electronically variable resistors called *optocouplers* or sometimes *photocouplers* are ideally suited for the phase shifter application. An optocoupler is a light-sensitive resistor, called a *photoresistor*, which is encapsulated with an LED or incandescent light source. When the LED is forward biased, the photoresistor is illuminated, causing its resistance to decrease.

Although they are available commercially, optocouplers are somewhat hard to find and also fairly expensive. If you would like to experiment with a phase shifter, you can make your own optocouplers using inexpensive *cadmium sulfide* (CdS) cells, LEDs, and heat-shrink tubing. Figure 5.30 shows how an optocoupler can be fabricated.

Fig. 5.30 Constructing an optocoupler using an LED and CdS photoresistor

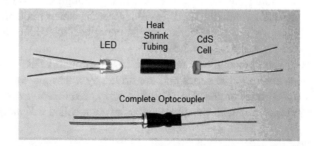

Cadmium sulfide photoresistors are specified in terms of minimum resistance R_{light} at a specific illumination level and maximum dark resistance R_{dark}. A good device for our purposes is the Perkin-Elmer model VT90N1, which is available from Digi-Key and Parallax, Inc. The parameters for the VT90N1 are $R_{light} \cong 10\,k\Omega$ and $R_{dark} \cong 200\,k\Omega$. Based on my experience with CdS photoresistors, the min and max values easily vary by $\pm 10\%$ or more, so these are definitely not precision devices.

CdS photoresistors are sensitive to visible light and may be used with just about any LED you might have lying around. I experimented with several different LED colors and found that high-brightness amber LEDs worked best, but high-brightness red LEDs worked ok as well.

Vintage optocouplers use incandescent light sources. Compared to LEDs, incandescent sources have slower response times and require much higher drive current but are sometimes preferred in order to more closely emulate the characteristics of vintage effects.

An Experimental Phase Shifter Circuit

An experimental phase shifter circuit is shown in Fig. 5.31. This circuit uses four cascaded all-pass filter sections, so the maximum phase shift approached at the asymptote for this circuit is 720°. The CdS photoresistors used in the prototype range from $R_{dark} \cong 150\,k\Omega$ and $R_{light} \cong 10\,k\Omega$, which allows the critical frequency to sweep between the following limits:

MinimumCritical Frequency	**MaximumCritical Frequency**
$f_{o(min)} = \dfrac{1}{2\pi R_{dark} C_1}$	$f_{o(max)} = \dfrac{1}{2\pi R_{light} C_1}$
$= \dfrac{1}{2\pi(150\,k\Omega)(0.01\,\mu F)}$	$= \dfrac{1}{2\pi(10\,k\Omega)(0.01\,\mu F)}$
$= 106\,Hz$	$= 1.6\,kHz$

The critical frequency range will vary depending on the exact values of R_{dark} and R_{light} for the photoresistors, which can vary significantly from one to another. The critical frequency range of the all-pass filters can be adjusted by reducing or increasing the values of capacitors C_2 through C_5. Using the next higher standard capacitor value (0.022 μF) would lower the center frequencies to $f_{o(min)} = 48$ Hz and $f_{o(max)} = 723$ Hz. Decreasing the capacitors to 0.0068 μF would give us $f_{o(min)} = 156$ Hz and $f_{o(max)} = 2.3$ kHz.

The optocouplers are labeled LDR_1 through LDR_4 in the schematic. The LEDs are connected in series and are driven by a triangle-wave low-frequency oscillator (LFO) built around U4. This is a form of *relaxation oscillator* in which U4A is a *comparator* with positive feedback (often called a *Schmitt trigger*) and U4B is an

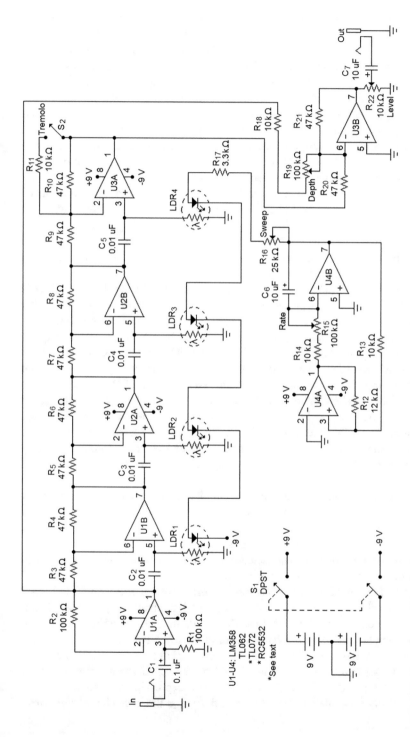

Fig. 5.31 Experimental 720° phase shifter

integrator. We will talk more about comparators in the sections that follow. The oscillation frequency is adjustable from about 0.2 Hz to 3 Hz using 100 kΩ linear potentiometer R_{15}. The amplitude of the oscillator output is approximately 12 V_{P-P}.

The phase-shift oscillator of Fig. 3.46 could also be adapted for use as an LFO with this circuit. This would be a good exercise in adapting an existing design to fit a different application. If you really want to tackle this project, the oscillation frequency range must be lowered significantly. And, in order to drive the LEDs sufficiently, the output voltage compliance of the oscillator should be increased. This would require a redesign of the phase-shift oscillator circuit to operate from ±9 V supplies, or perhaps a higher-compliance op amp buffer could be used. There is more than one correct solution to any design problem.

As mentioned previously, the classical whooshing effect of the phaser is achieved by summing the output of the last all-pass filter, the wet signal, with the original dry signal which is buffered by U1A. Mixing these signals causes dynamic cancellation and reinforcement of the harmonics present in the signal. The greater the number of all-pass sections, the more "teeth" the comb filter response has, and the more dramatic the phaser effect becomes.

Potentiometer R_{19}, labeled *Depth*, adjusts the amount of the dry signal that is summed with the wet signal produced by the all-pass network. Maximum comb filter notch depth is achieved when the wet and dry signals are equal in amplitude.

The *Sweep* potentiometer R_{16} adjusts the amount by which the critical frequencies of the all-pass stages are shifted by the LFO, by limiting the maximum brightness of the LEDs. Assuming the LEDs have forward voltage drops of about 1.5 V, when the depth pot is set to minimum resistance, the LED current reaches a maximum value of about 2.5 mA. Setting the depth pot to higher resistance limits the brightness of the LEDs and the maximum change of LDR resistance. When switch S_2 is closed, the feedback resistance of U3A drops to about 8.2 kΩ. This disrupts the normal operation of the fourth all-pass filter, causing the output amplitude to vary with frequency. The effect is an interesting combination of phase-shifting and tremolo.

When possible, I like to design effects circuits that use a single 9 V battery for power. I went with a bipolar power supply this time because the single-polarity version of the circuit would be much more complex, and because the total supply voltage would only be 9 V, the LDR LEDs would have to be wired in parallel since there would not be enough voltage headroom to drive them all in series. This would increase current drain, shortening battery life.

Using LM358 or TL062 op amps, total supply current is less than 10 mA, so it should be possible to use a MAX1044 (see Fig. 3.54) to create a negative supply rail and eliminate one of the 9 V batteries. Current drain using RC4558, RC5532, or TL072 op amps will exceed the 10 mA limit of the MAX1044.

If you feel up to building this circuit, you may need to experiment with the values of C_2–C_5 to get optimal performance, based on the actual characteristics of your LDRs. As mentioned earlier, using a sinusoidal sweep oscillator (such as a phase-shift oscillator) is another option that would change the character of the effect.

Although the circuit itself is very quiet, it will take up a lot of space on the protoboard, and you will probably have several fairly long wires running from place to place, so there's a good chance a prototype will pick up some noise.

Flangers

Flanging is an effect that occurs when a signal is subjected to a variable time delay and then recombined with the original, undelayed signal. It is commonly believed that the flanging effect has its origins in recording engineers playing identical tracks on two different reel-to-reel tape players. Very slightly slowing one of the reels (by placing a finger on the flange of the reel, hence the term flanging) will cause a variable time delay between the two signals, and the various frequency components in the signals will begin to sweep through modes of cancellation and reinforcement. Continuous slowing and resynchronizing the tracks results in the classic flanger sound. Flanging sounds somewhat similar to phase-shifting but is noticeably different.

Outside of a recording studio, tape player/recorder-based flangers are impractical. Today, the typical flanger is designed around an analog time-delay network, or the effect may be created using digital signal processing (DSP) techniques. Again, in keeping with the old-school approach I've taken here, we will not get into the DSP approach.

The basic operating principle of the flanger can be understood from examination of the block diagram shown in Fig. 5.32. The input signal is sampled and applied to the time-delay section. After some time (usually in the 1–10 ms range), the sample is available at the output. The time delay is continuously varied, determined by the clock frequency. The delayed "wet" signal is summed with the original "dry" signal, which produces a swept comb filter response giving us the flanging effect.

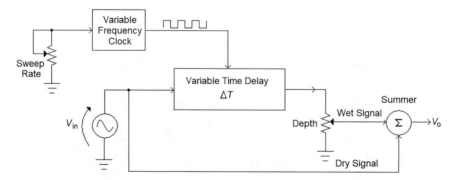

Fig. 5.32 Conceptual block diagram for a flanger

Flanging vs. Phase-Shifting

The typical phaser is built around an all-pass filter section, which has two definite phase limits at high and low frequencies. Consider a phaser made by cascading five all-pass stages of the form shown in Fig. 5.29a. Each section has a maximum phase shift of 180° at very low frequencies. The output of the last all-pass section has a phase angle of 180° × 5 = 900° as $f \rightarrow$ 0 Hz. As frequency increases, the output phase angle φ approaches zero degrees, following an arctangent curve. This is shown in Fig. 5.33a, where the phase angle is confined to a range of ±180 degrees.

Another way to think about the phase shift of the all-pass section is to imagine the unit circle on the complex plane. Again, starting at f = 0 Hz, the all-pass phase shift is 900° (or, equivalently, 180° on the unit circle). As frequency increases at a constant rate from 0 Hz, the phase vector of the all-pass section rotates clockwise

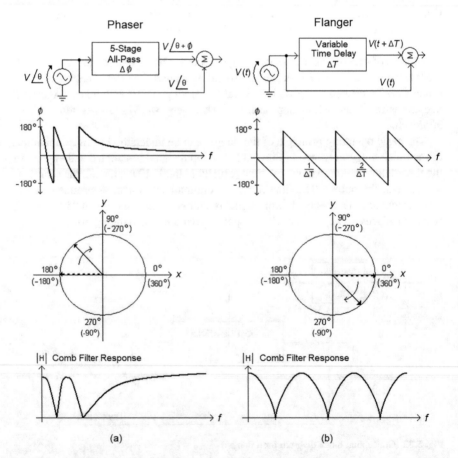

Fig. 5.33 (a) Phase response of the all-pass filter section and phaser frequency response. (b) Phase response of the flanger time-delay section and flanger frequency response

in the negative or lagging phase direction, starting at 180°. The phase angle changes rapidly at first but slows as frequency increases, finally approaching zero degrees asymptotically toward a total of 2.5 complete rotations (900° from the starting position). These phase cycles determine the number of "teeth" in the comb filter response and the spacing of the notches when the output of the all-pass section is summed with the original input signal.

In contrast to the phaser, the flanger is built around a time-delay section. The phase angle of the time-delayed output signal increases linearly with frequency (in the negative or lagging direction), ideally without any limit. As frequency increases continuously from 0 Hz, the output phase starts at $\varphi = 0°$ and rotates clockwise around the unit circle (lagging phase) at a constant rate. This causes the frequency response of the flanger to have equally spaced notches in its comb filter response, as shown in Fig. 5.33b.

The difference between the comb responses of the phase shifter and the flanger is what accounts for the difference in the sound of the two effects. Keep in mind that both the phase shifter and the flanger will have constantly varying response characteristics because the phaser all-pass critical frequency f_o and the flanger delay time Δt vary continuously with their LFO and clock signals. This causes the comb response curves to compress and stretch horizontally as the phaser LFO voltage and flanger clock frequency vary.

Bucket-Brigade Devices

The heart of the flanger is the time-delay network. The time-delay section is traditionally implemented using a specialized IC called a *bucket-brigade device* (BBD) or *sampled analog delay* (SAD) register as it is sometimes called. A BBD is an analog shift register, in which an analog input voltage is sampled, and that sample value is passed from stage to stage at a rate determined by a clock signal.

The basic structure of the BBD is illustrated in Fig. 5.34, where a 512-stage shift register is used as an example. The actual switches in the BBD chip are SPST MOSFET-based, *analog switches*, but I have taken some liberty by showing SPDT switches to make the main idea easier to understand (I hope). The analog switches are opened and closed by a digital clock signal. A two-phase clock is required for proper analog switch synchronization.

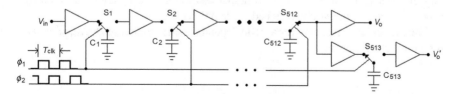

Fig. 5.34 Functional internal block diagram for a typical BBD

The input voltage is sampled on each falling edge of Φ_1. Small integrated capacitors serve as storage cells for the sampled voltages. It takes 512 clock cycles to shift a sample across to the upper output terminal V_o. The sampled voltage is present at V_o for half a clock cycle. An extra 513th stage V'_o can be summed with V_o to extend the output for an entire clock cycle. The total time delay for a sample to shift through register is $\Delta T = 512T_{CLK}$. The higher the clock frequency, the shorter the time delay will be.

The input signal applied to the BBD should have a DC offset of about $0.5V_{DD}$. This offset voltage is necessary because the BBD operates from a single-polarity power supply. We don't want the input signal peaks to exceed the supply rails (0 V and $+V_{DD}$), which would cause clipping and could possibly damage the BBD.

One of the most popular BBDs used in flanger designs over the years was the SAD-1024, which was manufactured by EG&G Reticon. The SAD-1024 is a 16-pin DIP that contains two independent, 512-stage N-channel MOSFET-based delay units. The 512-stage sections may be connected in series to form a 1024-stage device, doubling the maximum time delay. The SAD-1024 is designed to operate from a +15 V power supply and may be clocked from 1.5 kHz to 1.5 MHz. Minimum clock frequency is determined by the leakage of the internal storage capacitors. Maximum clock frequency is determined by how quickly the storage capacitors can be charged/discharged by the buffers in the SAD.

Unfortunately, the SAD-1024 has been out of production for many years, and the chips that are left tend to be expensive and hard to find. At the time this was being written, I had seen SAD-1024s selling online for as much as $50.00 each.

Alternative BBDs are the MN3xx series of devices produced by Panasonic/ Matsushita. These BBDs are out of production but still available from various sources. The PMOS devices (made with P-channel MOSFETs) operate from a negative power supply, while the NMOS devices use a positive power supply.

Pin-compatible replacements for a number of MN3xxx devices are being produced by a company called *CoolAudio*. These are all N-channel devices, designed to operate from a +5 V supply, but they can be used safely with supply voltages as high as +9 V. Table 5.2 highlights some major specifications for a sample of the MN3xxx and equivalent CoolAudio V3xx BBDs. The minimum clock frequency specified for these devices is 10 kHz, while the maximum specified clock frequency varies for different part numbers. In practice, these devices can be clocked at significantly higher frequencies with an increase in output distortion.

The maximum input signal voltages listed in Table 5.2 for the various BBDs are relatively low. These values are based on maximum harmonic distortion of 2.5%. Higher input voltages can be used if higher distortion is acceptable, but in no case should the peak input voltage exceed the supply rails.

The maximum time delay for a BBD depends on the number of stages, N, and the minimum clock frequency.

Table 5.2 Sample of MN3xx/V3xx series BBD specifications

Device	Type	V_{DD}	Stages	Taps	$f_{clk(max)}$	$V_{in(max)}$
MN3001	PMOS	−15 V	2 x 512	–	800 kHz	1.8 V_{rms}
MN3002	PMOS	−15 V	512	–	800 kHz	1.8 V_{rms}
MN3004	PMOS	−15 V	512	–	100 kHz	1.8 V_{rms}
MN3006	PMOS	−15 V	512	–	200 kHz	1.8 V_{rms}
MN3011	PMOS	−15 V	3328	396, 662, 1194, 1726, 2790, 3328	100 kHz	1.0 V_{rms}
MN3204	NMOS	+4 to 10 V	512	–	200 kHz	0.5 V_{rms}
MN3205 V3205SD	NMOS	+4 to 10 V	4096		100 kHz	
MN3206	NMOS	+4 to 10 V	512	–	200 kHz	0.36 V_{rms}
MN3207 V3207	NMOS	+4 to 10 V	1024	–	200 kHz	0.36 V_{rms}
MN3208 V3208	NMOS	+4 to 10 V	2048	–	100 kHz	0.36 V_{rms}
MN3214	NMOS	+4 to 10 V	1024	140, 379, 621, 798, 1024	200 kHz	0.36 V_{rms}

$$T_{D(max)} = \frac{N}{2f_{clk(min)}} \tag{5.16}$$

The minimum time delay is determined by the number of stages, N, and the maximum clock rate allowed by the BBD. This is given by

$$T_{D(min)} = \frac{N}{2f_{clk(max)}} \tag{5.17}$$

Based on the published clock frequency specs, the minimum and maximum delay possible for a few of the MN3xx devices and the SAD-1024 would be:

	MN3205	**MN3207**	**SAD-1024**
Stages	4096	1024	1024
$f_{clk(max)}$	100 kHz	200 kHz	1.5 MHz
$T_{D(min)}$	20.48 ms	2.56 ms	0.034 ms
$f_{clk(min)}$	10 kHz	10 kHz	10 kHz
$T_{D(max)}$	204.8 ms	51.2 ms	51.2 ms

Clock and LFO Generation

As shown in Fig. 5.34, BBDs require a two-phase clock signal. Specialized clock generator ICs are available, designed specifically for use with the MN3xx and V3xx chips. These clock generators are the MN3102 and the equivalent V3102D. The

Fig. 5.35 Clock generator
chip for MN3xx/
V3xx BBDs

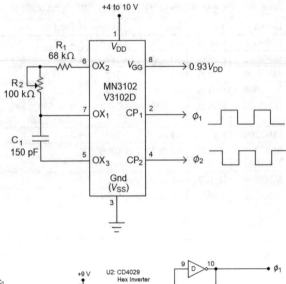

Fig. 5.36 Clock circuit for sweeping an NMOS BBD delay

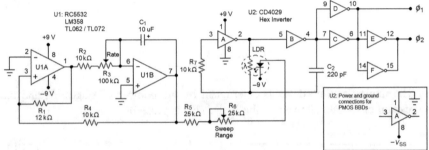

schematic for a variable frequency clock generator, compatible with the NMOS
BBDs, is shown in Fig. 5.35. The frequency is adjustable from approximately
50–120 kHz.

In order to electronically sweep the oscillation frequency of the clock generator,
an electronically variable resistor (or its equivalent) must be used in place of the
potentiometer.

You can build your own CMOS oscillator with an LDR variable resistor as shown
in Fig. 5.36. The clock inputs of BBDs have fairly high input capacitance (up to
700 pF). To help ensure that the CMOS oscillator has sufficient drive for the clock
inputs, unused inverters are connected in parallel which increases current drive
capability.

The value of R_5 may need to be adjusted to best suit the sensitivity of the LDR
that you are using in the circuit. Also note that for the PMOS BBDs the power
connections to the CMOS oscillator must be modified as a negative supply rail V_{SS} is
required. This modification produces clock pulses that vary from 0 V to $-V_{SS}$.

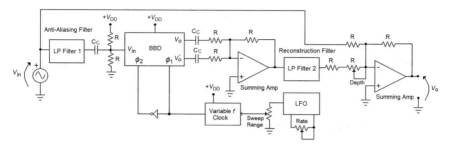

Fig. 5.37 Flanger using a BBD time delay

At the time of writing, the MN3xx and V3xx devices, and other stomp-box-related supplies, including photocells, optocouplers, and enclosures, are readily available from various sources, including www.smallbearelec.com, www.buildyourownclone.com, and www.coolaudio.com, among others.

A BBD-Based Flanger

The block diagram for a BBD-based flanger is shown in Fig. 5.37. The variable LFO sweeps the clock frequency that is applied to the BBD. The *Rate* pot determines the speed with which the time delay is varied. The *Sweep Range* pot controls the amount of deviation of the clock frequency. The *Depth* pot determines how much of the delayed wet signal is summed with the dry signal.

Sampling Frequency and Aliasing

An important result from the field of digital signal processing is called the *Nyquist-Shannon sampling theorem*, which says that the highest input signal frequency should be less than half the sampling frequency. In equation form, this is written as

$$f_{in(max)} \leq f_{clk}/2 \qquad (5.18)$$

The frequency $f_{clk}/2$ is often called the *Nyquist frequency*. If the input signal frequency exceeds the Nyquist frequency, *aliasing* will occur. Aliasing is a phenomenon that can occur in sampled signal systems in which signal frequencies higher than $f_{clk}/2$ are translated down to lower frequencies. If the input frequency exceeds the Nyquist frequency, but is lower than the clock frequency ($f_{clk}/2 < f_{in} < f_{clk}$), the frequency of the aliased signal will be

$$f_{\text{alias}} = f_{\text{clk}} - f_{\text{in}} \qquad\qquad (5.19)$$

Input frequencies greater than the clock frequency result in *folding*. Folding is another form of aliasing. I don't want to get too far into the weeds with this (I tend to do that enough already), so we'll leave the details of folding for the DSP books.

Anti-aliasing and Reconstruction Filters

In general, aliased frequency components are not harmonically related to the original signal and are usually undesirable. To help prevent aliasing, a low-pass input filter is used, with a corner frequency somewhat less than $f_{\text{clk}}/2$. This filter passes the desired input signal frequencies while attenuating higher frequencies which would be aliased. As you might expect, the low-pass input filter is often called an *anti-aliasing filter*, as shown in Fig. 5.37.

The output of the BBD will be a sampled approximation of the original input signal. This causes frequency components that are multiples of the clock frequency to be present in the delayed signal. These frequencies are removed by a low-pass filter at the BBD output, called a *reconstruction filter*. Basically, the reconstruction filter smooths the BBD output signal. Setting the reconstruction filter corner frequency slightly higher than the maximum signal frequency works well for this application. The phaser doesn't require these filters because it is a true analog design.

Oversampling

It's not a bad idea to sample at least three times faster than the highest signal frequency to be processed. Assuming $f_{\text{in(max)}} = 10$ kHz for convenience, a reasonable minimum clock frequency would be about 50 kHz, which oversamples by a factor of 5. Oversampling makes the design of the anti-aliasing and reconstruction filters less critical. We will revisit these concepts again in this chapter.

Chorus Effect

The chorus effect may be produced in much the same way as flanging, but there are some significant differences. The time delay for the chorus effect is typically longer, usually around 20–50 ms. The deviation of the time delay is typically not as large for the chorus as for the flanger, and in some cases, the clock frequency is not varied at all. In a flanger, the deviation of the time delay can be quite large, say a 10-to-1 ratio. The chorus effect is more subtle, with time-delay deviation of around ±20% or less. For example, if the time delay is $T_D = 20$ ms, then the LFO would be adjusted to cause the delay to vary about ±4 ms. The chorus effect can be made more full-sounding by using a BBD with multiple delay taps, such as the MN3214. The block diagram for a basic chorus effect is shown in Fig. 5.38.

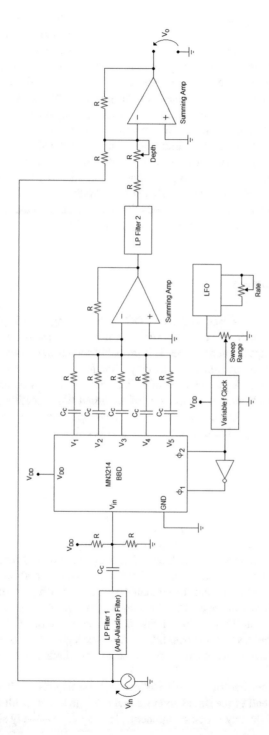

Fig. 5.38 Chorus effect using a multi-tap BBD

There are many opportunities for experimentation with the chorus effect circuit of Fig. 5.38. As with the phase shifter, the triangle-wave LFO could be replaced with a sinusoidal oscillator. Another potentially interesting circuit modification would be the replacement of the summing amp input resistors with potentiometers.

It is also possible to simulate a chorus effect using phase shifter-type circuits. Figure 5.39 shows a modified version of the phase shifter described earlier that produces a reasonable chorus effect. In this circuit, the outputs of each all-pass filter stage are summed at the output. This produces a much more complex form of frequency-dependent cancellation and reinforcement of the signal components. In addition, switch S_3 has been added to allow selection of fast or slow sweep speeds. In the fast position, the circuit adds a pitch modulation vibrato effect. This is really a very versatile and cool circuit to experiment with. With a bit of tweaking and some simple modifications, you can get sounds that range from chorus, to Leslie rotary speakers, to a Robin Trower "Bridge of Sighs" type of effect.

Envelope Followers

The *envelope follower* is used to derive a DC control voltage that is proportional to the trace of the envelope of a complex signal, which in our case is a signal produced by a guitar or perhaps a human voice. This probably sounds like a complex function, but basically, an envelope follower is a fancy name for a *precision rectifier* cascaded with a low-pass filter. A precision rectifier is a circuit that uses op amps to reduce barrier potential and other nonlinearities of a normal diode rectifier, producing a nearly ideal rectifier. As you will soon see, envelope followers are an essential building block in the design of many different effect circuits.

Signal Envelope

It is important to be clear on exactly what is meant by an envelope in the context of waveforms and signals. Complex waveforms like speech and musical instrument signals consist of a relatively wide range of spectral components that vary dynamically in overall amplitude at a much slower rate. This is a form of amplitude modulation. If we look at such a modulated signal over a time span that is long compared to the period of the average frequency content of the signal, we may see a waveform like that on the left side of Fig. 5.40. The envelope of the signal is the curve that traces the positive and negative peaks of the signal. Note that the envelope is symmetrical about the time axis. This is a common characteristic of speech and music signals.

In Fig. 5.40, the precision rectifier eliminates the negative half of the signal, leaving the upper half of the signal and its envelope. The low-pass filter removes the remaining high-frequency signal components, leaving the relatively slowly varying

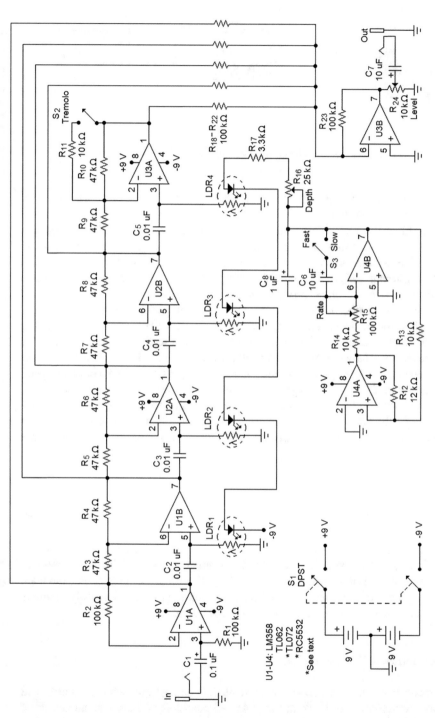

Fig. 5.39 Chorus/vibrato effect circuit adapted from a phaser

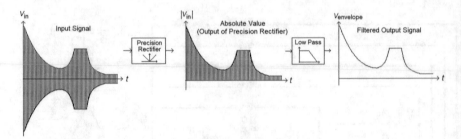

Fig. 5.40 Input signal processed by an envelope follower

envelope voltage. It is possible to remove the upper half of the envelope and retain the bottom if so desired.

The block diagram of Fig. 5.40 uses a precision full-wave rectifier, but a precision half-wave rectifier may be used here as well. The full-wave rectifier is usually preferable because the frequency doubling characteristic of this rectifier makes output filtering easier. The precision full-wave rectifier gives us an electronic equivalent of the *absolute value function*.

The low-pass filter is designed such that the envelope follower responds quickly enough to follow the envelope of the complex signal while still effectively filtering out the high-frequency components produced by the rectifier. Since the envelope varies slowly relative to the audio signal, the low-pass filter is designed for a corner frequency $f_{env} < f_C < f_{signal}$. A corner frequency of $f_C \cong 10$ Hz or so would be a good place to start if you were designing an envelope follower for a typical audio source.

Sometimes, rather than characterizing a first-order RC filter by its corner frequency f_C, the time constant $\tau = RC$ is used instead. The time constant approach is more commonly used when dealing with filters that have a very low corner frequency.

Precision Rectifier Circuits

The precision rectifier is the heart of the envelope follower. In addition to virtually eliminating diode nonlinearity and barrier potential effects, precision rectifiers can also provide gain if necessary. In terms of signal processing applications, the main disadvantages of any precision rectifier compared to passive, diode-only rectifiers are limited high frequency response and added complexity.

Precision Half-Wave Rectifiers

Two commonly used precision half-wave rectifier circuits and their typical input and output waveforms are shown in Fig. 5.41. All precision rectifiers take advantage of

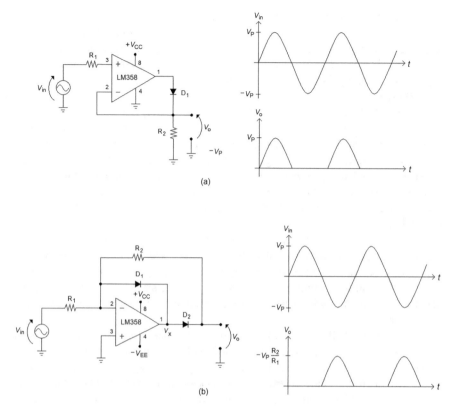

Fig. 5.41 Precision half-wave rectifiers and input/output waveforms. (**a**) Super diode. (**b**) Two-diode circuit

the fact that when a diode is placed inside the feedback loop of an op amp, its effective barrier potential is reduced to nearly zero ($V_F' \cong V_F/A_{OL}$). We first discussed this phenomenon when we examined the rail splitter of Fig. 4.17.

The circuit in Fig. 5.41a is sometimes called a *super diode*. For positive-going inputs, the op amp forward biases D_1, closing the feedback loop. We have 100% feedback, so $A_V = 1$ and the output equals the input signal. When V_{in} goes negative, the op amp no longer forward biases D_1, which now acts as an open circuit. The output is held at approximately ground potential by R_2. In a digital circuit, R_2 would be called a *pull-down resistor*. R_2 also provides a path to ground for the inverting input bias current. Typically, R_2 will have a value from 1 to 10 kΩ.

Resistor R_1 serves two purposes. First, it protects the input of the op amp. When v_{in} goes negative, the noninverting input goes more negative than the V_{EE} supply rail (which is ground here). This could damage the op amp, so R_1 then prevents the noninverting input of the op amp from drawing too much current under these conditions. A guitar pickup couldn't damage the op amp but some other source might.

The second function of R_1 is to prevent the source that drives the op amp from excessive loading during negative-going signal conditions. When v_{in} is more negative than V_{EE}, the noninverting input becomes a low resistance to ground (a forward-biased, base-emitter junction). Here, R_1 prevents the noninverting input from loading down the driving source. Typical values for R_1 range from 1 to 100 kΩ.

We could use a bipolar supply to power the circuit, but this would actually slow down the response of the rectifier. The reason for this is that when the input goes negative and D_1 reverse biases, the feedback loop is open, and the output of the op amp saturates heavily negative trying to drive current through the diode. When input goes positive, again it takes more time for the op amp transistors to recover from hard saturation. In the single-polarity circuit of Fig. 5.41a, the op amp does not actually drive the feedback loop for negative-going inputs, so this problem is avoided.

If a negative output rectifier is needed, we can't simply reverse the direction of the diode, as is done with a passive diode rectifier. We would also need to use a negative power supply so the op amp output can actually go negative. The LM358 is a good choice in this circuit because it is designed for single-supply operation and can drive its output to within 10 mV of the negative supply rail, which is ground in this circuit.

The two-diode half-wave rectifier in Fig. 5.41b blocks the positive-going half of the input signal while passing an inverted replica of the lower, negative-going portion. For positive input voltages, the op amp drives V_x negative. The op amp forces virtual ground at the inverting input via forward-biased diode D_1. Diode D_2 is reverse biased, which isolates the V_o terminal from the output of the op amp. As long as the input voltage is positive, V_o is pulled down to virtual ground via resistor R_2.

When the input voltage goes negative, V_x goes positive, which reverse biases D_1 and forward biases D_2. The op amp now drives the feedback loop via D_2 and R_2. The V_o terminal goes positive such that inverting input pin 2 maintains virtual ground. Putting these pieces together, the output voltage is given by

$$v_o = \begin{cases} -v_{in} \dfrac{R_2}{R_1} & , v_{in} < 0 \\ 0V & , v_{in} \geq 0 \end{cases} \tag{5.20}$$

The notation used in (5.20) is used to describe *piecewise continuous* functions. It really does look horrible and mysterious, but it's the most efficient way to be mathematically precise when working with these functions.

The advantage of the super diode is its simplicity and the fact that it can operate from a single-polarity supply. The two-diode rectifier of Fig. 5.41b can operate at higher frequencies than the super diode, it can provide gain, and the output polarity can be inverted by reversing the direction of the diodes, but it has the disadvantages of being more complex and requiring a bipolar power supply.

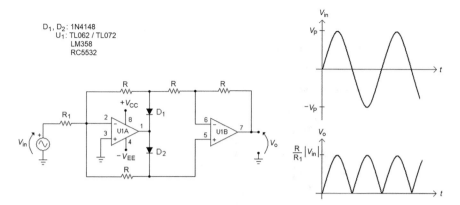

Fig. 5.42 Precision FW rectifier with input and output waveforms

Precision Full-Wave Rectifier

A precision full-wave rectifier is shown in Fig. 5.42. In this circuit, all resistors are equal in value except perhaps R_1, which is chosen to set the overall gain of the op amp. The output of the circuit is given by the following equation:

$$V_o = |V_{in}| \frac{R}{R_1} \tag{5.21}$$

Equal-value resistors R are usually set somewhere in the range from 10 to 100 kΩ. The full-wave rectifier of Fig. 5.42 has lower output resistance and higher current drive capability than either of the half-wave rectifiers of Fig. 5.41.

An Experimental Envelope Follower

An experimental envelope follower is shown in Fig. 5.43. A second dual op amp is used to allow buffering of the input and output of the envelope follower. Input buffer U1A is used because the input resistance of the precision rectifier varies from 1 to 10 kΩ, which is too low for a guitar to drive efficiently. Op amp U2B prevents loading of the low-pass filter. The gain of the rectifier is adjustable from about 0.9 to 10. I breadboarded this circuit, and its performance is discussed next.

In the following discussion, the gain of the rectifier was set to $A_V = 10$, and switch S_1 was set to the "No Filter" position. A scope capture of input and output signals is shown in Fig. 5.44. Here, the D string was picked used to produce the input signal. The markers labeled 1 and 2 on the left side of the display show ground (0 V) for each scope trace. The upper trace shows the open D string signal, which goes positive and negative, as usual.

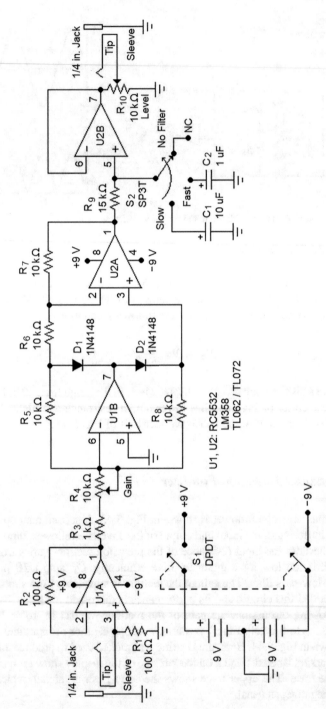

Fig. 5.43 Practical envelope follower circuit (positive output voltage)

Fig. 5.44 Open D string waveform (upper trace) and unfiltered output of precision rectifier (lower trace)

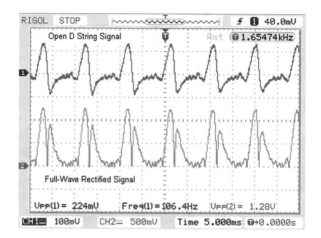

The lower trace is the unfiltered output of the envelope follower. Note that the negative portions of the signal are inverted and flipped up above ground. The frequency doubling property of the full-wave rectifier can be seen in the lower scope trace as well.

We will now look at the operation of the envelope follower with the low-pass filter enabled. In the slow response setting, the time constant of the filter is $\tau_{slow} = 150$ ms. In the fast response setting, the filter time constant is $\tau_{fast} = 15$ ms. Scope captures of the output of the envelope follower, using a slow sweep speed (500 ms/div) with the open D string ($f = 147$ Hz) being picked, are shown in Fig. 5.45. The slow sweep speed is necessary to view the envelope, which is a slowly varying signal. The instantaneous signal varies so quickly that it appears as a nearly solid trace at this sweep speed.

Notice that the slow response setting ($C = 10$ μF, $\tau = 150$ ms) provides the best filtering of the high-frequency noise but smooths out rapid changes in the envelope. The fast response setting ($C = 1$ μF, $\tau = 15$ ms) doesn't smooth out the envelope as much but also is less effective at removing high-frequency noise from the output signal. Circuit performance could be improved by using a second-order, active LP filter.

It is interesting to note that the envelope of the string waveform decays in a fairly exponential manner. This is a common characteristic of many natural phenomena. The envelope shows some nonlinearity, which is also common for this type of real-world signal.

The envelope follower in Fig. 5.43 operates from two 9 volt batteries. A charge pump like the MAX1044 of Fig. 3.54 could be used to realize a ±9 V bipolar supply using a single 9 V battery. This circuit will work well with power supplies ranging from ±5 to ±15 V. As a design exercise, you may want to try your hand at redesigning the circuit for single-polarity operation using the super diode rectifier.

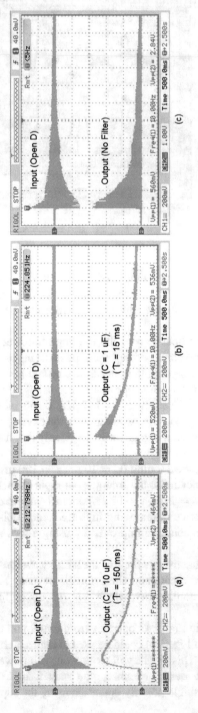

Fig. 5.45 Input and output waveforms for the envelope follower

String Frequency-to-Pulse Converter

While we are on the subject of envelope followers, here is a very cool application. Suppose you wanted to generate pulses at the same fundamental frequency as the string being picked. This would be useful for tuning purposes, for driving a digital circuit, or perhaps for synchronizing another signal with the guitar signal. Using a simple comparator as a *zero-crossing detector* would not work because noise and large harmonics in the signal can cause multiple unwanted zero-crossings. Setting the trigger level of the comparator to a fixed value other than 0 V is not a good solution either, because the signal will decay in amplitude rather quickly, dropping below the comparator threshold.

The solution to this dilemma is the circuit shown in Fig. 5.46a. This circuit uses an envelope follower to generate a reference voltage V_{REF} that changes dynamically with the signal envelope. This produces the output pulses shown in Fig. 5.46c. These nice, clean output pulses can be used to reliably trigger a frequency counter or other digital circuit.

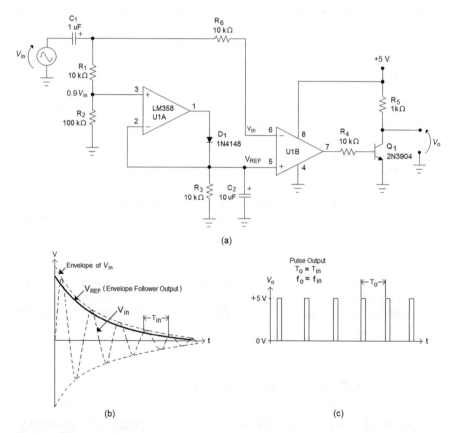

Fig. 5.46 Envelope follower pulse generator with typical input and output waveforms

Let's take a closer look at the operation of this circuit. Op amp U1A is a super diode, half-wave rectifier that passes the positive half of the signal envelope. A voltage divider, formed by R_1 and R_2, scales the output of the envelope follower by a factor of 0.9 so that V_{REF} is slightly lower than the amplitude of the original signal envelope. R_3 and C_2 filter the rectifier output, leaving the envelope voltage which is applied to the inverting input of U1B.

Op amp U1B is used as a comparator. We've seen comparators before, in the relaxation oscillator section of the phaser in Fig. 5.41. A comparator is a circuit that compares two analog input voltages and produces a high or low digital output, depending on which input voltage is greater. Since op amp U1B has no negative feedback, its gain is extremely high ($A_{OL} \cong 100,000$); therefore, its output will saturate high or low with a very small differential input voltage

$$\left(V_{(+)} - V_{(-)}\right) < 1\text{mV} \quad \text{(for output saturation)}$$

Refer to the waveforms in Fig. 5.46b. Here, v_{in} and its envelope are shown as dashed signals. The output of the envelope follower V_{REF} is the heavy, solid curve. When the instantaneous value of the input is less than the envelope follower output ($V_{in} < V_{REF}$), the op amp output is driven high. This forward biases the B-E junction of the NPN transistor, driving it into saturation, which produces $V_o = 0$ V.

When the instantaneous value of v_{in} exceeds the envelope follower output level, the op amp output is driven low, and transistor Q_1 goes into cutoff, producing $V_o =$ +5 V. The resulting output pulse train has the same frequency and polarity as the guitar signal. Also, because the output pulse train alternates between 0 and +5 V, the signal is compatible with many digital logic circuits. The circuit works well with a 9 volt battery as a supply voltage if 5 V logic compatibility is not a concern.

The circuit works best with a high-output pickups (humbuckers), producing a clean output pulse train for up to 10 s when the low E string is plucked. Performance could be improved if an input buffer with $A_V \cong 5$ is added. Also, a low-pass RC input filter could be added (after the buffer) to reduce noise sensitivity. Setting the corner frequency to $f_C \cong 300$ Hz would be a good place to start. If you are just learning electronic design, this is a great opportunity to try your hand at a little basic design work!

The LM358 was specified for this circuit again because it is designed to operate from a single-polarity supply voltage as low as +3 V. And as noted previously, the LM358 output terminal will drive to within about 10 mV of the negative supply rail. This guarantees that the transistor will be in cutoff when the output of the LM358 is low.

Compression, Sustain, and Dynamic Range

A compressor is a circuit that reduces the *dynamic range* of a signal. Dynamic range DR is defined as the decibel ratio of the maximum signal voltage to the minimum signal voltage that can be resolved. As an equation, this is written as

$$DR = 20 \ \log \left(\frac{V_{max}}{V_{min}} \right) \tag{5.22}$$

In principle, the dynamic range of an ideal analog signal is infinite since the minimum voltage is zero. In practice, the minimum signal level is determined by circuit noise, which swamps out signals that are very small.

Digital audio provides an easy example of dynamic range calculation. The least significant bit voltage V_{LSB} defines the smallest possible voltage change that can occur during playback of a digitized signal. The *full-scale output voltage* V_{FS} is the maximum output that can be produced. Though they are essentially obsolete, we will use the audio CD as our example of a digital audio source in this example. The resolution of a standard audio CD is 16-bits. That is, 16-bit binary numbers are used to represent digitized samples of the analog signal voltage. Let's assume our CD player has a maximum output voltage $V_{FS} = 10$ V. This gives us

$$
\begin{aligned}
V_{LSB} &= V_{FS}/2^n \\
&= 10V/2^{16} \\
&= 10V/65,536 \\
&= 152.59 \ \mu V
\end{aligned}
$$

The dynamic range of the audio CD is

$$
\begin{aligned}
DR &= 20 \log \left(\frac{V_{max}}{V_{min}} \right) \\
&= 20 \log \left(\frac{10V}{152.59 \ \mu V} \right) \\
&= 96 \ dB
\end{aligned}
$$

The dynamic range of digitized audio can also be calculated using $DR = 20 \log 2^n$, where n is the number of bits used to represent each sample of the signal. Try this formula on the digital audio example above to verify the relationship.

The dynamic range of human hearing varies from one person to another, especially with age and exposure to damaging loud sound levels, but a range of about 130 dB is often assumed. Setting the threshold of hearing as the 0 dB reference level, the threshold of pain is about 130 dB greater. Sound power levels greater than 130 dB are not perceived as being louder, even though increased energy is being transferred to the eardrum.

Sustain describes the decay characteristic of the signal produced by the guitar. As we have seen previously, the signal envelope produced by a guitar will tend to decrease exponentially with time. The physical construction of the guitar, the string tension, and even the interaction of the vibrating string with magnetic field of the pickups affect the rate of decay. In any case, however, it is usually desirable for a guitar to have very long sustain. Sometimes the terms sustain and compression are used synonymously, but they do not really mean the same thing. Also, sustain is defined differently in the context of electronic music, ala synthesizers and ADSR (attack, decay, sustain, release) modules. We are not using this interpretation here.

The logarithmic amplifier of Fig. 5.25 can be thought of as a compressor of sorts. Recall that the gain of the log amp decreases exponentially as the instantaneous value of the input signal increases. This effectively compresses high-amplitude peaks while expanding the signal near zero-crossings. This form of compression is not really useful to us because it severely distorts the guitar signal. I tried to build a compressor using this approach, long ago when I was in high school. At the time, I didn't have a deep enough understanding of these concepts, and needless to say, my compressor didn't work. I was able to figure out a solution when I was in college, and I had gained a much better understanding of signals and circuits (thanks, Dr. Schuler).

What we really are after in this case is not compression of the instantaneous guitar signal, but rather to hold the overall envelope of the signal relatively constant, without severely distorting the signal. That is, as the envelope of the signal decays, the gain of the system should compensate by increasing by the same proportion. As you can probably guess, envelope followers and voltage-controlled amplifiers are very useful in this application.

Voltage-Controlled Amplifiers

Back in Chap. 3, we saw that a voltage-controlled amplifier may be implemented using an operational transconductance amplifier or OTA. The input to a transconductance amplifier is a voltage, and the output is a current. The gain parameter is transconductance g_m, which is the same gain parameter for FETs. In this section, we will again work with the very popular CA3080 OTA.

The CA3080 has decent frequency performance specs, with $SR = 8$ V/μs and $GBW = 2$ MHz, and the CA3080 can operate from power supplies of ± 4 V to ± 18 V. The circuit design equations of Chap. 3 are repeated here for convenience.

$$I_{ABC} = \frac{V_{EE} - V_{BE}}{R_{ABC}} \tag{5.23}$$

$$g_m = 19.2 I_{ABC} \tag{5.24}$$

$$I_o = g_m V_{in} \tag{5.25}$$

$$V_o = I_o R_L \tag{5.26}$$

A CA3080 set up for noninverting operation is shown in Fig. 5.47. Remember, the OTA has a differential input like a standard op amp, but the output acts like a current source rather than a voltage source. The gain of the OTA is controlled by

Fig. 5.47 Noninverting
operational
transconductance amplifier

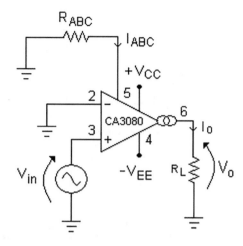

setting the amplifier bias current I_{ABC} via selection of resistor R_{ABC}. This is the key
to using the OTA as a voltage-controlled amplifier.

Experimental OTA-Based Compressor

An experimental compression/sustain circuit is shown in Fig. 5.48. The envelope
follower works as described previously, with gain adjustable for a maximum of
about 30. Note that diodes D_1 and D_2 are oriented to produce a negative envelope
voltage.

Under no-signal conditions, or when S_2 is in the "Normal" position, $I_{ABC} \cong$
50 µA, and the gain of the OTA is about $A_V \cong 2$. When an input signal is present and
S_2 is set for "Sustain," the envelope follower output voltage controls the OTA bias
current, I_{ABC}. The larger the signal, the more negative the envelope voltage goes,
reducing the gain of the OTA. A compression ratio of about 5 to 1 (14 dB) is possible
with this circuit. The "Gain Offset" adjustment introduces a variable offset into the
precision rectifier, which allows the no-signal gain of the OTA to be adjusted.

As with most of the other circuits presented, this circuit will work with supply
voltages of ±9 to ±15 V. This is definitely not a high-performance compressor, but
high degree of adjustability of the various sections allows some very interesting
behavior.

Experimental LDR-Based Compression/Sustain

The circuit of Fig. 5.49 is a simple but quite effective compression/sustain unit. The
optocoupler photoresistor acts as a variable feedback resistance for noninverting op

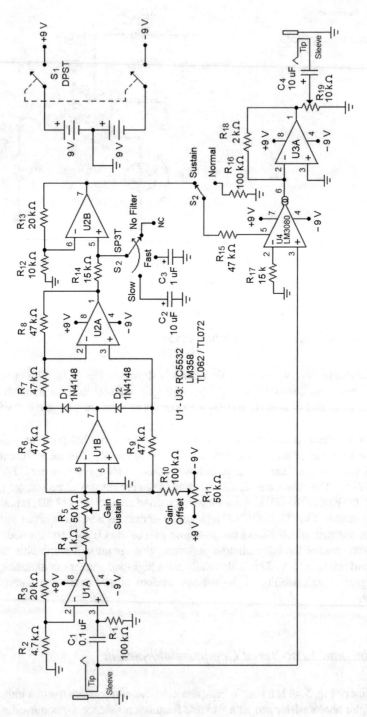

Fig. 5.48 Experimental OTA-based compression circuit

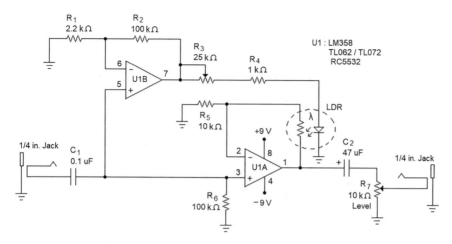

Fig. 5.49 LDR-based compression/sustain circuit

amp U1A. Under dark conditions, the feedback resistance is high which causes U1A to have high gain. Under light conditions, the gain of U1A is reduced.

The optocoupler LED is driven by op amp U1B, which is set for a relatively high gain of 46.5. When a signal is applied, U1B drives the LED into forward bias on positive input peaks. The higher the signal amplitude, the lower the average resistance of the LDR, reducing the gain of U1A. Potentiometer R_3 adjusts the compression level of the circuit.

Envelope follower action is provided by the LED, which acts as a rectifier as well as a light source that is proportional to signal amplitude. Low-pass filtering is realized by the inherently slow response of the CdS photoresistor, which effectively averages out the audio frequency light pulses from the LED. The slow response time of the photoresistor also gives the circuit an interesting momentary delay before compression occurs. OTA-based compressors work so quickly that this effect is not usually audible.

The LDR-based circuit takes a rather brute-force approach to achieving compression and sustain, but it is surprisingly effective if resistor values are adjusted for optimal performance. I designed this circuit because the OTA-based compressor was a bit too complex for less advanced experimenters. I also wanted to show that there are many alternative designs that can accomplish a given goal, and often the simplest solution is the best solution.

Tremolo

Tremolo is a periodic variation of the loudness of an instrument. The rate of loudness variation is usually adjustable, in the range of 1 to 10 Hz. Tremolo is actually a form of amplitude modulation. The term tremolo is often used incorrectly when applied to

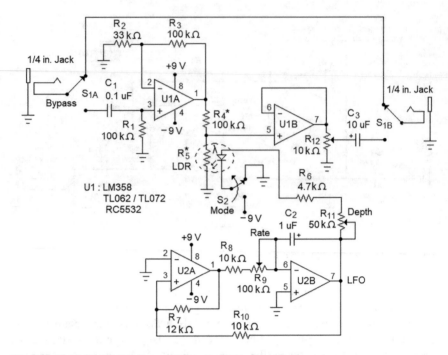

Fig. 5.50 Tremolo circuit using an LDR-controlled voltage divider

describe the so-called tremolo bar of a guitar. The tremolo bar actually varies the pitch of the guitar, which is called *vibrato*.

We can combine many of the concepts applied in previous applications to implement the tremolo effect. In Fig. 5.50, optocoupler R_5 is used to form an electrically adjustable voltage divider with R_4. The value of R_4 is chosen to be approximately the same as the maximum resistance of the LDR, which is assumed to be $R_{dark} = 100$ kΩ in this example. The low-frequency triangle-wave oscillator drives the optocoupler, which varies the attenuation of the voltage divider. Assuming that the light resistance of the LDR is $R_{light} = 10$ kΩ, the gain of the divider varies from 0.5 to 0.09. The lower the value of R_{light}, the more pronounced the tremolo effect becomes.

LFO oscillation frequency may be varied from about 1–10 Hz. Increasing the value of C_2 would shift the adjustment range lower in proportion to the capacitor size. Switch S1 was included to refresh your memory on true bypassing. Mode switch S_2 allows selection of a traditional smooth sounding tremolo and a more abrupt loud/soft transition.

This circuit works quite well, but there are many different implementations possible. For example, an alternative tremolo circuit could be designed using a VCA in place of the LDR voltage divider. The CA3080 could be used in this application.

Another alternative approach would be to eliminate the variable voltage divider and replace R_3 with the LDR photoresistor. In this approach, the LFO would directly vary the gain of U1A, much as is done in the compression/sustain circuit of Fig. 5.49. These are some great exercises if you are just learning to design your own effects circuits.

Reverberation

Reverberation, or *reverb* for short, is the simulation of the acoustic characteristics of the guitar being played in a room that reflects sound energy, usually multiple times. The delay time of the sound reflections depends mainly on the size of the room. The decay time depends on the sound absorption characteristics of the walls and objects in the room.

Delay Time

Typical reverb delay times range from a few milliseconds to hundreds of milliseconds. You can estimate a reverb delay time using the speed of sound (about 1080 ft/s) and room dimensions. Let's look at two examples. In the first case, assume you are at one end of a room that is 20 ft long, and in the second, you are in a large room such as a gymnasium that is 100 ft long. The time it takes for sound to leave your guitar, hit the wall, and return in each case would be

$$\text{20-foot-long room} \qquad \text{100-foot-long room}$$
$$T_{reverb} = \frac{2 \times 20\text{ft.}}{1080\text{ft.}/s} \qquad T_{reverb} = \frac{2 \times 20\text{ft.}}{1080\text{ft.}/s}$$
$$= 37\text{ms} \qquad = 185\text{ms}$$

This is a very oversimplified model of the actual situation, but the basic concept should make sense. As a guitar effect, the longer reverb time is probably more desirable, but too long of a delay results in a distinct echo rather than reverberation. There are no hard and fast rules for what delay sounds best however. It's really a matter of taste, and a delay that works well with one style of music or song may not work with another.

Decay Time

Reverb consists of multiple reflections of the sound that decays in amplitude over time. This results in an approximately exponential decay envelope, much like those shown back in the oscilloscope displays of Fig. 5.45. The exponential envelope is described by (5.27), where V_o is the initial amplitude of the signal, e is the base of the natural logarithms ($e \cong 2.718\ldots$), t is the time elapsed after the sound is initially created, and τ is the time constant of the decay envelope.

$$V_{envelope} = V_o e^{-t/\tau} \tag{5.27}$$

The value of the time constant τ depends on how quickly sound energy is absorbed in the room. Rooms that absorb sound have a very small time constant. Rooms that reflect sound efficiently have a larger time constant. A practical reverb time constant range is $0.1\ \text{s} \leq \tau \leq 1\ \text{s}$, where $\tau = 0.1\ \text{s}$ is typical of the average living room and $\tau = 1\ \text{s}$ is equivalent to a large cathedral.

Theoretically, an exponentially decaying signal takes infinitely long to reach zero amplitude. In most electrical engineering applications, we use the fact that after 5τ, the envelope has decayed by over 99%, and we consider the signal to have died out. In acoustic engineering, the time it takes for the reverb envelope to drop by 60 dB (a factor of 0.001) from its initial value, designated as T_{60}, is often used to specify the decay characteristic. This is roughly equivalent to seven time constants. That is, $T_{60} \cong 7\tau$.

Reverb Springs

A reverb spring unit, or tank, consists of a housing that contains movable input and output coils, mounted over magnetic cores, and two or more long coil springs that mechanically couple the two windings. For all practical purposes, the input winding can be considered to be a motor, and the output winding acts as a generator. When the input coil is driven with the guitar signal, its magnetic core moves. The resulting vibrations travel down the springs, moving the core of the output coil, inducing a time-delayed signal that is fed back into the amplifier. Vibrations are reflected back and forth through the springs which simulates the exponential decay of a room with good reverb characteristics. Typical reverb spring units are shown in Fig. 5.51.

Fig. 5.51 Reverb spring units

Long reverb springs simulate a larger room. Often, multiple springs of different lengths are used to simulate reflections from walls of different distances from the sound source. This produces a richer, more realistic reverb sound.

Reverb spring units come in a wide variety of sizes and input/output impedance variations. Since the springs are driven by coils, the input and output impedances are best modeled as inductors. Input impedance usually ranges from 8 to 2000 Ω, while output impedance usually ranges from around 500 Ω to 12 kΩ. Low-impedance input coils require higher drive current. Drive currents ranging from 2 to 30 mA are common.

The output of the reverb spring unit is fairly low in amplitude, usually around 10–50 mV$_{P-P}$. Low output voltage coupled with the rather low input impedance means that an input buffer and output amplifier are needed for proper reverb tank operation. An example is shown in Fig. 5.52, where a reverb spring has been added to the power amplifier covered previously (Fig. 4.12). This circuit uses a Belton model BS3EB2C1B, three-spring reverb unit. This reverb has the following specifications at a frequency of 1 kHz:

Input coil	Output coil	Decay characteristic
$Z_{in} = 800\ \Omega$	$Z_o = 2575\ \Omega$	$1.75\ \text{s} \leq T_{60} \leq 2\ \text{s}$
$L_{in} = 150\ \text{mH}$	$L_o = 400\ \text{mH}$	or
$R_{in} = 75\ \Omega$	$R_o = 230\ \Omega$	$0.25\ \text{s} \leq \tau \leq 0.29\ \text{s}$
$I_{in} = 3.1\ \text{mA}$		

In Fig. 5.52, it is assumed that R_1 is adjusted such that the signal at the wiper averages about 50 mV$_{P-P}$ in amplitude. In order to obtain $I_{in} = 3.1$ mA, the input coil requires about 2.5 V$_{P-P}$, which requires input buffer U2A to have $A_V \geq 50$.

Being conservative and assuming the output of the reverb unit is about 10 mV$_{P-P}$, op amp U2B requires a gain of $A_V \geq 50$ to bring the signal up to 500 mV$_{P-P}$. The reverb signal amplitude is adjusted with 10 kΩ pot R_{17}. The reverb signal is summed with the clean signal at the inverting input of U1B. Note that for single-supply operation, the usual biasing components and coupling capacitors will have to be used.

A Digital Reverb

Digital reverb is one of the few digital effects that I have included, simply because these modules are inexpensive and easy to use. One such module is the BTDR-1H digital reverb from Belton, shown in the photo of Fig. 5.53. The device pins are hidden under the package but are identified below. A standard 9 V battery is shown in the photo for scale.

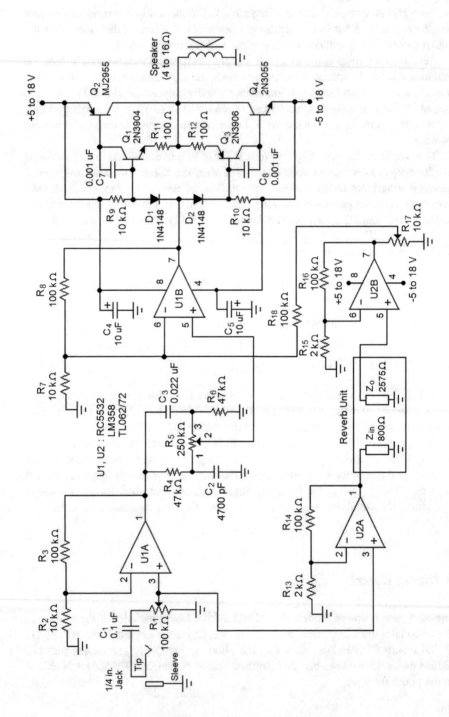

Fig. 5.52 Power amplifier with reverb spring

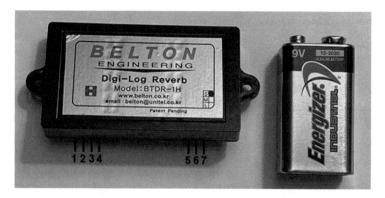

Fig. 5.53 Digital reverb module

The BTDR-1H reverb module is available in three different decay time versions: short decay ($T_{60} = 2$ s), medium decay ($T_{60} = 2.5$ s), and long decay ($T_{60} = 2.85$ s). These modules are designed to operate from a single-polarity, regulated 5 V power supply. Primary design parameters are $Z_{in} = 10$ kΩ, $Z_o = 220$ Ω, and $A_v = 1$. Maximum input signal voltage is 1.5 V_{pk}.

The pins of the reverb module are assigned as follows:

Pin Function

1 V_o

2 V_o

3 Signal Ground

4 V_{in}

5 Power Ground(Use pin 5 alone, if separate signal and power grounds are not used.)

6 N.C.

7 $V_{CC}(+5$ V$)$

A reverb circuit using the BTDR-1H is shown in Fig. 5.54. A low-power regulator, the 78L05, is used here because the reverb module requires a 5 volt supply for its internal logic circuitry. This circuit is designed to be used as an external effect that is connected between the guitar and the main amp. The BTDR-H1 has internal coupling capacitors, so external input and output coupling caps are not required.

Op amp U1A is set up for $A_v = 2$, while op amp U1B acts as a differential amplifier with approximately unity gain. The output of U1A sits at a DC level of 4.5 V, which biases amplifier U1B for proper operation. The reverb level adjustment has a slight effect on the gain and the bias point of op amp U1B, but the variation is not significant.

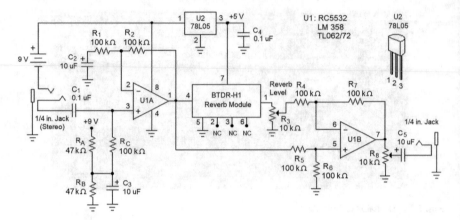

Fig. 5.54 Digital reverb circuit

Modulation and Pitch Shifting

Modulator circuits are used in many electronic communication applications, including, of course, guitar effects. Modulation is a process by which some parameter of a signal v_1 is controlled or varied by another signal, which we will call the modulating signal v_m. In the mathematically simplest case, we can assume the signal to be modulated, v_1, is a sinusoid. The basic sinusoidal signal has the form

$$v_1 = V_1 \sin(2\pi f_1 t + \phi_1) \tag{5.28}$$

There are three possible parameters that we can vary or modulate:

V_1 Varying the peak amplitude V_P produces amplitude modulation.
f_1 Varying frequency produces frequency modulation.
Φ_1 Varying phase produces phase modulation.

We have actually seen some examples of these forms of modulation already. The voltage-controlled amplifier is really an amplitude modulator, as is the compressor, which controls the peak amplitude of a guitar signal by electronically varying the gain of a VCA. In the case of the tremolo, the amplitude of the guitar signal is modulated by a low-frequency oscillator.

Amplitude Modulation

The block diagram shown in Fig. 5.55 is a generic amplitude modulation circuit. If no modulating signal is applied ($v_m = 0$ V), the gain of the VCA is determined by the DC bias voltage. If we have a DC bias voltage present and apply a complex signal $v_m(t)$, the output $v_o(t)$ will have an envelope that matches the complex modulating signal, as shown in the block diagram.

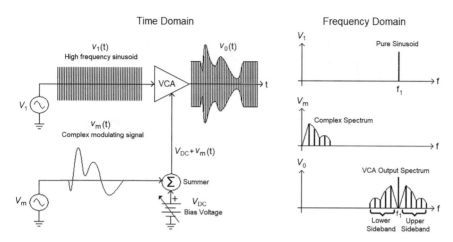

Fig. 5.55 Amplitude modulation of a pure sinusoid with a complex signal, using a VCA

The right side of Fig. 5.55 shows the amplitude modulator input and output signals in the frequency domain. Modulation of a sinusoidal signal v_1 with a complex signal v_m results in an output signal that contains replicas of the original modulating signal spectra, translated up in frequency, symmetrically located on each side of f_1. These are called *sidebands*.

The math behind this amplitude modulation circuit is expressed in the following equation:

$$v_o(t) = V_1 \sin\left(2\pi f_1 t\right) \times \left[V_{DC} + v_m(t)\right] \tag{5.29}$$

In the case of a complex modulating signal, $v_m(t)$ consists of many sinusoidal components and may be expressed as the sum of these sinusoids—if we happen to know what they are. However, in the case of a guitar or just about any musical signal for that matter, the signal spectrum of v_m will be constantly changing and difficult if not impossible to quantify at this level of detail.

The main thing to notice in (5.29) is that amplitude modulation really boils down to the multiplication of one signal by another. As we have seen before, trigonometric identities like (5.6) are very useful in the analysis of modulator circuits. And you thought those horrible trig identities you were tortured with in school had no practical purpose!

Balanced (Ring) Modulation

If we reduce the DC term in (5.29) to zero, by turning the DC voltage in Fig. 5.56 down to zero ($V_{DC} = 0$ V), something interesting happens to the output of the VCA; the original sinusoidal signal v_1 goes away, and all that is left are the two frequency-shifted, mirror-image spectra of v_m. This is shown in Fig. 5.56a. The circuit is now what we call a *balanced modulator* or *ring modulator*.

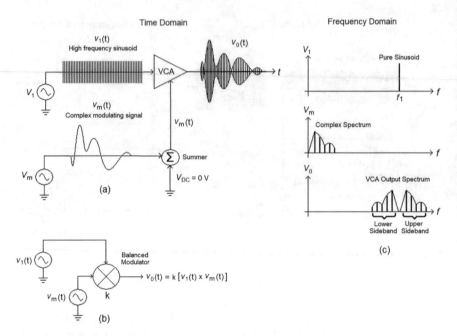

Fig. 5.56 VCA balanced or ring modulator. (a) The circuit. (b) General symbol for a balanced modulator. (c) Typical input and output signal spectra

Since the output of the balanced modulator is proportional to the product of the two input signals, the simplified block diagram symbol of Fig. 5.56b is often used. The equation for the output of the balanced modulator is given by (5.30), where k is the gain of the balanced modulator.

$$v_o(t) = k v_1(t) v_m(t) \tag{5.30}$$

In guitar effects applications, most of the time, the guitar signal is mixed with a pure sinusoidal tone, where the guitar signal is v_m and v_1 is generated by an adjustable frequency oscillator. This causes translation of the guitar signal spectrum and is a form of pitch shifting. Because the modulator produces sum and difference frequencies, in general, the spectrum of the output signal will not be harmonically related to the original signal.

An Experimental Ring Modulator

Balanced modulation is widely used in many areas of electronics, and there are chips designed specifically to perform this task. An example is the LM1496. This venerable device may be used in applications ranging from audio to around 50 MHz. A circuit using the LM1498 as a balanced modulator is shown in Fig. 5.57.

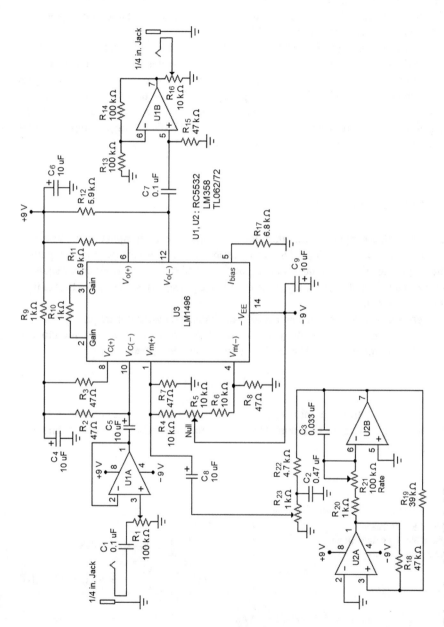

Fig. 5.57 Balanced modulator/pitch shifter

The triangle-wave oscillator used in several previous circuits is used again here. The output of the oscillator is low-pass filtered to smooth off the peaks of the waveform and to reduce the amplitude of the harmonics that make up the triangular waveform. To minimize distortion, both the audio input signal and oscillator signal applied to the LM1496 should be held below 0.5 V_{P-P}.

The input resistance of the balanced modulator is only 47 Ω, so the input signal is buffered by U1A to prevent severe loading of the guitar pickup or the preceding effect circuit. Potentiometer R_5 is adjusted to minimize output of the oscillator signal when no instrument signal is present. Op amp U1B buffers the output of the balanced modulator and provides additional gain, if desired. The current drain from the negative supply rail is less than 10 mA, so a MAX1044 charge pump converter could be used to operate the circuit from a single 9 V battery.

Frequency Doubling

Consider what happens if we apply the same signal to both inputs of the balanced modulator, as shown in Fig. 5.58. Since the circuit multiplies the two inputs together, the output signal is the square of the input v_m^2, scaled by a factor k. When we square a sinusoidal signal, the frequency doubles. The output signal is one octave higher than the original source.

Here's an example of the math behind the balanced modulator frequency doubler. To keep things reasonably straightforward (well, sort of), let's assume the input to the balanced modulator in Fig. 5.58 is a pure sine wave given by

$$v_m = V_m \sin(\omega_m t)$$

I am using ω to represent frequency in rad/sec to simplify the notation. Remember, $\omega = 2\pi f$. Using the trigonometric identity for the product of two sine functions (5.6), we have

Fig. 5.58 Balanced modulator squaring circuit/ frequency doubler

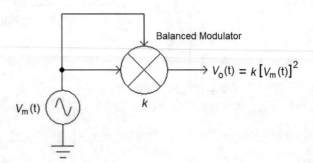

$$v_o = k[V_m \sin(\omega_m t)]^2$$
$$= k[V_m \sin(\omega_m t)][V_m \sin(\omega_m t)]$$
$$= \frac{kV_m^2}{2} \cos[(\omega_m - \omega_m)t] - \frac{kV_m^2}{2} \cos[(\omega_m + \omega_m)t]$$
$$= \frac{kV_m^2}{2} \cos[0] - \frac{kV_m^2}{2} \cos[2\omega_m t] \qquad (5.31)$$
$$= \frac{kV_m^2}{2} - \frac{kV_m^2}{2} \cos[2\omega_m t]$$

The first term in (5.31) is simply a DC level that we can disregard in this application, as it could be blocked by using a coupling capacitor on the output. The second term is the original input signal, doubled in frequency and phase shifted by +90°.

A Deeper Dive

The calculation you just endured definitely shows that the squaring circuit doubles the frequency of a simple sinusoid. However, a real-world audio signal will be a complex, dynamic mixture of sinusoids of different frequencies. As we've seen from the previous discussion of harmonic distortion, and from Figs. 5.55 and 5.56, some crazy stuff can happen when more complicated signals are processed. Likewise, when we apply a complex signal to the squaring circuit, things get a little messy. Consider an input signal that consists of two sinusoidal components

$$v_m(t) = 4\sin(2t) + 2\sin(8t)$$

Without showing all of the steps in the calculation, if we apply this signal to the circuit of Fig. 5.58, with $k = 1$, the resulting output is

$$v_o(t) = [4\sin(2t) + 2\sin(8t)]^2$$
$$= [4\sin(2t) + 2\sin(8t)] \times [4\sin(2t) + 2\sin(8t)]$$
$$= 10 - 8\cos(4t) + 8\cos(6t) - 8\cos(10t) - 2\cos(16t)$$

We obtain the doubled frequencies ($\omega = 4$, $\omega = 16$) in the second and last terms, but we also get the sum and differences of the input frequencies ($\omega = 6$, $\omega = 10$) as well as a DC offset. Note that the sum and difference frequencies are not harmonically related to the original input signals. These non-harmonically related components tend to make the output sound dissonant.

An experimental squaring circuit is shown in Fig. 5.59. The circuit operates much like the previous ring modulator, except that the oscillator is not required. Potentiometer R_5 is adjusted for minimum distortion.

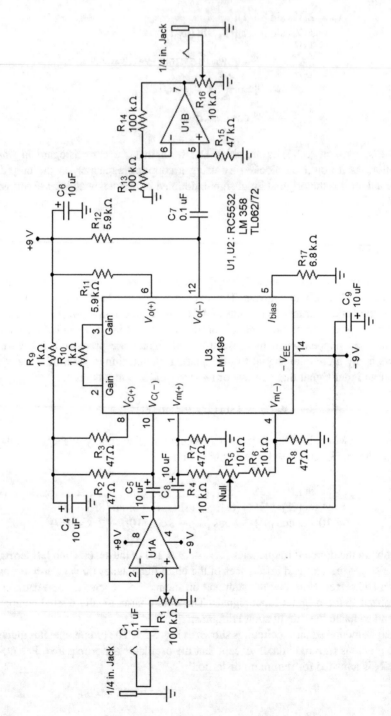

Fig. 5.59 Squarer (frequency doubler) using a balanced modulator

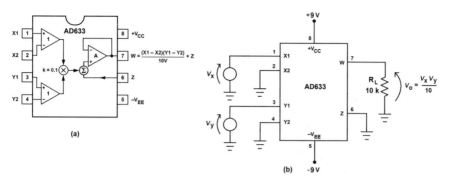

Fig. 5.60 (a) AD633 analog multiplier. (b) A simple application circuit

Analog Multipliers

Analog multipliers are another class of linear ICs that may be used to implement VCAs, amplitude modulation and pitch shifting. The AD633 shown in Fig. 5.60a is an example of a common analog multiplier, with a fixed scaling factor of $k = 0.1$. The AD633 has a slew rate of 20 V/µs and can deliver up to ±30 mA to a load. The AD633 works best with a supply of ±15 V but can be operated with a supply voltages from ±8 to ±18 V, so we can use two 9 volt batteries or a single 9 V battery and MAX1044 charge pump to power the chip.

A basic multiplier circuit using the AD633 is shown in Fig. 5.60b. Using a ±9 V supply, the peak input voltage of v_x and v_y should be limited to a range of about $\pm$ 2.5 V_{pk}, giving an output signal in the range of ±6.25 V_{pk}.

Though simpler to use, analog multiplier ICs are more expensive than balanced modulators like the LM1496. At the time of writing, in single-unit quantities, the AD633 costs about \$17 USD, while the LM1496 costs \$0.45. As mentioned before, if you are new to circuit design and you'd like to learn more, a good way to start is to experiment with modifying existing designs. Redesigning the circuits of 5.57, 5.58, and 5.59 using the AD633 in place of the LM1496 would be a great learning exercise and fun too.

Vocoders

One of my favorite musicians is Peter Frampton. When I was in college, *Frampton Comes Alive*, one of the best-selling albums of all time, was released. On several of his more popular songs, he used a device called a *talk box*. A talk box is just a small box that holds a speaker, or speaker driver (technically, it's a linear motor, but anyway...), which is driven by the guitar amplifier. The box is sealed except for a hole into which a length of hose is inserted. The hose is routed up a microphone stand, and the end is placed in the corner of your mouth. When you play the guitar,

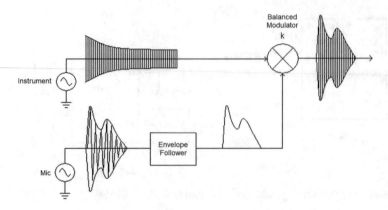

Fig. 5.61 Block diagram for a simple vocoder

the sound waves from the talk box enter your mouth via the hose, where they substitute for the sound waves normally produced by your vocal cords. The mouth is then used to modulate the guitar signal, and the microphone/amp system reproduces the vocalized guitar sound.

Naturally, I had to build one of these things, and I had a lot of fun with it. Eventually, it was lost, but it has since been replaced by a nice commercial unit, and every once in a while, I still like to use it and play along with "Do You Feel Like We Do?"

A *vocoder* is the electronic equivalent of a talk box. The term vocoder is a contraction of *vo*ice and en*coder*. The block diagram for a simple vocoder is shown in Fig. 5.61. In this circuit, the filtered output of an envelope follower modulates the amplitude of the guitar signal via a balanced modulator. The balanced modulator function could be implemented using an LM1496 or an analog multiplier like the AD633. This circuit operates such that when the vocal envelope voltage increases, the gain of the balanced modulator increases. This impresses the voice signal envelope onto the guitar signal. This is essentially the same function that is performed by a traditional talk box. A few songs that use a vocoder are "Robot Rock" by Daft Punk and "The Raven" by The Alan Parsons Project.

A primitive, experimental vocoder is shown in Fig. 5.62. The input buffer provides a gain of about $A_V = 22$ in order to provide a reasonably large signal for the precision rectifier when using a typical microphone. The input sensitivity is adjusted with audio taper pot R_1. The switchable filter section was left in place to allow for experimentation. Potentiometer R_{10} allows adjustment of the envelope voltage to prevent overdriving of the balanced modulator.

With no vocal signal present, the balanced modulator should produce no output signal even if the guitar signal is present. Null potentiometer R_{15} is adjusted for minimum output with only one signal present. When a vocal signal is applied, the envelope of this audio source is mixed with the guitar signal. Effectively, the gain of the balanced modulator varies with the amplitude of the envelope.

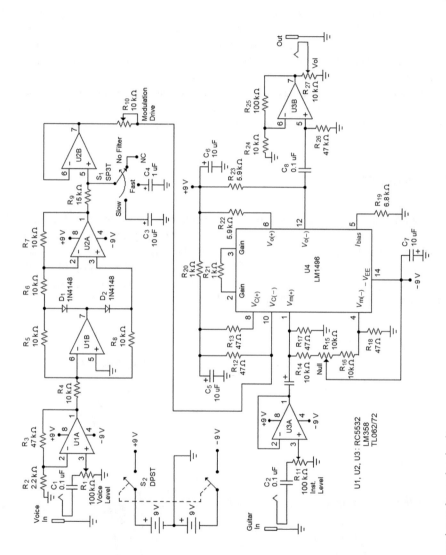

Fig. 5.62 Basic experimental vocoder circuit

This is really a bare-bones vocoder circuit that is fun to experiment with, but it is definitely not a high-performance circuit. Commercially available vocoders usually use banks of multiple bandpass filters (often implemented using DSP techniques these days) to divide the vocal and guitar signals into frequency bands that are applied to individual envelope followers. This approach produces richer, high-quality sound but requires a lot more circuitry.

As with all of the effects circuits covered here, I encourage you to think of different ways to implement this design. For example, redesigning the vocoder to use an analog multiplier in place of the LM1496 would be a good exercise. You could also experiment with using different precision rectifier circuits as well. Here's another idea that you could try. Rather than a guitar, a synthesizer or simple oscillator may also be used to supply the instrument input signal as well. The phase-shift oscillator of Fig. 3.46 would be a good circuit to use in this application.

Wah-Wah Circuits

The traditional *wah-wah* effect is typically implemented using a variable *bandpass* (BP) filter, where the center frequency of the filter is swept up and down with a foot pedal. Variable *bandstop* or *notch* filters may also be used, but are not as common. Examples of passive, second-order, series RLC bandpass and bandstop filter networks and typical response curves are shown in Fig. 5.63. The *center* or *resonant frequency* of either filter is given by

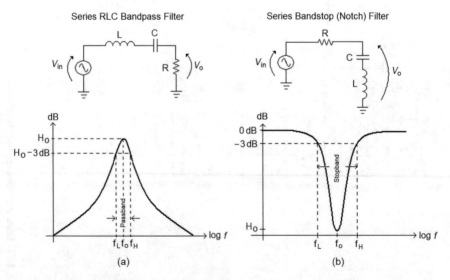

Fig. 5.63 Passive RLC filters. (a) Bandpass filter. (b) Bandstop (notch) filter

$$f_o = \frac{1}{2\pi\sqrt{LC}} \tag{5.32}$$

These filters are classics, and definitely worth to study, but they are not practical for direct application as wah-wah circuits. The problem here is that in order to vary the center frequency, we must be able to easily vary the inductance L or capacitance C. While variable capacitors and inductors are available, they are generally not suitable for audio frequency applications and inductors especially are inconvenient to adjust.

Active filter techniques allow us to easily vary the effective values of either L or C in a circuit. Active filters also make it possible to eliminate inductors altogether. This is an advantage because at audio frequencies, suitable inductors can be physically large and heavy, expensive, and hard to find.

IGMF Bandpass Filter

One of the most commonly used active BP filter circuits is called an *infinite-gain, multiple-feedback* bandpass filter, which is shown in Fig. 5.64. We will simply refer to this filter as an *IGMF*. The response of the IGMF will resemble the BP response curve shown in Fig. 5.63a.

The bandwidth of the BP filter response is given by

$$BW = f_H - f_L \tag{5.33}$$

The maximum passband response H_o occurs at the center frequency f_o. The corner frequencies of the response are defined as the points at which the passband response H_o drops by 3 dB.

When working with bandpass and notch filters, an important parameter is the *quality factor Q*. In practical terms, Q is a measure of the sharpness of the passband

Fig. 5.64 IGMF bandpass active filter

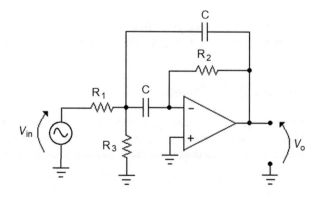

of the filter. High-Q filters have a narrow passband. The Q of a bandpass filter is given by

$$Q = \frac{f_o}{BW} \tag{5.34}$$

The IGMF circuit is good for Q up to about 25 using a decent op amp such as a TL071/72 or RC5532. There are other circuits that are better suited for obtaining higher Q values.

In order to simplify the design of the IGMF, we normally use equal-value capacitors. The design equations for this circuit are

$$f_o = \frac{\sqrt{\left(\frac{1}{R_2 C^2}\right)\left(\frac{1}{R_1} + \frac{1}{R_3}\right)}}{2\pi} \tag{5.35}$$

$$R_1 = \frac{Q}{2\pi f_o A_o C} \tag{5.36}$$

$$R_2 = \frac{Q}{\pi f_o C} \tag{5.37}$$

$$R_3 = \frac{Q}{2\pi f_o C\left(2Q^2 - A_o\right)} \tag{5.38}$$

$$A_o = -\frac{R_2}{2R_1} \tag{5.39}$$

The lowest guitar string frequency is about 84 Hz, while the highest frequency at the 23rd fret is about 1.3 kHz. Not including harmonics, this is only about 4 octaves, which is a practical range of frequencies over which we might want to sweep the center frequency of a bandpass filter to create a wah-wah effect.

An IGMF Design Example

There are quite a few free parameters that we can choose when designing an IGMF filter. A reasonable design starting point is to set f_o to the geometric center (*geometric mean*), between the upper and lower frequency limits, which is

$$f_o = \sqrt{f_L f_H}$$
$$= \sqrt{84 \times 1300} \qquad (5.40)$$
$$\cong 330 \text{Hz}$$

It doesn't hurt to build in a bit of gain, so we will set $A_o = 2$ for this design. A reasonable value of Q to start with is $Q = 5$, but you can experiment with any value you choose up to about 25.

Finally, a capacitor value must be chosen. You can choose just about any standard value, but something in the range from 0.1 μF to 0.001 μF will usually work in audio filter applications. Choosing $C = 0.01$ μF and inserting these values into the design equations produces

$$C = 0.01 \mu\text{F}$$
$$R_1 = 120.6 \text{ k}\Omega(\text{use}120 \text{ k}\Omega)$$
$$R_2 = 482.2 \text{ k}\Omega(\text{use}470 \text{ k}\Omega)$$
$$R_3 = 5 \text{ k}\Omega(\text{use}4.7 \text{ k}\Omega)$$

The standard resistor values are close enough to the calculated values that the filter will work well. If the capacitor value chosen initially had resulted in insanely large or small resistor values, or if the resistances were not even close to standard values, we simply choose a new capacitor value and recalculate.

Varying f_o of the IGMF

A very convenient characteristic of the IGMF is that the center frequency of the filter can be changed without affecting the values of Q or A_V simply by varying the value of R_3. We must not let R_3 go to zero ohms, or the circuit will become unstable, most likely causing oscillation. If we use a 10 kΩ, audio taper potentiometer for R_3, with a series resistor of 100 Ω to keep the filter stable, the frequency range is

$$f_{o(\min)} = 242 \text{ Hz}, f_{o(\max)} = 2.3 \text{ kHz}$$

I experimented with the circuit and found that using a higher-resistance pot to sweep the filter to lower frequencies had little effect on the sound, so a 10 kΩ pot works well here for R_3. If you would like to calculate different values of R_3 based on various new center frequencies, the following formula can be used:

$$R_{3(\text{new})} = R_{3(\text{old})} \left(\frac{f_{o(\text{old})}}{f_{o(\text{new})}} \right)^2 \qquad (5.41)$$

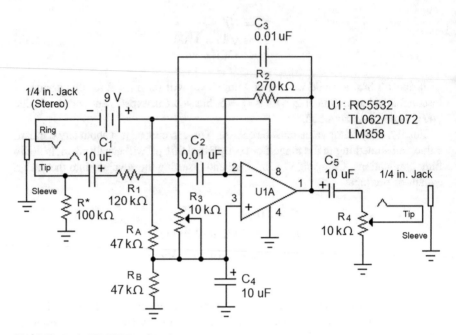

Fig. 5.65 Basic IGMF BP wah-wah

Experimental IGMF Wah-Wah Circuits

A basic wah-wah designed around the IGMF BP filter is shown in Fig. 5.65. For convenience, this circuit operates from a single 9 V battery. The circuit in Fig. 5.66 expands the basic wah-wah by adding a second BP filter with a lower center frequency. The outputs of the bandpass filters are summed by op amp U2A. The overall response of this filter will have two peaks. Sweeping both filters requires either the use of a dual 10 kΩ potentiometer in the pedal or you could use a single pot to vary a pair of photocoupler LDRs, like those used in the circuits studied previously. Because this circuit requires R_3 to be reduced to a very low value to work well, an LDR that has $R_{light} \cong 100$ Ω should be used.

You can experiment with different capacitor values to change the wah-wah filter parameters. The effect of these changes will be most pronounced in the two-stage wah-wah. You could also experiment with connections to the frequency sweep potentiometers of the two-stage wah-wah. For example, reversing the connections relative to the ends of R_{3A} and R_{3B} will cause the BP filters to sweep in opposite directions, which might produce interesting results.

Eliminating Switching Pop

An optional resistor R* is included in both Figs. 5.65 and 5.66 to help prevent popping from occurring when the effect is switched in and out of the signal chain.

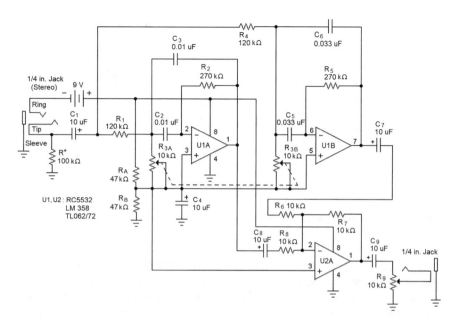

Fig. 5.66 Two-stage wah-wah circuit

This resistor provides a charging path for input coupling capacitor C_1. If this resistor is not used, the left-side terminal of the capacitor floats until it is connected to a source. When the effect is connected to a signal source, the cap begins charging, which causes a pop or thump when charging current initially flows. The value of this resistor is not critical, but it should be kept relatively high, as it does reduce the input resistance of the circuit somewhat.

A Gyrator-Based Wah-Wah Circuit

A gyrator is a network that has the electrical characteristics of an inductor but uses no actual inductors in its design. In other words, a gyrator is a *synthetic inductor*. The schematic for a gyrator and its equivalent inductor is shown in Fig. 5.67.

The gyrator circuit is quite similar to that of the all-pass filter used in the design of the phase shifter. A somewhat unusual feature of this circuit is that the output of the op amp does not serve as an output terminal for the gyrator.

The effective inductance of the gyrator is

$$L = R_1 R_2 C_1 \tag{5.42}$$

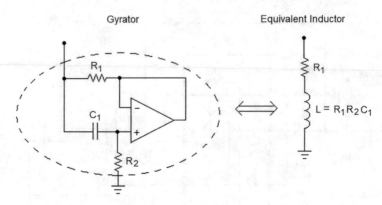

Fig. 5.67 Op amp gyrator and equivalent circuit element

Note that the gyrator does not simulate an ideal inductor, but rather an inductor with a series resistance. This resistance is similar to the actual winding resistance that would be present in a real inductor. A very cool thing about the gyrator is that we can change the inductance very easily by replacing either R_1 or R_2 with a potentiometer.

The Q of an inductor tells us how ideally the inductor behaves. The more ideal the inductor is, the higher its Q will be. An ideal inductor would have infinite Q. The Q of a real inductor is given by $Q = X_L/r_S$, where r_S is the series resistance of the winding. For the gyrator, $r_S = R_1$, so we can define the Q of the gyrator as follows:

$$Q = \frac{X_L}{R_1} = \frac{2\pi f L}{R_1} = \frac{2\pi f R_1 R_2 C_1}{R_1}$$
$$= 2\pi f R_2 C \qquad (5.43)$$

This is an interesting result because R_1 cancels from the equation. This means that we can change the value of inductance by varying R_1 without affecting Q. The inductance of the gyrator can be also adjusted by varying R_2, but the value of Q increases or decreases in direct proportion to L.

An important thing to notice is that one terminal of the gyrator must be connected to ground. This limits the usefulness of the gyrator as a simulated inductor. For example, while the gyrator could be used to directly implement the bandstop filter of Fig. 5.63b, the bandpass filter of Fig. 5.63a could not be implemented in its current form because the inductor is not grounded on one end.

A wah-wah circuit, using a gyrator, is shown in Fig. 5.68a. The equivalent RLC circuit representation is shown in Fig. 5.68b. The gyrator section and capacitor C_2 form an equivalent series-resonant circuit, which behaves as a variable notch filter.

Op amp U1B is configured as a noninverting amplifier. At resonance, $X_L = X_{C2}$, cancelling to zero, so the impedance Z of the gyrator notch filter is minimum with $Z_o = R_1$. This causes op amp U1B to have maximum voltage gain at f_o. The notch filter response is inverted, producing the desired bandpass response.

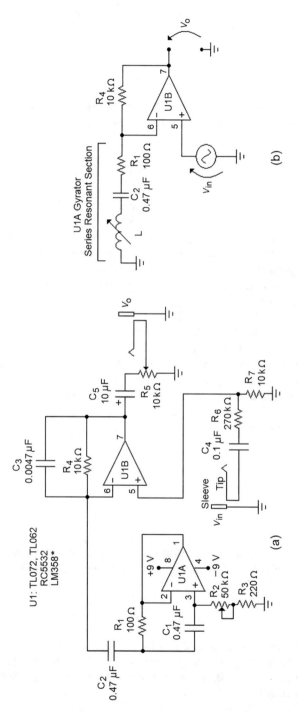

Fig. 5.68 (a) Experimental gyrator-based wah-wah and (b) equivalent circuit

The center frequency of a passive RLC bandpass filter is given by (5.32), which is repeated here for convenience.

$$f_o = \frac{1}{2\pi\sqrt{LC}} \qquad (5.32)$$

The center frequency of the gyrator BP filter is given by (5.44), which corresponds exactly with (5.32) because $L = C_1 R_1 (R_2 + R_3)$.

$$f_o = \frac{1}{2\pi\sqrt{C_2 C_1 R_1 (R_2 + R_3)}} \qquad (5.44)$$

Using the component values given in the schematic, the sweep range of the filter extends from $f_{o(min)} = 150$ Hz to $f_{o(max)} = 2.3$ kHz.

Capacitor C_3 is used to limit high frequency response but has negligible effect over the intended frequency range of the circuit. Without this capacitor, the filter exhibits a large peak at about 100 kHz, which could result in instability. Resistor R_3 is used to prevent the denominator of (5.44) from going to zero.

The gain of the filter is quite high at f_o, being approximately $H_o = 100$ (40 dB). To prevent the filter from being overdriven, voltage divider R_6-R_7 attenuates the guitar signal by a factor of 0.036. This effectively reduces the peak gain of the filter to $H_o = 3.6$. The output level is adjusted by R_5.

The structure of the equivalent circuit in Fig. 5.68b allows for easy derivation of a useful frequency response equation, which is given in (5.45). This is a slightly modified version of the usual gain equation for noninverting op amp U1B. Here, $R_4 = 10 \angle 0°$ kΩ, $X_L \angle 90° + R_1 \angle 0°$ is the equivalent inductive reactance and series resistance of the gyrator, and $X_{C2} \angle -90°$ is the capacitive reactance of C_2.

$$H = 1 + \frac{R_4 \angle 0°}{X_L \angle 90° + R_1 \angle 0° + X_{C2} \angle -90°} \qquad (5.45)$$

The fact that the variables in (5.45) are complex numbers complicates the math a bit, but the important thing to realize is that at resonance, $|X_L| = |X_{C2}|$, and because inductance and capacitance have opposite phase angles, they cancel. This explains how the circuit transforms the notch response into a bandpass response. At resonance, the gain is given by (5.46) where all variables are real numbers.

$$H_o = 1 + \frac{R_4}{R_1} \qquad (5.46)$$

The circuit was simulated, resulting in the frequency response curves shown in Fig. 5.69. The value of R_2 was stepped from 0 to 50 kΩ. The overall gain of the circuit is attenuated by the R_6-R_7 voltage divider; otherwise, the peak response would be about 40 dB.

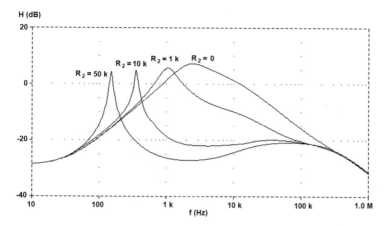

Fig. 5.69 Frequency response of the gyrator wah-wah circuit

The gyrator and active filters in general require higher bandwidth op amps than most other applications. For this reason, though they will still work, the LM358 and TL062 are not the best choices for the gyrator wah-wah. The higher bandwidth of the TL072 and RC5532 will give better performance.

As in previous effect circuits, the gyrator wah-wah provides lots of potential for experimentation and customization. For example, it would be interesting to set R_2 to a fixed value and replace R_1 with a potentiometer (and a resistor to prevent $R_1 = 0 \, \Omega$ from occurring). This would allow the Q of the bandpass filter to vary with f_o.

Envelope-Controlled Filter (Auto-Wah)

The traditional wah-wah effect is controlled by a foot pedal that varies the center frequency of a BP filter. An interesting effect can be obtained by controlling some filter parameter, typically the center frequency of a bandpass filter, with the envelope of the instrument. This forms what is called an *envelope-controlled filter* or *auto-wah*.

The schematic diagram for one possible implementation of an envelope-controlled filter is shown in Fig. 5.70. This circuit uses the same brute-force envelope follower approach that was used in the compression/sustain circuit of Fig. 5.49. Positive input signal peaks are sufficiently amplified by U1B to drive the LED into conduction. The slow response of the photoresistor averages out the effect of the light pulses. Op amp U1A forms an IGMF bandpass filter. The value of the photoresistor varies with the signal envelope, which in turn controls the center frequency of the BP filter.

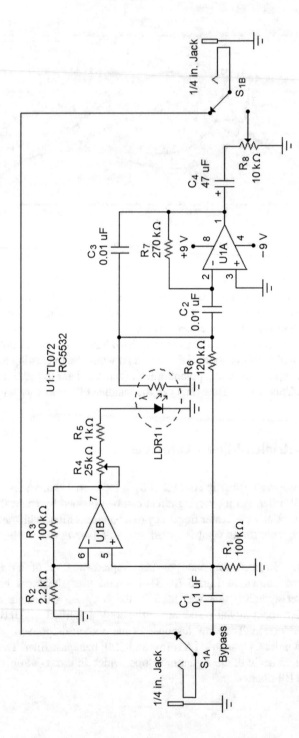

Fig. 5.70 Envelope-controlled filter auto-wah

Noise Gates

In response to requests from readers of the previous editions, noise gate operation is now discussed. A noise gate shorts the instrument signal to ground when the input signal envelope is less than some threshold value. Typically, the threshold voltage is set slightly above the noise floor of the guitar or the effect box preceding the noise gate.

A simple but very effective noise gate circuit is shown in Fig. 5.71a. The output is switched either by a normally-open (NO), CMOS analog switch or an NPN transistor. Op amp U1A is a super diode half-wave rectifier that passes the positive half of the signal envelope. Resistor R_1 provides a DC path to ground for the noninverting input bias current of U1A when the output switching device is an open circuit. As usual, R_2 and C_2 form a low-pass filter that removes high-frequency signal components, leaving the envelope voltage. The filtered envelope voltage is applied to the inverting input of op amp U1B.

The noise-floor threshold voltage V_{th} is set by *sensitivity* potentiometer R_5. The threshold voltage V_{th} is applied to the noninverting input of op amp U1B, which acts as a comparator. When no input signal is present, $+V_{env} < V_{th}$ and the output of U1B goes high. This closes the analog switch, shorting the output to ground, squelching output noise.

When an input signal is present, $+V_{env} > V_{th}$ and the output of comparator U1B goes low. The control input of the analog switch is driven low, opening the switch, which allows the signal to pass to the output. As long as the envelope voltage is greater than the threshold voltage, the audio signal will pass through to the output of the noise gate. Threshold voltage V_{th} can be set between 0 and 860 mV for the component values shown. If a greater range of adjustment is needed, a 10 kΩ potentiometer can be used.

The TS12A4514 is inexpensive (about \$0.45) and provides very high performance, with an on-state resistance of $R_{on} \cong 10\ \Omega$. This provides a noise reduction of 40 dB, but it's not the kind of part most people are likely to have sitting around. If you'd rather not order a TS12A4514, an NPN transistor can be substituted, though with some decrease in performance. Figure 5.71c shows the pin connections for a 2N3904, when used in place of the TS12A4514.

Waveforms from a PSpice simulation of the noise gate using the 2N3904 BJT switch, with a 1 V_{P-P}, 200 Hz sine input signal, are shown in Fig. 5.72. The upper waveform shows the input signal, amplitude modulated to produce an envelope that is fairly typical of an actual audio signal. For this simulation, the threshold voltage was set to $V_{th} = 370$ mV. The output signal shown in the lower part of the figure clearly shows the gating effect of the circuit.

Notice that there is slight attenuation of the output signal in Fig. 5.72. This is due to the voltage divider formed by R_7 and R_L. This simulation was run with $R_7 = 10$ kΩ initially, which caused this loss. Using $R_7 = 1$ kΩ as shown in the schematic will result in negligible attenuation when driving a typical amplifier or effects circuit.

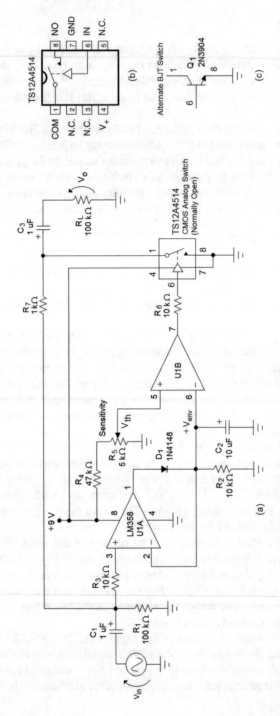

Fig. 5.71 (a) Noise gate circuit. (b) CMOS analog switch pinout. (c) Alternate BJT switch connections

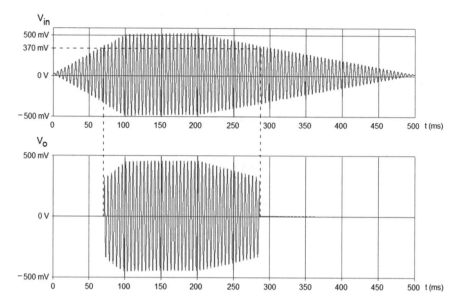

Fig. 5.72 Input and output waveforms for the noise gate

An input buffer with a low adjustable gain, say $A_{V(\text{max})} \leq 2$, could be used to compensate for possible attenuation due to loading.

A Little Deeper Look at the LM358

You might also be wondering about the purpose of R_3. The input stage of the LM358 is a PNP Darlington differential pair, which is shown in Fig. 5.73. Here, we are driving the inverting input directly with a sinusoidal voltage. Because the LM358 is operating from a single-polarity supply, negative-going input signals will be shorted to ground when the C-B junction of Q_{1A} becomes forward biased. This would clip the input voltage at about –0.7 V, distorting the audio signal. Adding resistor R_3, as in Fig. 5.71, helps prevent the noninverting input U1A from clipping negative-going inputs. Resistor R_1 back in Fig. 5.41a serves the same purpose.

Sampling, Quantization, DACs, and ADCs

The last effect that we will cover in this chapter is an experimental circuit that samples the instrument signal and outputs a quantized, 8-bit reconstruction of the original signal. The circuit is shown in Fig. 5.74. There is a lot going on in this circuit, and we need to cover a few new concepts, so let's dive right in.

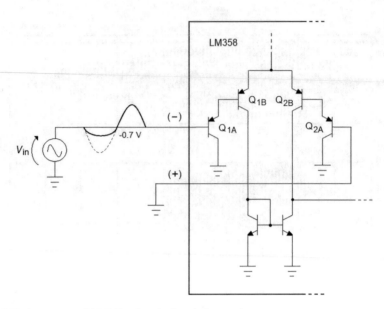

Fig. 5.73 Input stage of LM358 using single-polarity supply

Starting with the familiar sections, U4B is a unity gain buffer. Op amp U4A buffers the input with gain $A_V = 4.3$. Note that because no output coupling capacitor is present, the output of U4A, labeled v'_{in} in the schematic, has a DC offset of $V_{CC}/2$ volts. This serves an important function that we will soon see.

The CD40106 U1 is a hex, CMOS Schmitt trigger inverter. Inverters U1A and U1B form a relaxation oscillator that provides an adjustable frequency clock signal. The frequency is adjustable over a range of approximately $f_{CLK} = 6$ to 20 kHz.

U2 and U3 are CMOS 4-bit counters that are cascaded to form an 8-bit binary up/down counter. Output Q_0 of U3 is the least significant bit (LSB) of the counter, while output Q_3 of U2 is the most significant bit (MSB). Maximum count is 1111 $1111_2 = 255_{10}$. When counter mode control pin 10 is high ($U/\overline{D} = 1$), an up-count results (the counter *increments*). A low input ($U/\overline{D} = 0$) causes down-counting (the counter *decrements*). When a given Q output is high, we will assume $V_{OH} = +V_{CC}$. When Q is low, $V_{OL} = 0$ V.

The R-2R Ladder

Resistors R_9 through R_{24} comprise an interesting structure called an *R-2R ladder*. In this circuit, the R-2R ladder serves as a *digital-to-analog converter (DAC)*. A DAC accepts an *N*-bit binary number input and produces a proportional analog output voltage.

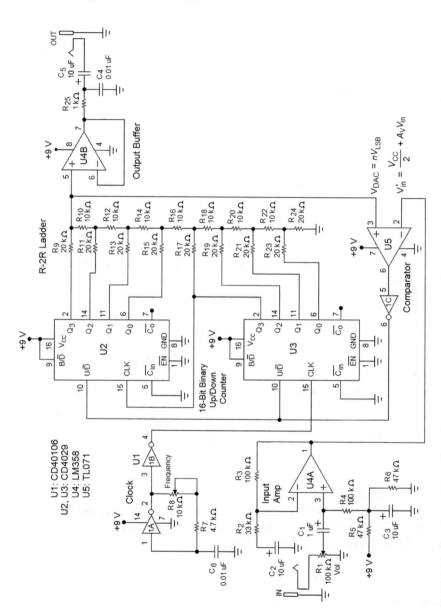

Fig. 5.74 An R-2R ladder-based quantizing effect circuit

The output of the 8-bit counter serves as the input to the R-2R ladder. If we start at zero and apply an increasing count to a DAC, the output V_{DAC} will increase in steps from 0 V to maximum output voltage. The smallest step is 1 LSB, which produces a voltage given by

$$V_{LSB} = \frac{V_{CC}}{2^N} \tag{5.47}$$

where N is the number of input bits. For our circuit, $V_{CC} = 9$ V and $N = 8$, so

$$V_{LSB} = \frac{9 \text{ V}}{256}$$
$$\cong 35.16 \text{ mV}.$$

The maximum output voltage is called the *full-scale voltage* V_{FS} which is given by

$$V_{FS} = V_{CC} - V_{LSB} \tag{5.48}$$

In this circuit, the full-scale voltage only differs from V_{CC} by about 35 mV, and the 9 V battery is not very precise anyway, so we will assume $V_{FS} \cong V_{CC} = 9$ V.

The ladder produces an output voltage V_{DAC} that is given by

$$V_{DAC} = nV_{LSB} \tag{5.49}$$

where n is the value of the binary input to the R-2R ladder. As an example, let's assume that the output of the 8-bit counter is $n = 1011\ 0101_2 = 181_{10}$. The resulting output of the ladder should be

$$V_{DAC} = nV_{LSB}$$
$$= 181 \times 35.16 \text{ mV}$$
$$= 6.36 \text{ V}$$

The output of U4A (v'_{in}) and V_{DAC} are applied to comparator U5. A TL071 was chosen for the comparator because of its high slew rate (20 V/µs). Schmitt trigger inverter U1C speeds up the comparator output even more, assuring a fast, clean signal for the $U/\overline{D}$ inputs of the CD4029s.

DAC Operation

Let's assume the counter is reset to zero, and no input signal is applied yet. The Q-outputs of the counter drive all R-2R ladder input bits low, and we get $V_{DAC} = 0$ V. With no signal present, the output of U4A is just the DC offset, and

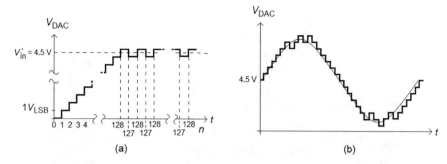

Fig. 5.75 (a) DAC output with a DC input. (b) DAC tracking a sinusoidal signal

$V'_{in} = V_{CC}/2 = 4.5$ V. Since $V_{DAC} < V'_{in}$, the comparator/inverter drives $U/\overline{D}$ high, so the counter increments. V_{DAC} increases by $1 V_{LSB}$ per clock cycle.

Eventually, the count increases until $V_{DAC} > V'_{in}$, and the comparator drives $U/\overline{D}$ low, causing the counter to decrement. After the next clock cycle, $V_{DAC} < V'_{in}$; $U/\overline{D}$ goes high, and the counter increments. Now, $V_{DAC} > V'_{in}$, $U/\overline{D}$ goes low, and the counter decrements. This up/down process repeats over and over, as shown in Fig. 5.75a.

If a sinusoidal signal is applied to the input, V'_{in} will vary as shown in the continuous waveform in Fig. 5.75b. The comparator senses differences between V_{DAC} and V'_{in} causing up/down counting, forcing V_{DAC} to track V'_{in} as closely as it can. This forms what is called a *tracking analog-to-digital converter (ADC)*.

No, the previous sentence was not a typo. Although the R-2R ladder is a DAC (it accepts a digital input and produces an analog output), feedback via the comparator transforms the ladder back into an ADC. Most analog-to-digital converters are built around DACs.

The distinction between DAC and ADC in this circuit is somewhat blurred. To be technically precise, this circuit accepts an analog signal with a continuous range (voltage) and a continuous domain (time) and outputs an approximation of that signal that has a discrete range and a discrete domain. The output voltage can only change in discrete steps, at discrete times.

Effect on an Audio Signal

Two main effects will occur when an audio signal is applied to this circuit. First, the output is an 8-bit approximation of the original signal. This introduces a form of distortion called *quantization noise* into the signal. R_{25} and C_4 form a low-pass filter which reduces the harshness of this noise on the output. This would be considered to be a reconstruction filter, like we saw used in the flanger in Fig. 5.37.

The input signal is sampled once every clock cycle. Since the clock operates at a relatively low frequency (6–20 kHz), aliasing will occur. With $f_{CLK} = 6$ kHz, signal

Fig. 5.76 Prototype construction of the quantizer circuit

frequencies greater than 3 kHz (the Nyquist frequency) will be aliased to lower frequencies. When $f_{CLK} = 20$ kHz, the Nyquist frequency is 10 kHz. Aliasing is the second form of distortion, which varies with adjustment of the clock frequency.

This circuit is highly experimental and is based on an idea I had when presenting the tracking ADC to my students at the college. If you want to experiment a bit, changing the clock frequency range by using a different value for capacitor C_6 is the most obvious place to start. Using a lower clock frequency will produce more radical aliasing and could be cool. Increasing the maximum clock frequency will reduce aliasing effects but make the circuit more sensitive to wiring and component placement-induced effects. A photo of the breadboarded circuit is shown in Fig. 5.76.

Final Comments

The subject of guitar effects falls under the more general area of signal processing, which is one of the most interesting subjects in electronics. We have really only touched the surface of the theory and design of guitar effects circuits, but hopefully you now have a better understanding of how these circuits work, which might help you to use them more effectively. The circuits presented here are a great starting point for experimentation with your own effects designs. I encourage you to try them out and modify them.

The math used to explain many of the circuits we've examined can be intimidating, but most of the time, basic algebra and trigonometry are all that you need. Keep in mind that the concepts used here are also applicable to many other areas of electronics and even mechanical system analysis.

As in previous editions, we've concentrated mainly on analog effects in this chapter. Bucket-brigade devices are right on the border between digital and analog domains, having characteristics that belong to both. The addition of the ADC/quantization experimental effect is about as far into digital signal processing as I could go, without adding hundreds of pages to the book. Maybe this book will inspire some reader to write a similar text on DSP-based effects?

Summary of Equations

Frequency and Period Relationships for Sinusoids

$$f = \text{Cycles/second (Hertz, Hz)} \tag{5.1}$$

$$f = \frac{1}{T} \tag{5.2}$$

$$T = \frac{1}{f} \tag{5.3}$$

$$v(t) = V_P \sin\left(2\pi f t + \Phi\right) \tag{5.4}$$

$$\omega = 2\pi f \tag{5.5}$$

Trig Identity: Product of Two Sine Functions

$$A \sin\left(\omega_1 t\right) B \sin\left(\omega_2 t\right) = \frac{AB}{2}\left[\cos\left(\omega_1 - \omega_2\right)t - \cos\left(\omega_1 + \omega_2\right)t\right] \tag{5.6}$$

BJT, JFET, MOSFET, and Triode Transconductance Equations

$$I_C \cong I_S e^{V_{BE}/\eta V_T} \tag{5.7}$$

$$I_D = I_{DSS}\left(1 - \frac{V_{GS}}{-V_P}\right)^2 \tag{5.8}$$

$$I_D = k(V_{GS} - V_T)^2 \tag{5.9}$$

$$I_P = k(V_P + \mu V_G)^{3/2} \tag{5.10}$$

PN Junction Diode Forward Current

$$I_F \cong I_S e^{V_{AK}/\eta V_T} \tag{5.11}$$

Log Amp Output Voltage

$$V_o = \eta V_T \ln\left(\frac{V_{in}}{R_1 I_S}\right) \tag{5.12}$$

All-Pass Filter Phase and Center Frequency Equations

$$\varphi = 180^0 - 2\,\tan^{-1}\left(\frac{f_{in}}{f_o}\right) \tag{5.13}$$

$$\varphi = -2\,\tan^{-1}\left(\frac{f_{in}}{f_o}\right) \tag{5.14}$$

$$f_o = \frac{1}{2\pi R_1 C_1} \tag{5.15}$$

Maximum and Minimum BBD Time Delay

$$T_{D(\max)} = \frac{N}{2f_{clk(\min)}} \tag{5.16}$$

$$T_{D(\min)} = \frac{N}{2f_{clk(\max)}} \tag{5.17}$$

Nyquist Sampling Theorem Relationships

$$f_{clk} \geq 2f_{in(\max)} \tag{5.18}$$

$$f_{alias} = f_{clk} - f_{in} \tag{5.19}$$

Precision Half-Wave Rectifier Output Voltage

$$v_o = \begin{cases} -v_{in}\dfrac{R_2}{R_1} & ,v_{in} < 0 \\ 0\,V & ,v_{in} \geq 0 \end{cases} \tag{5.20}$$

Precision Full-Wave Rectifier Output Voltage

$$V_o = \frac{R}{R_1}|V_{in}| \tag{5.21}$$

Dynamic Range

$$DR = 20\log\left(\frac{V_{max}}{V_{min}}\right) \tag{5.22}$$

LM3080 OTA Analysis Equations

$$I_{ABC} = \frac{V_{EE} - V_{BE}}{R_{ABC}} \tag{5.23}$$

$$g_m = 19.2 I_{ABC} \tag{5.24}$$

$$I_o = g_m V_{in} \tag{5.25}$$

$$V_o = I_o R_L \tag{5.26}$$

Exponential Decay

$$V_{envelope} = V_o e^{-1/\tau} \tag{5.27}$$

Instantaneous Form for Phase-Shifted Sine Wave

$$v_1 = V_1 \sin\left(2\pi f_1 t + \phi_1\right) \tag{5.28}$$

Amplitude Modulator Output Voltage Equation

$$v_o(t) = V_1 \sin\left(2\pi f_1 t\right) \times [V_{DC} + v_m(t)] \tag{5.29}$$

Balanced (Ring) Modulator Output Voltage Equation

$$v_o(t) = kv_1(t)v_m(t) \tag{5.30}$$

Trigonometric Identity for the Square of a Sine Function

$$[V_m \sin(\omega_m t)]^2 = \frac{kV_m^2}{2} - \frac{kV_m^2}{2} \cos(2\omega_m t) \tag{5.31}$$

Series RLC Resonant Frequency

$$f_o = \frac{1}{2\pi\sqrt{LC}} \tag{5.32}$$

IGMF Bandpass Filter Equations

$$BW = f_H - f_L \tag{5.33}$$

$$Q = \frac{f_o}{BW} \tag{5.34}$$

$$f_o = \frac{\sqrt{\left(\frac{1}{R_2 C^2}\right)\left(\frac{1}{R_1} + \frac{1}{R_3}\right)}}{2\pi} \tag{5.35}$$

$$R_1 = \frac{Q}{2\pi f_o A_o C} \tag{5.36}$$

$$R_2 = \frac{Q}{\pi f_o C} \tag{5.37}$$

$$R_3 = \frac{Q}{2\pi f_o C(2Q^2 - A_o)} \tag{5.38}$$

$$A_o = -\frac{R_2}{2R_1} \tag{5.39}$$

$$f_o = \sqrt{f_H f_L} \tag{5.40}$$

$$R_{3(new)} = R_{3(old)} \left(\frac{f_{o(old)}}{f_{o(new)}}\right)^2 \tag{5.41}$$

Op Amp Gyrator Inductance

$$L = R_1 R_2 C_1 \tag{5.42}$$

Q-Factor of a Gyrator Inductor

$$Q = 2\pi f R_2 C \tag{5.43}$$

Center Frequency of Gyrator Wah-Wah Filter

$$f_o = \frac{1}{2\pi\sqrt{C_1 C_2 R_1 (R_2 + R_3)}} \tag{5.44}$$

Gyrator Wah-Wah Response Equations

$$H = 1 + \frac{R_4 \angle 0°}{X_L \angle 90° + R_1 \angle 0° + X_{C2} \angle -90°} \tag{5.45}$$

$$H_o = 1 + \frac{R_4}{R_1} \tag{5.46}$$

Analog-to-Digital and Digital-to-Analog

$$V_{LSB} = \frac{V_{CC}}{2^N} \quad (N \text{ is the number of bits}) \tag{5.47}$$

$$V_{FS} = V_{CC} - V_{LSB} \tag{5.48}$$

$$V_{DAC} = n V_{LSB} \quad (n \text{ is the value of the binary input}) \tag{5.49}$$

Chapter 6
Low-Power Vacuum Tube Amplifiers

Introduction

The invention of the transistor has made vacuum tubes obsolete in virtually all applications, except for one major exception. Musical instrument amplifiers in general and guitar amps in particular have proven to be one of the last refuges for tube-based designs. No one can argue that modern solid-state amplifiers are efficient, inexpensive, and capable of sounding great too. However, vacuum tube amplifiers, which we will usually just call tube amps from now on, have a certain mojo that really can't be duplicated with transistors or op amps.

Compared to previous chapters, this one starts out a bit slowly, covering a lot of background material. Once we get in the groove though, the main thrust of this chapter is basic vacuum tube design and analysis methods used in low-power, class A amplifier applications. These circuits typically make up the stages of the typical guitar amplifier up to the power output stage. Based on suggestions from readers of the second edition of this book, additional design examples are presented, which should allow you to design your own basic audio amplifier stages with just about any tube you happen to have on hand.

This chapter assumes a decent general knowledge of electronics. Serious study and experimentation with the circuits covered in the preceding chapters is sufficient preparation, though if you aren't familiar with vacuum tube basics, additional review material on basic tube operating principles may be found in Appendix C.

Commonly Used Vacuum Tubes

There are still hundreds of different tubes available, and complete data sheets from many different manufacturers can be found online. The tubes that we will use in design examples in this chapter are the 12AT7, 12AU7, and 12AX7 dual triodes, as well as the dual triode/pentode 6AN8. These tubes are still very commonly used in many guitar amplifier applications and are readily available at reasonable cost.

Parts Sources and Availability

A good source for tubes, sockets, power transformers, chassis boxes, speakers, and just about any other vacuum tube-related supplies you can think of is Antique Electronic Supply (www.tubesandmore.com). Another good source of tubes and tube-related supplies is Tube Depot (www.tubedepot.com). You can also usually find lots of useful tube stuff on eBay and at electronic flea markets called hamfests. This name derives from the fact that amateur radio operators (hams) and other electronics hobbyists originated these gatherings. You never know what you are going to find at the typical hamfest.

The 12Axx tubes are still in production in China and Eastern Europe and are also readily available as *new old stock* (NOS) for reasonable prices. NOS tubes are no longer being produced by the original manufacturers but are available in brand-new, unused condition. A few of the more common NOS tube manufacturers are General Electric (GE), RCA, Sylvania, Amperex, Tung-Sol, Raytheon, Mullard, Phillips, Motorola, and Telefunken. The 6AN8 and the popular 6FQ7 are not currently in production, but they are still readily available as NOS.

In general, the quality of NOS tubes is very high. Specific variants of some tubes are reputed to have superior performance and sell for outrageously high prices. I have tried to be objective and unbiased here, but having experimented with many different tubes in many different designs over the years, I really can't say that I prefer the sound of one 12AX7 over another. Of course, if a given tube is defective or has deteriorated in performance because of long use, there will definitely be a difference in performance. I have had great success using NOS and new Chinese, Russian, and other Eastern European manufactured tubes. In all honesty, I really don't pay any attention to brand names. I usually just use whatever brand tube I happen to have on hand, and things work just fine. My suggestion is to experiment and see what sounds best to you.

Vacuum Tube Parameters and Data Sheets

By current standards, much of the terminology and notation used in older vacuum tube literature is somewhat inconsistent and not well defined. Where possible, I use the original terminology and notation, but in order to make things more consistent with current electrical engineering conventions, I sometimes prefer to use transistor-like notation, such as V_{PP} instead of B+.

Tube data sheets frequently list ratings based on specific applications such as vertical or horizontal oscillators, vertical deflection amplifiers, class A amplifiers, push-pull amplifiers, as well as others. Eventually you get used to the old-school approach to data sheets, and it begins to make sense, but it does take some getting used to.

Tube data sheets may list parameters under the headings of "Design Center Values" and/or "Design Maximum Values." These are both maximum DC and signal-related values that should not be exceeded. The term Design Center Values was used prior to 1957. The term Design Maximum Values was adopted in 1957. Regardless of which notation is used, these sections list the maximum values for tube currents and voltages under various operating conditions. Usually these maximums refer to average DC parameter values, but sometimes, they specify maximum allowable instantaneous values. Determining which interpretation is correct often requires some practical experience.

Many tube voltage and current ratings vary with different biasing arrangements (class A, B, grid leak bias, fixed bias, etc.) and between triode/pentode operation as well. Generally, the class A design center and design maximum values are the most conservative, so if you want to maximize tube reliability, use these limits. Regardless of the other ratings for a tube, the maximum plate power dissipation limit should not be exceeded.

Absolute Maximum Ratings

One thing that I found particularly frustrating was finding the absolute maximum plate current and plate-to-cathode voltage ratings for a given tube. You can search tube data sheets all day long and not find this information.

A rare example of a data sheet that explicitly lists some absolute maximum plate current and voltage values is GE data sheet ET-T880A, for the 12AU7. The design center maximum and absolute maximum values are:

GE ET-T880A data sheet: 12AU7 design center maximum ratings		
	Class A1 amplifier	Vertical deflection amplifier
DC plate voltage	300 V	300 V
Peak positive pulse plate voltage	–	1200 V
DC cathode current	20 mA	20 mA
Peak cathode current	–	60 mA

The peak values listed are the absolute maximum instantaneous values the tube is guaranteed to handle, so we want to make sure that our designs never exceed these ratings. Typically, the instantaneous maximum (peak) ratings for plate current and plate voltage are at least three times the design center maximums. So, if peak values are not listed for a given tube, a reasonable estimate would be to assume that absolute maximum instantaneous plate current and voltage ratings are three times the design center maximums.

The DC values listed are the maximum allowable average DC voltage and current values. Basically, these are the maximum allowable Q-point current and voltage values. The designs presented in this chapter won't exceed these limits, but you may find vintage amplifier designs that operate tubes at much higher voltages than would be allowed, according to the data sheets. For example, the Fender AA1164 Princeton Reverb amp operates its 12AT7 reverb driver tube at a plate voltage $V_P = 400$ V. The published 12AT7 design center maximum DC plate voltage is 300 V. There are many other similar examples, but I recommend designing circuits that fall within the manufacturer's maximum ratings.

As a point of interest, for the parameters listed above, class A1 means class A biasing where the grid of the tube never goes positive with respect to the cathode ($V_{GK} \leq 0$ V). This is the biasing that we will be using in this chapter, but we will simply call it class A.

Vertical deflection amplifiers are found in old CRT (cathode-ray tube) televisions. This is a particularly demanding application that subjects tubes to very high peak voltages and currents, which is why the absolute maximum ratings are usually encoded into this portion of the data sheet, if at all.

Other Data Sheet Parameters

Here's an example of one of the more unfamiliar ways that some data sheets present voltage and current values. A particular GE data sheet (ET-T1515A, March 1959) lists the typical class A, triode-mode, zero-signal plate current for the 6L6GC as 40 mA, with $V_{PP} = 250$ V. The max signal plate current is listed as being 44 mA. At first glance, this seems to indicate that if we increase the plate current by only 4 mA, we will exceed the tube current limits. This is not the case though. What GE is actually telling us is that we can bias the 6L6G up for $I_{PQ} = 40$ mA, and an input signal can increase the plate current by 44 mA which gives us

$$
\begin{aligned}
I_{PM} &= I_{PQ} + I_{P(signal)} \\
&= 40 \text{ mA} + 44 \text{ mA} \\
&= 84 \text{ mA}.
\end{aligned}
$$

As I said before, compared to current IC and transistor data sheets, vacuum tube data sheets take some getting used to.

Table 6.1 lists some of the major parameters for the tubes that are used in this book, as well as some others that may be useful. I compiled this data from many manufacturers' data sheets. Some parameters are estimates, based on values obtained from multiple data sheets and characteristic curves. I show you how to determine some of these parameters from tube characteristic curves in the last section of this chapter. Where two voltage or current values are listed, the upper value is for operation in the pentode mode, and the lower value is for triode-mode operation.

Table 6.1 Selected tube data

Tube	Type	μ	g_m (mS)	r_P (Ω)	$I_{P(DC)}$ (mA)	$V_{P(DC)}$ (V)	$P_{P(max)}$ (W)	R_K (Ω)	Heater (V/A)	Notes
EL34	Pentode	160	10.6	1.5 k	150	500	30	130	6.3 V	6CA7
		8	4.7	1.7 k		450			1.5 A	$R'_L = 3.5$ k (P) = 3.5 k (T)
EL84	Pentode	418	11	38 k	60	300	12	130	6.3 V	6BQ5
		19	5.4	3.5 k	40	250			760 mA	$R'_L = 5.2$ k (P) = 3.5 k (T)
KT88	Beam power	165	11	15 k	140	800	42	–	6.3 V	
	Tetrode	8	12	670	160				1.6 A	
PCL86	Pentode	21	10.5	63 k	55	250	9	220	12.6 V	Dual T/P
	Triode	100	1.6	–	4	–	0.5	–	300 mA	Similar to 12AX7
5AR4GT	Dual diode	–	–	–	825	1550	–	–	5 V	20 V at 275 mA
	Indirect heat				3.7 A$_{(pk)}$				3 A	RMS 450-0-450
5U4GB	Dual diode	–	–	–	225	1550	–	–	5 V	50 V at 275 mA
	Direct heat				4.6 A$_{(pk)}$				3 A	RMS 450-0-450
5Y3GT	Dual diode	–	–	–	125	1400	–	–	5 V	50 V at 275 mA
	Direct heat				440 mA$_{(pk)}$				2 A	RMS 350-0-350
6AN8	Pentode	1241	7.3	170 k	12	330	2.3	56	6.3 V	Dual T/P
	Triode	21	4.5	4.7 k	15	330	2.8	–	450 mA	Similar to 12AU7
6AQ5	Beam power	213	4.1	52 k	50 mA	275	12 W	–	6.3 V	
	Pentode	9.5	4.8	2 k	50 mA	250	10 W		450 mA	
6BF5	Beam power	90	7.5	12 k	36	250	5.5	–	6.3 V	
	Pentode								1.2 A	
6BQ7	Dual triode	39	6.4	6.1 k	9	150	–	–	6.3 V	
									400 mA	

(continued)

Table 6.1 (continued)

Tube	Type	μ	g_m (mS)	r_P (Ω)	$I_{P(DC)}$ (mA)	$V_{P(DC)}$ (V)	$P_{P(max)}$ (W)	R_K (Ω)	Heater (V/A)	Notes
6BX7	Dual triode	10	7.7	1.3 k	60	500	10 (Ea.)	390	6.3 V, 1.5 A	12 W total power, Both triodes
6BQ7	Dual triode	20	2.6	7.7 k	10	330	4 (Ea.)	–	6.3 V, 600 mA	5.7 W total power, Both triodes
6K6GT	Beam power	220	2.2	100 k	75	315	8.5	400	6.3 V	
	Pentode	6.7	2.7	2.5 k	50	250	7	–	400 mA	
6L6GC	Beam power	172	5.2	33 k	70	500	30	300	6.3 V	KT66, $R'_L = 4.2$ k (P)
	Pentode	8	4.7	1.7 k	54	450	30	–	900 mA	= 5 k (T)
6V6GT	Beam power	250	3.8	65 k	45	350	14	–	6.3 V	$R'_L = 5$ k (P)
	Pentode	9.8	5	2 k	–	315	10	–	450 mA	= 5 k (T)
12AT7	Dual triode	61	5.5	11 k	15	300	2.5	–	6.3 V, 300 mA	ECC81, 6021
12AU7	Dual triode	21	3.0	6.5 k	20	300	2.75	–	6.3 V, 300 mA	ECC82, 5814
12AV7	Dual triode	40	7.3	5.5 k	18	300	2.7	60	6.3 V, 300 mA	5965
12AX7	Dual triode	100	1.5	65 k	8	300	1	–	6.3 V, 300 mA	ECC83, 7025
5963	Dual triode	21	3.2	6.6 k	20	250	2.5	–	6.3 V, 300 mA	Computer version of 12AU7

Equivalent tube numbers are given for some of the tubes listed. Some are identical in all respects, such as 12AT7 = ECC81, 12AU7 = ECC82, and 12AX7 = ECC83. Some other tubes may be very similar, with only slight differences in some parameters. For example, the 12AU7 and the 5963 are almost identical electrically, with the primary difference being the lower DC plate voltage limit of 250 V for the 5963 versus 300 V for the 12AU7. It is best not to make tube substitutions without carefully studying the circuit in which the tube will be used to make sure the substitute tube will work properly.

Unlike IC and transistor circuit design techniques, tube amplifier design is very much a graphical procedure, and tube characteristic curves are used extensively in the design process. Characteristic curves for the tubes used in most of the design examples are included in Appendix B. Alternatively you can obtain a few tube data books (GE, RCA, Sylvania, etc.) to find these curves, or you can download them from online. A great source for tube data sheets is Frank Filipse's site http://frank. pocnet.net/.

Tube Pin Numbering

Regardless of the size of the tube envelope (medium shell octal, small-button 9-pin, etc.), all tubes use the same pin numbering system, which is shown in Fig. 6.1. The basing diagrams are bottom views with pins numbered going clockwise from the index. The tube on the left of the figure is a 12AU7. The right side of the figure shows a 6L6GC.

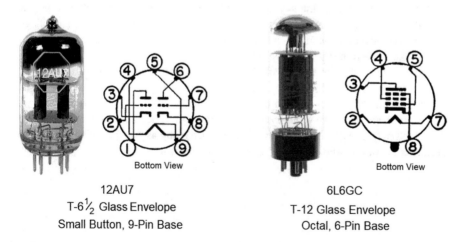

12AU7
T-6½ Glass Envelope
Small Button, 9-Pin Base

6L6GC
T-12 Glass Envelope
Octal, 6-Pin Base

Fig. 6.1 Common vacuum tube packages and basing diagrams

General Amplifier Design Principles

I feel fairly confident in saying that most readers of this book are creative, technically minded people that probably find designing circuits to be far more interesting and enjoyable than analyzing them. In general, the ability to design electronic circuits (synthesis) is at least one step higher in understanding and abstraction than analysis. The design procedures outlined here will certainly not transform you into a tube design guru overnight. But hopefully you will gain additional insight into the design process and deepen your understanding of vacuum tube circuits. If you are already knowledgeable about tube circuits, there may not be much new to you here.

By necessity, the design approach used here is somewhat formulaic and procedural, but I try to point out places where a given approach can be modified to suit different goals. It is likely that once you have worked through the designs presented, you will be able to adapt the design procedures to suit your needs.

There are many different ways we can approach the design of an amplifier, but it is important to keep the overall goal of the design in mind. In the context of this book, there are actually two general goals that need to be met: designing useful, cool guitar amplifiers and explaining the theoretical and practical techniques used in basic vacuum tube circuit design.

It's difficult to say which goals are most significant to the readers of a book such as this one. To hobbyists on a tight budget, keeping costs low is important. If you must work with lots of parts that have been scavenged from old equipment, then your designs will necessarily be centered on this limitation. For example, if you only have a power transformer that has a 250 V_{rms}, 50 mA high-voltage secondary, and a 6.3 V, 1 amp, low-voltage secondary, then this will essentially define the maximum capabilities of your amplifier, as well as limiting power supply circuit variations. If money is no object, then the sky is the limit as far as circuit complexity and power output is concerned. If, by some chance, this book is being used in a formal college course, then technical accuracy and some degree of rigor is important. Hopefully, I've walked this tightrope successfully.

Most of the time, there is no single correct solution to a given design problem. Factors to be considered include performance, complexity, cost, efficiency, reliability, parts availability, and aesthetics. All of these factors must be weighed against one another, and there are virtually an infinite number of design solutions that satisfy these factors to one degree or another. In the end, as long as you are satisfied with the performance of the amplifier, the design was successful.

Cathode Feedback Biasing

Similar to JFETs, vacuum tubes are normally-on devices. Although the device physics are different, from a circuit analysis standpoint, vacuum tube cathode feedback bias works using the same principle as JFET source feedback bias. Cathode feedback biasing is also sometimes called *self-biasing*, *automatic biasing*, or simply *cathode biasing*.

A cathode feedback biased triode is shown in Fig. 6.2. Plate current I_P flows down through the tube and R_K, causing a voltage drop which raises the cathode voltage V_K to

$$V_K = I_P R_K \tag{6.1}$$

The resistance seen looking into the grid is practically infinite so $I_G \cong 0$ A. Because of this, there will be no voltage drop across R_G, and the grid voltage is $V_G = 0$ V. Applying Kirchhoff's voltage law (KVL) around the grid-cathode loop, the resulting grid-to-cathode bias voltage is given by

$$V_{GK} = -I_P R_K \tag{6.2}$$

The cathode supplies free electrons that can flow up to the positively charged plate. If the grid was floating, and not connected to ground, it would have no effect on the flow of electrons through the tube. Since the grid is at ground potential ($V_G = 0$ V), it is negative with respect to the cathode. The grid repels some of the

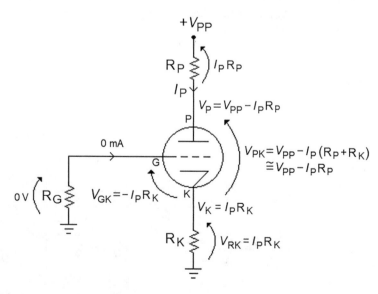

Fig. 6.2 Cathode feedback biasing

electrons that would normally flow from the cathode up to the plate, and current I_P is reduced. The more positive the cathode is with respect to the grid, the smaller the current flow becomes, and the triode is biased more toward cutoff. The plate-to-cathode voltage V_{PK} is found using

$$V_{PK} = V_{PP} - I_P(R_P + R_K)$$ (6.3)

It turns out that very often $R_P >> R_K$, so we can sometimes make the following approximation:

$$V_{PK} \cong V_P = V_{PP} - I_P R_P$$ (6.4)

This approximation will be used in some of the upcoming design examples because it simplifies the design procedure and results in a slightly more conservative design as well.

Cathode feedback is a form of negative feedback, which increases biasing stability. The Q-point of a cathode feedback biased tube will tend to be fairly insensitive to parameter variations as the tube ages, or if it is replaced by a new tube. This is a very desirable characteristic.

Fixed Biasing

Fixed biasing (also called *grid biasing*) is not commonly used in low-power, class A amplifier designs, so we won't spend much time discussing it here. In fixed biasing, a separate bias voltage V_{GG} is applied directly to the grid of the tube, as shown in Fig. 6.3. Notice that the grid voltage V_G is a negative voltage, with respect to ground. Since the grid current is zero, the grid voltage is

Fig. 6.3 Fixed biasing or grid biasing

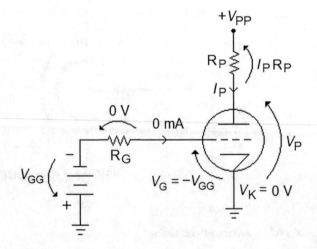

$$V_G = -V_{GG} \qquad (6.5)$$

The advantage of fixed biasing is that because a cathode resistor is not needed, a greater percentage of the supply voltage is available to drop across the tube. This allows a fixed bias amplifier to produce a larger output signal swing than an equivalent cathode feedback amplifier.

The need for a separate, negative grid biasing voltage is a disadvantage of fixed biasing. Fixed biasing is also more sensitive to tube parameter variations than cathode feedback biasing. Care must be taken when replacing fixed bias tubes, especially in power amplifier stages. Often, fixed bias stages need to be *rebiased* when a tube is replaced.

Sometimes a low-value cathode resistor will be present, even though grid biasing is used. This resistor provides a bit of negative feedback which stabilizes the amplifier. This also allows us to determine I_P indirectly by measuring the low voltage dropped across the resistor.

Class A, Resistance-Coupled, Common Cathode Amps

Class A amplifiers are used extensively in tube amp designs in voltage gain/driver stages and in power output stages as well. *Resistance coupling* is old-school tube terminology for what is now usually referred to as capacitive coupling in modern transistor circuit analysis and design. As with the transistor amplifiers we covered previously, the fundamental job of coupling and bypass capacitors is to act as short circuits for AC signals while blocking DC.

The basic resistance-coupled, common cathode, class A amplifier stage is shown in Fig. 6.4, with all pertinent voltages and currents shown. Cathode feedback is used here to establish the Q-point. This circuit is very similar to the source-feedback biased JFET, common source amplifier studied back in Chap. 3. We start by examining the DC equivalent circuit, which is formed by treating all capacitors as open circuits and reducing the input signal voltage to zero. Note that the DC equivalent is identical to Fig. 6.2.

A positive-going input signal makes the grid become less negative with respect to the cathode, and plate current increases. When the input signal drives the grid more negative, plate current is reduced. Like the BJT CE and FET CS amplifiers, the common cathode amplifier inverts the output signal phase, as shown by the waveforms in Fig. 6.4. In order to keep the input resistance of the circuit high and to prevent any significant grid current from flowing, we must keep $V_{GK} \leq 0$ V (class A1 bias).

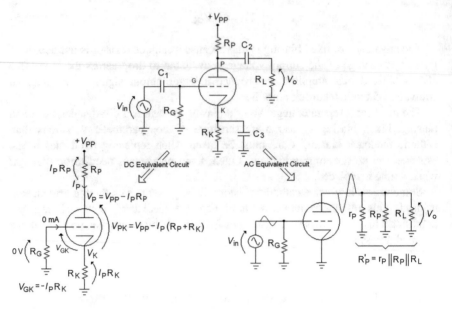

Fig. 6.4 Resistance-coupled, class A, common cathode amplifier

12AU7 Low-Power Amp Design Example

Let's get right into the interesting stuff by designing a practical amplifier based on the circuit of Fig. 6.4. We will use a 12AU7 dual triode for this example. Here are a few of the main 12AU7 parameters taken from Table 6.1, along with brief explanations.

12AU7 basic parameter summary	
$P_{D(max)} = 2.75$ W	Maximum DC plate power dissipation
$I_{P(max)} = 20$ mA	Maximum DC plate current
$V_{P(max)} = 300$ V	Maximum DC plate-to-cathode voltage
$g_m = 3$ mS	Typical transconductance
$r_P = 6.5$ kΩ	Typical internal dynamic plate resistance (similar to r_{CE} or r_o for a BJT)
$\mu = 20$	Amplification factor (maximum available voltage gain)

There are many different ways to approach the design of this amplifier. The design procedure used in this example will be presented in numbered steps to help keep things organized. Since this is our first design, there is also quite a bit of explanatory material included. Future design examples will be more streamlined.

1. Determine power supply voltage V_{PP}

    One thing that we usually know at the outset of the design process is the power supply voltage, V_{PP}. For low-power, resistance-coupled amplifiers using tubes such as the 12AU7 or 12AX7, the supply voltage typically ranges from 100 to 300 V. For this example, let's assume that $V_{PP} = 250$ V.

2. Determine suitable Q-point (V_{PQ}, I_{PQ})

There are many different ways to select a Q-point. A reasonable way to choose the Q-point plate voltage for most low-power, resistance-coupled, class A amplifiers is to start by setting the plate voltage to about half the power supply voltage

$$
\begin{aligned}
V_{PQ} &= \frac{V_{PP}}{2} \\
&= \frac{250 \text{ V}}{2} \\
&= 125 \text{ V.}
\end{aligned}
\tag{6.6}
$$

Keep in mind that we are going to use the approximation $V_{PK} \cong V_{PQ}$ (Eq. 6.4) during the design procedure. Also, we will see later that depending on the load being driven, we may wish to choose a different V_{PQ} in order to obtain more symmetrical output clipping levels.

We can choose any value for I_{PQ} as long as the resulting Q-point is located within the *safe operating area (SOA)*. Keeping I_{PQ} low reduces power dissipation and operates the tube on a more nonlinear portion of its characteristic curves. High I_{PQ} increases power dissipation and results in more linear tube operation. Somewhat arbitrarily, we will choose

$$
I_{PQ} = 5 \text{ mA.}
$$

The Q-point power dissipation of the 12AU7 plate is given by

$$
\begin{aligned}
P_{DQ} &= I_{PQ} V_{PQ} \\
&= 5 \text{ mA} \times 125 \text{ V} \\
&= 625 \text{ mW.}
\end{aligned}
\tag{6.7}
$$

Here is a summary of the various parameters calculated so far. The Q-point is well within the SOA.

$$
V_{PQ} = 125 \text{ V}, I_{PQ} = 5 \text{ mA}, P_{DQ} = 625 \text{ mW}
$$

Using the plate characteristic curve graph for the 12AU7 (from Appendix C, or download one from online), place a dot at the Q-point location as shown in Fig. 6.5.

3. Determine maximum plate voltage V_{PM}

Place a dot on the plate voltage axis at the supply voltage, $V_{PM} = V_{PP} = 250$ V. This is Point 1 in Fig. 6.5. We will call this maximum plate voltage V_{PM}, which occurs when the tube is in cutoff.

4. Determine maximum plate current I_{PM}

Since we know that $V_{PM} = 2V_{PQ}$, then by definition, $I_{PM} = 2I_{PQ} = 10$ mA. Mark this as Point 2 on the plate curves.

5. Plot the DC load line

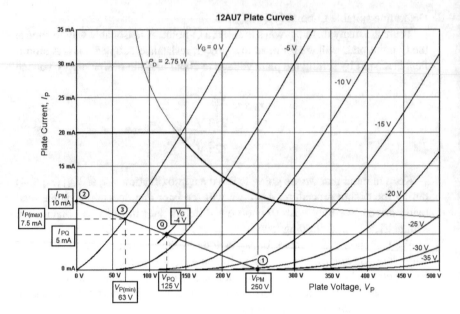

Fig. 6.5 Location of the Q-point and construction of DC load line

Draw a line from Point 1 to Point 2. This line is called the *DC load line*. Notice that the load line intersects the Q-point. If this doesn't happen when you plot the load line, it's an indication that something is wrong with your calculations.

6. Calculate required plate resistance R_P

The required plate resistance R_P is the reciprocal of the slope of the load line. We can use any convenient interval on the load line to find this value using (6.8).

$$R_P = \frac{\Delta V_P}{\Delta I_P} \tag{6.8}$$

Here, it is most convenient to use the endpoints of the load line in (6.9), giving us

$$\begin{aligned} R_P &= \frac{V_{PM}}{I_{PM}} \\ &= \frac{250\ \text{V}}{10\ \text{mA}} \\ &= 25\ \text{k}\Omega\,(\text{use either } 22\ \text{k}\Omega \text{ or } 27\ \text{k}\Omega). \end{aligned} \tag{6.9}$$

We will choose $R_P = 27\ \text{k}\Omega$.

Because we used $R_P = 27\ \text{k}\Omega$, which is larger than the theoretical value (25 kΩ), the actual plate voltage is found using Ohm's law and KVL to be

$$V_{PQ} = V_{PP} - I_{PQ}R_P$$
$$= 250 \text{ V} - (5 \text{ mA} \times 27 \text{ k}\Omega)$$
$$= 115 \text{ V}.$$

This is slightly less than the original Q-point goal of 125 V, but the difference is not significant. There will often be differences of 5% or so between theoretical and actual operating parameters because we are restricted to using standard resistor values and there are normal tube parameter and supply voltage variations as well.

The actual plate voltage (115 V) is shown in Fig. 6.7, but during the design process, I will often show the theoretical, design-goal Q-point values on the schematics.

7. Determine minimum allowable plate voltage $V_{P(min)}$

Locate the intersection of the load line with the $V_G = 0$ V plate curve (Point 3 in Fig. 6.5). The voltage coordinate of this point is the actual minimum useful plate voltage $V_{P(min)}$ for the amplifier. In this example, we have

$$V_{P(min)} \cong 63 \text{ V}.$$

If the amplifier is driven too hard, the grid will go positive with respect to the cathode, causing grid current to flow. This will cause the grid to dissipate significant power, which may damage the tube. It also causes the input resistance of the tube to decrease dramatically. Neither occurrence is a good thing. To prevent this, we keep $V_P > V_{P(min)}$.

Point 3 also locates the *maximum allowable* plate current $I_{P(max)}$. If the plate current exceeds this value, the grid will be more positive than the cathode. Maximum allowable plate current is always less than the *maximum possible* plate current I_{PM}. We don't really need to know $I_{P(max)}$, but it's easy to determine. In this example, we have

$$I_{P(max)} \cong 7.5 \text{ mA}.$$

8. Determine grid voltage V_G

The Q-point is located to the left of the $V_G = -5$ V curve. We must visually approximate the grid voltage that would intersect the Q-point. In this case, a good estimate is

$$V_G \cong -4 \text{ V}.$$

Note: Because we are using cathode feedback, we are actually interested in the grid-to-cathode voltage V_{GK}, but plate curves are always given in terms of V_G. This can be confusing at first because we are abusing the notation somewhat using $V_G = -I_P R_K$ (see Fig. 6.2).

9. Determine required cathode resistance R_K

   We now determine the value of the cathode resistor required to bias the tube up at the Q-point. We solve Eq. (6.2) for R_K which gives us

$$R_K = \frac{-V_G}{I_{PQ}}$$
$$= \frac{4\,V}{5\,mA}$$
$$= 800\,\Omega \text{ (use } 820\,\Omega\text{)}. \tag{6.10}$$

Alternative Determination of Cathode Resistance R_K

We can also determine the required cathode resistance using the tube transconductance curves. First, locate I_{PQ} on the plate current axis, and then draw a horizontal line to the right until it intersects the curve with the desired plate voltage V_{PQ}. In this example, since there is no curve for $V_{PQ} = 125$ V, we can graphically estimate the appropriate curve as shown in Fig. 6.6.

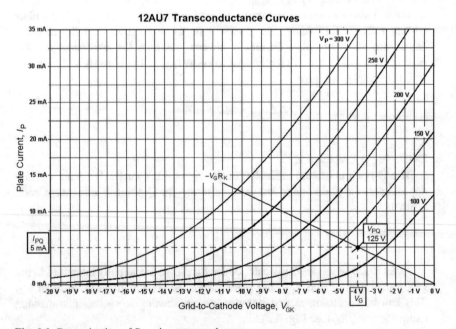

Fig. 6.6 Determination of R_K using transconductance curves

Draw a line down from the intersection of the $I_{PQ} = 5$ mA line and the $V_{PQ} = 125$ V curve (this is the Q-point). In this example, we find $V_G = -4$ V, which matches the value found in Step 8 using the plate curves. We use (6.10) again to find $R_K = -V_G/I_{PQ} = 4$ V/5 mA $= 800$ Ω.

Although the use of the transconductance curves was not necessary here, it is nice to see that the results are consistent. The line drawn from the origin through the Q-point is sometimes called the *bias line*. This is a graph of the equation $I_P = -V_G R_K$, which is just another form of (6.10). You don't need to plot this line, but it can serve to show how I_P will vary with changes in plate voltage V_P, for a given cathode resistor R_K.

10. Determine grid resistor R_G.

The final step in the design is to choose a grid resistor. Under normal conditions, the grid current is extremely small (ideally, $I_G = 0$ A), so a very large grid resistor may be used to keep the input resistance of the amplifier high. The maximum allowable self-bias grid resistance listed on the General Electric 12AU7 data sheet is $R_{G(max)} = 1$ MΩ. We will be a bit conservative and choose $R_G = 470$ kΩ.

DC and AC Load Lines

When an external load is capacitively (or transformer) coupled to the plate, the *AC load line* determines the large-signal limits of the amplifier output, that is, the AC load line maximum plate current and voltage and the output voltage clipping points. As a general rule of thumb, if $R_L \gg R_P$, the DC and AC load lines will be negligibly different, and we can use the DC load line for large-signal analysis.

In this example, let's assume that the load being driven is $R_L = 470$ kΩ, which is typical of when we are driving another amplifier stage. The complete amplifier is shown in Fig. 6.7. Because $R_L \gg R_P$ (i.e., 470 kΩ $\gg$ 27 kΩ), the AC load line would be virtually indistinguishable from the DC load line, so we do not need to construct the AC load line. We will soon see other examples where this is not the case.

Amplifier AC Performance

The AC equivalent plate resistance R'_P is given by (6.11). As noted at the start of this example, the internal dynamic plate resistance for the 12AU7 is $r_P = 6.5$ kΩ.

$$R'_P = r_P \parallel R_P \parallel R_L$$
$$= 6.5 \text{ kΩ} \parallel 27 \text{ kΩ} \parallel 470 \text{ kΩ} \quad\quad (6.11)$$
$$\cong 5.2 \text{ kΩ}.$$

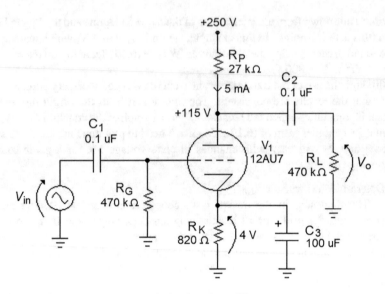

Fig. 6.7 Complete 12AU7 common cathode, class A amplifier stage

The voltage gain can be approximated by using an equation that is virtually identical to that used to find the gain of the JFET common source amplifier of Chap. 3.

$$\begin{aligned} A_V &= -g_m R'_P \\ &= -3 \text{ mS} \times 5.2 \text{ k}\Omega \\ &= -15.6. \end{aligned} \tag{6.12}$$

Incidentally, another equation that may be used to calculate voltage gain in place of (6.12) is (6.13), where μ is the amplification factor of the tube and $R'_L = R_P \parallel R_L$. I will leave it to you to verify that both methods give approximately the same voltage gain.

$$A_V = \mu \frac{-R'_L}{r_P + R'_L} \tag{6.13}$$

Removing bypass capacitor C_3 will reduce the voltage gain of the amplifier. Without cathode bypass, the gain is given by

$$A_V = \frac{-g_m R'_L}{1 + g_m R_K} \tag{6.14}$$

The input resistance of the amplifier is equal to the value of the grid resistor.

$$\begin{aligned} R_{in} &= R_G \\ &= 470 \text{ k}\Omega \end{aligned} \tag{6.15}$$

Output Voltage Compliance

If we really drive the amplifier hard, the Q-point will move along the load line all the way to clipping at the V_{PM} and I_{PM} limits. In order to keep from driving the grid too far positive, the Q-point should not be allowed to move to the left of Point 3 in Fig. 6.5. Back in Step 7, we graphically estimated the voltage coordinate of Point 3 to be $V_{P(min)} = 63$ V.

The maximum allowable change in plate voltage ΔV_{PL} moving to the left of the Q-point is

$$\begin{aligned} \Delta V_{PL} &= V_{PQ} - V_{P(min)} \\ &= 125\,V - 63\,V \\ &= 62\,V \end{aligned} \tag{6.16}$$

The operating point of the tube can move to the right on the load line to $V_{PM} = 250$ V. The maximum change in plate voltage ΔV_{PR} moving to the right along the load line is

$$\begin{aligned} \Delta V_{PR} &= V_{PM} - V_{PQ} \\ &= 250\,V - 125\,V \\ &= 125\,V \end{aligned} \tag{6.17}$$

These are the plate voltages at which the amplifier begins clipping. If the Q-point was perfectly centered, the clipping points would be equal, but this is difficult to achieve in practice.

Note the smaller of ΔV_{PL} and ΔV_{PR}, which we will designate as $V_{o(max)}$, the maximum unclipped output voltage, or *output compliance*. In this example, we have $\Delta V_{PL} < \Delta V_{PR}$, so

$$V_{o(max)} = \Delta V_{PL} = 62\,V_{pk}$$

If you prefer, the output voltage compliance can be given in peak-to-peak volts.

$$V_{o(max)} = 124\,V_{P-P}$$

Because we don't want to let the grid go positive with respect to the cathode, the maximum allowable peak input voltage is given by

$$\begin{aligned} V_{in(max)} &= I_{PQ}R_K \\ &= 5\,mA \times 820\,\Omega \\ &= 4\,V_{pk} = 8\,V_{P-P} \end{aligned} \tag{6.18a}$$

We can also calculate the maximum input voltage based on maximum output voltage (clipping) and the gain of the amplifier. The relationship is

$$V_{in(max)} = \frac{V_{o(max)}}{|A_V|}$$

$$= \frac{62\ V}{16.8} \tag{6.18b}$$

$$\cong 3.7\ V_{pk} = 7.4\ V_{P-P}$$

Equation (6.18a) tells us the input voltage that will cause the grid to go more positive than the cathode. This allows grid current to flow and drastically reduces input resistance. Equation (6.18b) tells us the input voltage that will result in clipping based on the load line. In this example, these values are close enough that the difference is negligible, but this doesn't always happen. Generally, we don't want to drive the grid positive, so (6.18a) gives the limit that is more significant.

Experimental Test Results

The circuit was constructed using several different 12AU7s and a common equivalent tube, the 5963. The resulting measurements are listed in Table 6.2. The grid voltages listed were based on measuring the drop across R_K using $V_{GQ} = -V_{RK}$ (see Fig. 6.2). The experimental voltage gain values vary the most from the theoretical values. It is common for actual gain to be lower than predicted. Keep this in mind for future reference.

Q-Point Location and Distortion

Looking at the load line for our circuit in Fig. 6.5, we see that when the Q-point is driven toward cutoff (Point 1), the plate curves begin flattening out. This results in a reduction of gain as V_o goes more positive (v_{in} is going negative). When the Q-point moves left, toward saturation (Point 2), the curves get steeper and gain increases. Large Q-point movement results in significant gain nonlinearity.

In general, biasing the tube at a high I_{PQ} and low V_{PQ} results in more linear amplification. This would be preferred in a high-fidelity amplifier. Biasing at low I_{PQ} and higher V_{PQ} results in more nonlinearity and higher distortion. Since we are

Table 6.2 Experimental values for 12AU7 design example

	Theoretical 12AU7	Ruby Tubes 12AU7	GE (NOS) 5963	RCA (NOS) 5963	Chinese 12AU7
I_{PQ}	5.0 mA	4.7 mA	4.8 mA	4.6 mA	4.7 mA
V_{GQ}	−4.0 V	−3.9 V	−3.9 V	−3.8 V	−3.9 V
V_{PQ}	125 V	123 V	119 V	125 V	122 V
A_V	−16.8	−13.4	−12.7	−12.7	−12.1

$V_{PP} = 250$ V, $R_P = 27$ kΩ, $R_K = 820$ Ω, $R_G = 470$ kΩ, $R_L = 470$ kΩ, $V_{GQ} = -V_{RK}$

designing a guitar amplifier, a little bit of distortion is normally a desirable characteristic.

If we are designing the first stage of an amp, a centered Q-point and the resulting symmetrical clipping is not usually too important because stages further down the line will normally be driven into clipping first. We want the later stages, especially the output stage, to have high output compliance. Of course, we can design the first stage with a Q-point very close to saturation or cutoff (I'm borrowing transistor terminology here) if we want distortion or clipping to occur in this section of the amp.

6AN8 Triode, Low-Power Amp Design Example

In this example, we will use the triode section of a 6AN8 dual triode/pentode to design an amplifier that will be used to drive a 50 kΩ load. This relatively low resistance is representative of the input resistance of a Baxandall tone control circuit, for example. This will require us to plot both DC and AC load lines for the amplifier.

The main parameters for the 6AN8 triode section are listed below. This is a medium-mu triode, very similar to the 12AU7.

6AN8 triode basic parameter summary	
$P_{D(max)} = 2.8$ W	Maximum DC plate power dissipation
$I_{P(max)} = 15$ mA	Maximum DC plate current
$V_{P(max)} = 330$ V	Maximum DC plate-to-cathode voltage
$g_m = 4.5$ mS	Typical transconductance
$r_P = 4700\ \Omega$	Typical internal dynamic plate resistance (similar to r_{CE} or r_o for a BJT)
$\mu = 21$	Amplification factor (maximum available voltage gain)

1. Determine power supply voltage.
    For this example, let's assume that $V_{PP} = 300$ V.
2. Determine suitable Q-point (V_{PQ}, I_{PQ}).
    In this example, we will choose the plate current first. Since $I_{P(max)} = 15$ mA, a reasonable and convenient quiescent plate current is $I_{PQ} = 5$ mA.
    The Q-point voltage will be chosen using (6.6).

$$V_{PQ} = \frac{V_{PP}}{2}$$
$$= \frac{300\ V}{2}$$
$$= 150\ V$$

We plot the Q-point on the 6AN8 triode plate curves as shown in Fig. 6.8. The Q-point is located within the SOA. The amplifier Q-point values are summarized below.

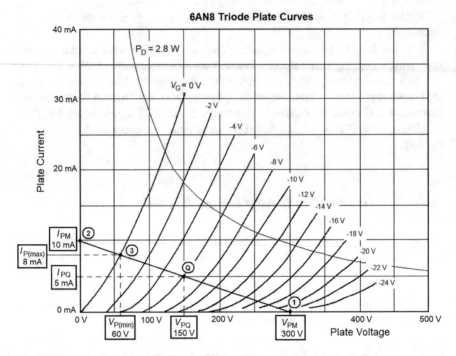

Fig. 6.8 Construction of the 6AN8 triode DC load line

$$V_{PQ} = 150\,\text{V}, I_{PQ} = 5\,\text{mA}, P_{DQ} = 750\,\text{mW}$$

3. Determine maximum plate voltage V_{PM}

   Place a dot on the plate voltage axis at the supply voltage, $V_{PM} = V_{PP} = 300$ V (Point 1 in Fig. 6.8).

4. Determine maximum plate current I_{PM}

   Because $V_{PM} = 2V_{PQ}$, we know $I_{PM} = 2I_{PQ} = 10$ mA (Point 2 in Fig. 6.8).

5. Plot DC load line (Point 1, through the Q-point, to Point 2).

6. Calculate plate resistance R_P.

$$R_P = \frac{V_{PM}}{I_{PM}}$$
$$= \frac{300\,\text{V}}{10\,\text{mA}}$$
$$= 30\,\text{k}\Omega \ (\text{use either } 27\,\text{k}\Omega \text{ or } 33\,\text{k}\Omega\ )$$

   We will choose $R_P = 33$ kΩ.

7. Determine minimum allowable plate voltage $V_{P(min)}$

   The intersection of the load line with the $V_G = 0$ V plate curve occurs at Point 3 in Fig. 6.8.

$$V_{P(\min)} \cong 60 \text{ V}$$

The maximum allowable plate current can also be estimated, which is $I_{P(\max)} \cong 8$ mA.

8. Determine grid voltage V_G.

The Q-point happens to fall right on the $V_G = -6$ V curve. Sometimes we get lucky.

$$V_G = -6 \text{ V}$$

9. Determine cathode resistance R_K.

Using (6.10), we are again fortunate to come up with a standard resistor value.

$$R_K = \frac{-V_G}{I_{PQ}}$$
$$= \frac{6 \text{ V}}{5 \text{ mA}}$$
$$= 1.2 \text{ k}\Omega$$

10. Determine grid resistor R_G.

The maximum allowable self-bias grid resistance listed on the Tung-Sol 6AN8 data sheet is $R_{G(\max)} = 1$ MΩ. We will choose $R_G = 470$ kΩ. The complete amplifier is shown in Fig. 6.9.

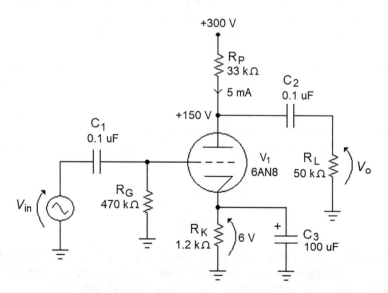

Fig. 6.9 Complete 6AN8 common cathode, class A amplifier stage

Amplifier AC Performance

The amplifier is driving a load resistance $R_L = 50$ kΩ. The total AC equivalent plate resistance is found using (6.11).

$$
\begin{aligned}
R'_P &= r_P \parallel R_P \parallel R_L \\
&= 4.7 \text{ k}\Omega \parallel 33 \text{ k}\Omega \parallel 50 \text{ k}\Omega \\
&= 3.8 \text{ k}\Omega
\end{aligned}
$$

Equation (6.12) gives the voltage gain.

$$
\begin{aligned}
A_V &= -g_m R'_P \\
&= -4.5 \text{ mS} \times 3.8 \text{ k}\Omega \\
&= -17.1
\end{aligned}
$$

The input resistance of the amplifier is

$$
\begin{aligned}
R_{in} &= R_G \\
&= 470 \text{ k}\Omega.
\end{aligned}
$$

The AC Load Line

The AC load line gives us a graphical representation of the large-signal limitations of amplifier operation. As mentioned in the previous example, when a load is capacitively (or inductively) coupled to the amplifier, an input signal will cause the operating point to move along the AC load line.

The AC and DC load lines always intersect at the Q-point. The AC load line maximum values take precedence and will be used to determine limits such as output compliance. In the present example, the load resistance value is not so large that it can be ignored (R_L is not much larger than R_P), so we will construct an AC load line to evaluate amplifier large-signal AC characteristics.

The first step in creating the AC load line is to determine the AC equivalent load resistance R'_L.

$$
\begin{aligned}
R'_L &= R_P \parallel R_L \\
&= 33 \text{ k}\Omega \parallel 50 \text{ k}\Omega \\
&= 19.9 \text{ k}\Omega
\end{aligned}
\tag{6.19}
$$

The AC load line for the amplifier is plotted in Fig. 6.10, along with the DC load line. The DC load line values have been removed from the plot to reduce clutter. The maximum plate voltage endpoint of the AC load line is given by

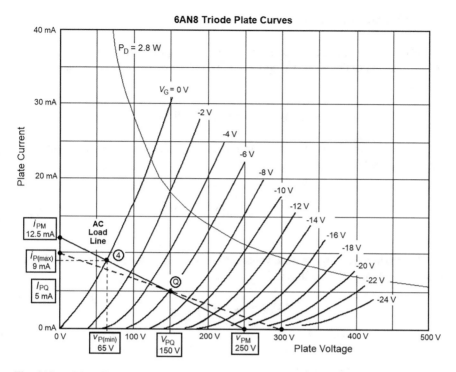

Fig. 6.10 AC load line for the 6AN8 amplifier

$$v_{PM} = V_{PQ} + I_{PQ}R'_L$$
$$= 150\,\text{V} + (5\,\text{mA} \times 19.9\,\text{k}\Omega) \qquad (6.20)$$
$$\cong 250\,\text{V}.$$

Note that lowercase italic letters are used here to represent points of interest on the AC load line. The left endpoint of the AC load line is given by

$$i_{PM} = I_{PQ} + \frac{V_{PQ}}{R'_L}$$
$$= 5\,\text{mA} + \frac{150\,\text{V}}{19.9\,\text{k}\Omega} \qquad (6.21)$$
$$\cong 12.5\,\text{mA}.$$

When we connect endpoints v_{PM} and i_{PM}, the AC load line intersects the DC load line at the Q-point. As noted before, if this does not occur, it is a good indication that a mistake has been made somewhere in the design process.

Output Voltage Compliance

To keep from driving the grid too far positive, the Q-point should not be allowed to move to the left of the $V_G = 0$ V curve (Point 4). The approximate coordinates of this point are.

$$v_{P(min)} \cong 65V, i_{P(max)} \cong 9 \text{ mA}.$$

The maximum allowable change in plate voltage ΔV_{PL} moving to the left from the Q-point to Point 4 on the AC load line is.

$$\begin{aligned}
\Delta V_{PL} &= V_{PQ} - v_{P(min)} \\
&= 150 \text{ V} - 65 \text{ V} \\
&= 85 \text{ V}
\end{aligned} \tag{6.22}$$

The maximum change in plate voltage ΔV_{PR} moving to the right from the Q-point to v_{PM} along the AC load line is.

$$\begin{aligned}
\Delta V_{PR} &= v_{PM} - V_{PQ} \\
&= 250 \text{ V} - 150 \text{ V} \\
&= 100 \text{ V}.
\end{aligned} \tag{6.23}$$

Since $\Delta V_{PL} < \Delta V_{PR}$, we have.

$$V_{o(max)} = \Delta V_{PL} = 85 \text{ V}_{pk} = 170 \text{ V}_{P-P}.$$

Applying (6.18a), the highest input signal voltage that won't overdrive the grid is.

$$\begin{aligned}
V_{in(max)} &= I_{PQ}R_K \\
&= 5 \text{ mA} \times 1.2 \text{ k}\Omega \\
&= 6 \text{ V}_{pk} = 12 \text{ V}_{P-P}.
\end{aligned}$$

Using (6.18b), the input voltage that just causes the output to clip is

$$\begin{aligned}
V_{in(max)} &= \frac{V_{o(max)}}{|A_V|} \\
&= \frac{85}{17.1} \\
&\cong 5 \text{ V}_{pk} = 10 \text{ V}_{P-P}.
\end{aligned}$$

The amplifier output should clip before the grid is driven positive, but in any case, we should limit V_{in} to less than 12 V_{P-P}.

12AX7 Low-Power Amp Design Example

The 12AX7 is a high-mu, dual triode and is probably the most commonly used vacuum tube in the low-level and driver sections of guitar amplifiers. Here is a summary of the major 12AX7 parameters:

12AX7 parameter summary	
$P_{D(max)} = 1$ W	Maximum DC plate power dissipation
$I_{P(max)} = 8$ mA	Maximum DC plate current
$V_{P(max)} = 300$ V	Maximum DC plate-to-cathode voltage
$g_m = 1.5$ mS	Typical transconductance
$r_P = 65$ kΩ	Typical internal dynamic plate resistance (similar to r_{CE} or r_o for a BJT)
$\mu = 100$	Amplification factor (maximum available voltage gain)

In this example, we will again assume that the amplifier is driving a load $R_L = 470$ kΩ.

1. Determine power supply voltage
   For this example, let's assume that $V_{PP} = 200$ V.
2. Determine suitable Q-point (V_{PQ}, I_{PQ})
   The 12AX7 is normally biased for operation at relatively low plate current ($I_{PQ} < 1$ mA). Somewhat arbitrarily, let's go for the following Q-point coordinates in this design:

$$V_{PQ} = 130 \text{ V}, I_{PQ} = 500 \text{ }\mu\text{A}, P_{DQ} = 65 \text{ mW}.$$

Mark the Q-point on the plate characteristic curves as shown in Fig. 6.11a.
3. Determine maximum plate voltage V_{PM}
   Mark the plate voltage axis at the supply voltage, $V_{PM} = V_{PP} = 200$ V.
4. Determine maximum plate current I_{PM}
   We can graphically determine I_{PM} by simply drawing the DC load line from V_{PM} through the Q-point, to the vertical axis of Fig. 6.11a, where we find

$$I_{PM} \cong 1.5 \text{ mA}.$$

Proportionality could also be used, which gives $I_{PM} \cong 1.43$ mA, but the difference is negligible.
5. Plot DC load line (already drawn in Step 4).
6. Calculate required plate resistance R_P (use DC load line endpoints).

$$\begin{aligned} R_P &= \frac{V_{PM}}{I_{PM}} \\ &= \frac{200 \text{ V}}{1.5 \text{ mA}} \\ &= 133 \text{ k}\Omega \text{ (use either } 120 \text{ k}\Omega \text{ or } 150 \text{ k}\Omega) \end{aligned}$$

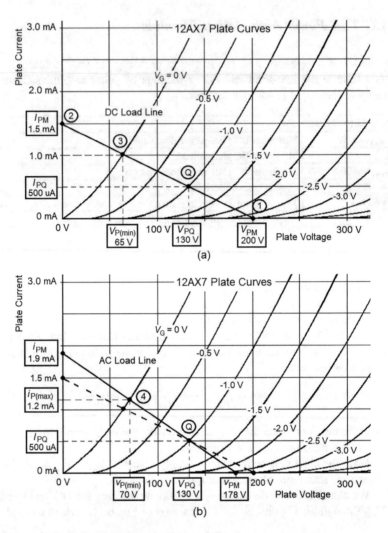

Fig. 6.11 12AX7 amplifier load lines. (**a**) DC. (**b**) AC

Let's use $R_P = 120\,k\Omega$ for this example. We were going for $V_{PQ} = 130$ V, but this R_P results in a slightly higher actual plate voltage $V_{PQ} = 140$ V. The difference isn't significant, and we will use $V_{PQ} = 130$ V in subsequent calculations.

7. Determine minimum allowable plate voltage $V_{P(min)}$

The intersection of the load line with the $V_G = 0$ V curve has the coordinates

$$V_{P(min)} \cong 65\,V, I_{P(max)} \cong 1\,mA.$$

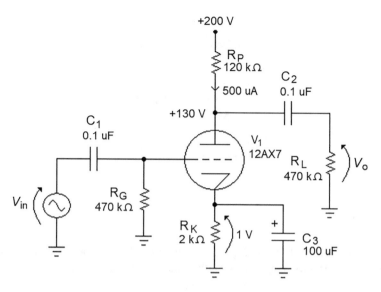

Fig. 6.12 Complete 12AX7 common cathode, class A amplifier

8. Determine grid voltage V_G

   We got lucky again. The Q-point falls right on the $V_G = -1$ V curve, so

$$V_G = --1V$$

9. Determine cathode resistance R_K (use (6.10))

$$R_K = \frac{-V_G}{I_{PQ}}$$
$$= \frac{1\,V}{500\,\mu A}$$
$$= 2\,k\Omega$$

   You are encouraged to use the 12AX7 transconductance curves to verify that $V_G = -1$ V and $R_K = 2$ kΩ are indeed the desired values for this circuit.

10. Determine R_G

   The maximum grid resistance specified for the 12AX7 is 1 MΩ. As in the last example, we use $R_G = 470$ kΩ. The complete 12AX7 amplifier is shown in Fig. 6.12.

Amplifier AC Performance

Using $R_L = 470$ kΩ and the 12AX7 data sheet parameters, the analysis equations give us

$$R'_P = r_P \| R_P \| R_L$$
$$= 65\,k\Omega \| 120\,k\Omega \| 470\,k\Omega$$
$$= 38.7\,k\Omega$$

$$A_v = -g_m R'_P$$
$$= -1.5\ mS \times 38.7\ k\Omega$$
$$= -58$$

$$R_{in} = R_G$$
$$= 470\ k\Omega$$

The AC Load Line

The AC load line is plotted in Fig. 6.11b, with the DC load line (dashed) shown for comparison purposes. We start by finding the AC equivalent load resistance.

$$R'_L = R_P \| R_L$$
$$= 120\ k\Omega \| 470\ k\Omega$$
$$= 96\ k\Omega$$

The endpoints of the AC load line are now found.

$$v_{PM} = V_{PQ} + I_{PQ} R'_L$$
$$= 130\ V + (500\,\mu A \times 96\ k\Omega)$$
$$= 178\ V$$

$$i_{PM} = I_{PQ} + \frac{V_{PQ}}{R'_L}$$
$$= 500\,\mu A + \frac{130\ V}{96\ k\Omega}$$
$$\cong 1.9\ mA$$

We graphically estimate the coordinates of Point 4.

$$v_{P(min)} \cong 70\ V, i_{P(max)} \cong 1.2\ mA$$

Output Voltage Compliance

The output clipping points are found as follows, using the AC load line:

$$\Delta V_{PL} = V_{PQ} - v_{P(\min)}$$
$$= 130 \text{ V} - 70 \text{ V}$$
$$= 60 \text{ V}$$

$$\Delta V_{PR} = V_{PM} - V_{PQ}$$
$$= 178 \text{ V} - 130 \text{ V}$$
$$= 48 \text{ V}$$

We see that $\Delta V_{PR} < \Delta V_{PL}$, so the output voltage compliance is

$$V_{o(\max)} = \Delta V_{PR} = 48 \text{ V}_{pk} = 96 \text{ V}_{P-P}$$

Using (6.18a), the maximum input voltage is

$$V_{in(\max)} = I_{PQ}R_K$$
$$= 500 \,\mu\text{A} \times 2 \text{ k}\Omega$$
$$= 1 \text{ V}_{pk} = 2 \text{ V}_{P-P}$$

Equation (6.18b) gives us approximately the same maximum input voltage.

$$V_{in(\max)} = \frac{V_{o(\max)}}{|A_V|}$$
$$= \frac{48 \text{ V}}{58}$$
$$\cong 0.83 \text{ V}_{pk} = 1.7 \text{ V}_{P-P}$$

Experimental Results

The circuit was constructed using several different 12AX7s where Q-point values and voltage gain were measured. As before, the listed grid voltage is related to the drop across the cathode resistor by $V_G = -V_{RK}$ (see Fig. 6.2). The resulting measurements are listed in Table 6.3, where again we find the experimental results in good agreement with theory.

12AT7 Low-Power Amp Design Example

The 12AT7 has specifications that are a compromise between the high-mu 12AX7 and the medium-mu 12AU7. In this example, we will use a relatively low supply voltage $V_{PP} = 125$ V, and we will drive a 470 kΩ load ($R_L = 470$ kΩ). The main 12AT7 parameters are given as follows:

Table 6.3 Experimental values for 12AX7 design example

	Theoretical 12AX7	Electro-Harmonix 12AX7	GE (NOS) 12AX7	RCA (NOS) 12AX7	Chinese 12AX7
I_{PQ}	500 μA	525 μA	550 μA	500 μA	555 μA
V_{GQ}	−1.0 V	−1.05 V	−1.10 V	−1.00 V	−1.11 V
V_{PQ}	130 V	137 V	134 V	139 V	133 V
A_V	−58	−52	−55	−51	−47

$V_{PP} = 200$ V, $R_P = 120$ kΩ, $R_K = 2$ kΩ, $R_G = 470$ kΩ, $R_L = 470$ kΩ, $V_{GQ} = -V_{RK}$

Fig. 6.13 Plotting the 12AT7 amplifier load line

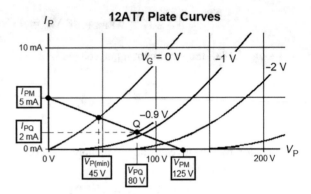

12AT7 parameter summary	
$P_{D(max)} = 2.5$ W	Maximum DC plate power dissipation
$I_{P(max)} = 15$ mA	Maximum DC plate current
$V_{P(max)} = 300$ V	Maximum DC plate-to-cathode voltage ($V_{PK(max)}$)
$g_m = 5.5$ mS	Typical transconductance
$r_p = 11$ kΩ	Typical internal dynamic plate resistance (similar to r_{CE} or r_o for a BJT)
$\mu = 61$	Amplification factor (maximum available voltage gain)

1. Determine power supply voltage

$$V_{PP} = 125 \text{ V}$$

2. Determine suitable Q-point (V_{PQ}, I_{PQ})
   This time we will set $V_{PQ} = 80$ V, which is greater than our usual choice of $V_{PP}/2$. This will produce a more centered Q-point between the $V_G = 0$ V plate curve and $V_{P(min)}$. A reasonable plate current is $I_{PQ} = 2$ mA, so we have

$$V_{PQ} = 80 \text{ V}, I_{PQ} = 2 \text{ mA}, P_{DQ} = 160 \text{ mW}.$$

   Mark the Q-point on the plate characteristic plot as shown in Fig. 6.13. A small segment of an approximate plate curve is sketched through the Q-point. It looks like this curve would represent $V_G \cong -0.9$ V.

3. Determine maximum plate voltage V_{PM}

$$V_{PM} = V_{PP} = 125 \text{ V}.$$

4. Determine maximum plate current I_{PM}

We find I_{PM} graphically by drawing the DC load line from V_{PM} through the Q-point, which gives us

$$I_{PM} \cong 5\text{mA}.$$

5. Calculate the required plate resistance R_P using the load line endpoints.

$$\begin{aligned} R_P &= \frac{V_{PM}}{I_{PM}} \\ &= \frac{125 \text{ V}}{5 \text{ mA}} \\ &= 25 \text{ k}\Omega \text{ (we will use } 22 \text{ k}\Omega \text{)} \end{aligned}$$

We don't need to plot an AC load line this time because $R_L >> R_P$ (i.e., 470 kΩ $>> 22$ kΩ).

6. Determine minimum allowable plate voltage $V_{P(min)}$

The coordinates of the intersection of the load line with the $V_G = 0$ V curve are

$$V_{P(min)} \cong 45 \text{ V}, I_{P(max)} \cong 3.5 \text{ mA}.$$

7. Determine cathode resistance R_K

$$\begin{aligned} R_K &= \frac{-V_G}{I_{PQ}} \\ &= \frac{0.9 \text{ V}}{2 \text{ mA}} \\ &= 450 \text{ }\Omega (\text{use } 470 \text{ }\Omega) \end{aligned}$$

8. Determine R_G

The maximum grid resistance specified for the 12AT7 is 1 MΩ. As you might have guessed, we will go with $R_G = 470$ kΩ.

Fig. 6.14 12AT7 triode
amplifier

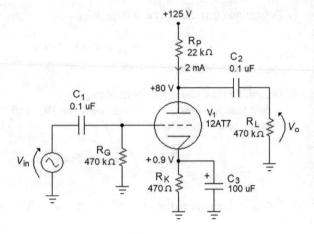

Amplifier AC Performance

The complete amplifier is shown in Fig. 6.14. The AC characteristics are

$$R'_P = r_P \| R_P \| R_L$$
$$= 11\,\text{k}\Omega \| 22\,\text{k}\Omega \| 470\,\text{k}\Omega$$
$$= 7.2\,\text{k}\Omega$$

$$A_V = -g_m R'_P$$
$$= -5.5\,\text{mS} \times 7.2\,\text{k}\Omega$$
$$\cong -40$$

$$R_{in} = R_G$$
$$= 470\,\text{k}\Omega$$

Output Voltage Compliance

The Q-point excursion limits are

$$\Delta V_{PL} = V_{PQ} - V_{P(min)}$$
$$= 80\,\text{V} - 45\,\text{V}$$
$$= 35\,\text{V}$$

$$\Delta V_{PR} = V_{PM} - V_{PQ}$$
$$= 125\,\text{V} - 80\,\text{V}$$
$$= 45\,\text{V}$$

$\Delta V_{PL} < \Delta V_{PR}$, so the output voltage compliance is

$$V_{o(max)} = \Delta V_{PL} = 35\,\text{V}_{pk} = 70\,\text{V}_{P-P}$$

Using (6.18a), the maximum input voltage is

$$V_{in(max)} = I_{PQ}R_K$$
$$= 2 \text{ mA} \times 470 \text{ }\Omega$$
$$= 0.94 \text{ V}_{pk} = 1.9 \text{ V}_{P-P}$$

Again, (6.18b) gives us approximately the same value as (6.18a).

$$V_{in(max)} = \frac{V_{o(max)}}{|A_V|}$$
$$= \frac{35 \text{ V}}{40}$$
$$\cong 0.9\text{V}_{pk} = 1.8\text{V}_{P-P}$$

Pentodes

The same design techniques we used with triodes will be used when working with pentodes. In fact, pentodes can be wired to behave exactly like triodes. However, pentodes are more complex devices, and there are many more circuit design variations possible, so before we begin an amplifier design example, let's run through a quick review of basic pentode operation.

In many ways, pentodes behave a lot more like transistors than triodes do. In particular, pentode plate characteristic curves look almost identical to those for depletion mode MOSFETs. Compared to triodes, pentodes exhibit much higher mu and dynamic plate resistance. These characteristics favor the pentode in many applications, especially high-power amplifier output stages. However, pentodes tend to have overdrive and distortion characteristics that sound more transistor-like than triodes. For this reason, many people prefer the distortion produced by triodes.

The various terminals of the pentode are identified in Fig. 6.15. Most of the time, the suppressor grid (G3) is connected to the cathode, so we will use the simplified symbol shown in Fig. 6.15b. A pentode can be used as a triode simply by connecting the screen grid (G2) to the plate, which is shown in Fig. 6.15c.

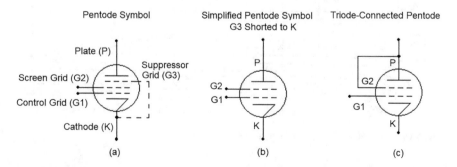

Fig. 6.15 (**a**) Pentode schematic symbol. (**b**) Simplified symbol. (**c**) Triode-connected pentode

Screen Grid (G2)

The primary function of the screen grid is to electrostatically shield the control grid (G1) from the plate. This greatly increases dynamic plate resistance r_P and reduces interelectrode capacitance. In normal operation, the screen grid current will be about 5 to 10% of the plate current. We will normally assume $I_{G2} \cong 0.1I_{PQ}$.

The screen grid is normally connected to a supply voltage point V_{SS} that is lower than the plate voltage V_{PQ}. Typical V_{SS} values range widely from around 10 to 100 V less than V_{PQ}. The value of V_{G2} is not critical, but as you can see in Fig. 6.16, these particular plate curves are most accurate when $V_{G2} = 150$ V. Lower voltage taps on the power supply rail are easily implemented in multi-stage amplifiers.

A resistor will usually be connected in series with the screen grid, shown as R_{G2} in Fig. 6.17. This resistor acts to limit screen grid current, should the tube be driven into clipping. The screen grid resistor will typically range from 470 Ω to 10 kΩ. The value is not critical, but generally high-power tubes (6 L6, EL34, etc.) will have smaller values ($R_{G2} \leq 1$ kΩ) because their normal screen grid current may be around 5 to 10 mA. Lower-power pentodes such as the 6AN8 can use a higher screen grid resistance, up to about 10 kΩ.

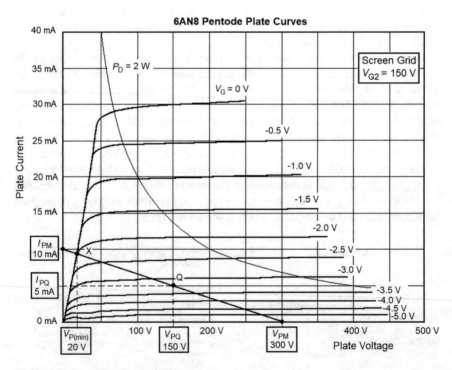

Fig. 6.16 DC load line for the 6AN8 pentode

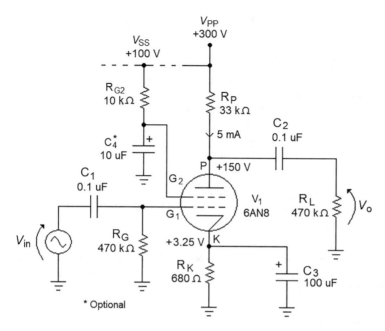

Fig. 6.17 Complete 6AN8 pentode amplifier stage

The tube will function as a triode if the screen grid is connected directly to the plate. This is commonly done in power amplifier output stages because the variety of true power triodes available is quite limited.

Suppressor Grid (G3)

Because of the large potential difference between the plate and the cathode, when electrons flow through the tube, they gain a lot of speed and kinetic energy. When these electrons strike the plate, secondary electrons are knocked loose and ejected from the plate. This is called *secondary emission*. These secondary emission electrons would normally be captured by the screen grid, resulting in high grid current and power dissipation, as well as other undesirable effects such as negative plate resistance. The suppressor grid repels the secondary emission electrons back to the plate, eliminating these problems.

The suppressor grid should always be connected to the cathode. The 6AN8 and 6L6GC, for example, are pentodes that have their suppressor grids internally connected to the cathode. Some pentodes, such as the EL34, require this connection to be made externally.

6AN8 Pentode, Low-Power Amp Design Example

6AN8 pentode parameter summary	
$P_{D(max)} = 2$ W	Maximum DC plate power dissipation
$I_{P(max)} = 12$ mA	Maximum DC plate current
$V_{P(max)} = 330$ V	Maximum DC plate-to-cathode voltage
$g_m = 7.3$ mS	Typical transconductance
$r_P = 170$ kΩ	Typical internal dynamic plate resistance (similar to r_{CE} or r_o for a BJT)
$\mu = 1241$	Amplification factor (maximum available voltage gain)

1. We will assume $V_{PP} = 300$ V and $V_{SS} = 100$ V.
2. Let's set the Q-point location to $V_{PQ} = V_{PP}/2 = 150$ V and $I_{PQ} = 5$ mA. This gives a quiescent power dissipation of $P_{DQ} = 750$ mW.
3. Maximum plate voltage: $V_{PM} = V_{PP} = 300$ V.
4. Maximum plate current: $V_{PM} = 2V_{PQ}$; therefore, $I_{PM} = 2I_{PQ} = 10$ mA.
5. The DC load line is plotted on the 6AN8 pentode plate curves in Fig. 6.16.
6. Determine required plate resistance

$$
\begin{aligned}
R_P &= \frac{V_{PM}}{I_{PM}} \\
&= \frac{300\ \text{V}}{10\ \text{mA}} \\
&= 30\ \text{k}\Omega\,(\text{use either } 27\ \text{k}\Omega \text{ or } 33\ \text{k}\Omega).\ \text{We will choose } R_P = 33\ \text{k}\Omega.
\end{aligned}
$$

7. Minimum allowable plate voltage (point x on load line) $V_{P(min)} \cong 20$ V.
8. Grid voltage is graphically estimated as $V_G \cong -3.25$ V.
9. Determine cathode resistance.

$$
\begin{aligned}
R_K &= \frac{-V_G}{I_{PQ}} \\
&= \frac{3.25\ \text{V}}{5\ \text{mA}} \\
&= 650\ \Omega\,(\text{use } 680\ \Omega).
\end{aligned}
$$

10. Grid resistor $R_G = 470$ kΩ.

The complete amplifier is shown in Fig. 6.17. The G2 resistor is connected to a 100 V supply node labeled V_{SS}. An optional screen grid bypass capacitor C_4 is shown in Fig. 6.17. The function of the screen grid bypass is to increase high frequency response and to prevent power supply voltage fluctuations from modulating the plate current, but this is not usually required in guitar amplifier applications.

Amplifier AC Performance

The amplifier is driving a sufficiently high load resistance $R_L = 470\,k\Omega$, so there is no need to construct an AC load line. The AC equivalent plate resistance is found as usual using (6.11).

$$R'_P = r_P \| R_P \| R_L$$
$$= 170\,k\Omega \| 33\,k\Omega \| 470\,k\Omega$$
$$= 26.1\,k\Omega$$

Again, we use (6.12) which gives the voltage gain

$$A_V = -g_m R'_P$$
$$= -7.3\ mS \times 26.1\ k\Omega$$
$$= -191$$

High voltage gain is typical of pentode, common cathode amplifiers. This is one of the advantages of the pentode over the triode.

The input resistance of the amplifier is

$$R_{in} = R_G$$
$$= 470\,k\Omega$$

Output Voltage Compliance

Another advantage pentodes have over triodes is maximum output voltage swing. Examination of the load line in Fig. 6.16 shows that the Q-point can move to the left until it intersects point x on the load line, where $V_{P(min)} = 20$ V. This is much lower than the typical value for triodes, which are around 50 V and higher.

The Q-point excursion limits are:

$$\Delta V_{PL} = V_{PQ} - V_{P(min)}$$
$$= 150\ V - 20\ V$$
$$= 130\ V$$

$$\Delta V_{PR} = V_{PM} - V_{PQ}$$
$$= 300\,V - 150\,V$$
$$= 150\,V$$

$\Delta V_{PL} < \Delta V_{PR}$, so the output voltage compliance is

$$V_{o(max)} = \Delta V_{PL} = 130\ V_{pk}$$
$$= 260\ V_{P-P}$$

This example illustrates that all things being equal, in addition to having higher voltage gain, pentodes also are capable of higher output voltage compliance than triodes.

Using (6.18a), the input voltage that will drive the grid positive with respect to the cathode is

$$\begin{aligned} V_{in(\,max\,)} &= I_{PQ}R_K \\ &= 5\text{ mA} \times 680\ \Omega \\ &= 3.4\text{ V}_{pk} = 6.8\text{ V}_{P-P} \end{aligned}$$

Because the pentode has much higher voltage gain than the triode, the signal voltage that will drive the output into clipping is usually much smaller than the input voltage that will drive the grid positive with respect to the cathode. Equation (6.18b) gives us

$$\begin{aligned} V_{in(\,max\,)} &= \frac{V_{o(\,max\,)}}{|A_V|} \\ &= \frac{130\text{ V}}{191} \\ &\cong 0.7\text{ V}_{pk} = 1.4\text{ V}_{P-P} \end{aligned}$$

Cathode Followers and Phase Splitters

A cathode follower is formed if the cathode terminal of the tube is used as the output. The cathode follower is analogous to the BJT emitter follower. The most useful characteristic of the cathode follower is its relatively low output resistance. Like the emitter follower, the output of the cathode follower has the same phase as the input voltage, and the voltage gain is always somewhat less than unity.

One of the most common applications for the cathode follower is as one half of a *phase splitter*. A phase splitter is an amplifier that provides two output signals that are 180° out of phase with each other (antiphase). A typical class A, resistance-coupled, phase splitter with output voltage waveforms is shown in Fig. 6.18. This circuit is sometimes called a *cathodyne*.

Think of the phase splitter as two amplifiers in one; load resistance R_{L1} is driven by an unbypassed, common cathode amplifier, while R_{L2} is driven by a cathode follower. Looking at the circuit as a common cathode amplifier, we find that the voltage gain will be very low (less than 1) because the cathode is unbypassed, and the DC equivalent plate and cathode resistances are nearly equal ($R_{K1} + R_{K2} \cong R_P$). Ideally, the voltage gain for both halves of the phase splitter will be equal, so that the output signals are exactly the same amplitude with a relative phase of 180°.

Phase splitters are usually used to drive push-pull output stages in power amplifier designs. We will examine this application in Chap. 7.

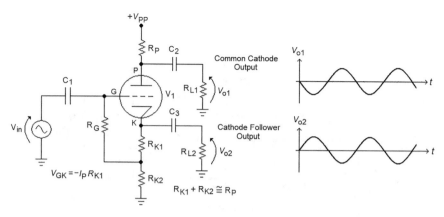

Fig. 6.18 Typical phase splitter and output voltage waveforms

Cathodyne Phase Splitter Design

Most of the cathodyne phase splitter design procedure is identical to that used in the design of the common cathode amplifiers of the previous examples. The first thing we need to do is choose a tube for the circuit. There are a few reasons that make the12AU7 a good choice for the phase splitter. The output resistance of the cathode follower is inherently low, suitable for driving lower-impedance loads. The low r_P of the 12AU7 allows the common cathode output to drive lower-impedance loads as well. The gain of the cathodyne is low because we are using a large amount of negative feedback, which can cause oscillation. Being a low-mu tube, the 12AU7 will be less prone to instability compared to a high-mu tube like a 12AX7.

We can determine the necessary R_P, R_{K1}, R_{K2}, and R_G values for the phase splitter by applying the same procedure used previously. In fact, we can use any of the previously designed common cathode amplifiers as the basis for an equivalent phase splitter.

We will use the 12AU7-based amplifier of Fig. 6.7 as our starting point for this design. The equivalent phase splitter is shown in Fig. 6.19.

Cathode Resistor Determination

Plate current I_P is mainly dependent on grid-to-cathode voltage. Since R_G is connected across R_{K1}, it is the voltage $V_{GK} = -I_P R_{K1}$ that biases the tube for the desired I_{PQ}. We set $R_{K1} = 820\,\Omega$ as previously determined, which should still give us $I_{PQ} \cong 5$ mA. The second cathode resistor R_{K2} reduces the common cathode voltage gain to about unity while at the same time lifting the gate and cathode to a higher voltage.

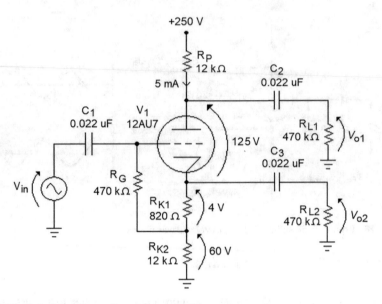

Fig. 6.19 Cathodyne phase splitter using a 12AU7

In the original design of Fig. 6.7, the plate resistor is $R_P = 27$ kΩ. We split this resistance between plate and cathode ends of the circuit. The new plate resistance R'_P is

$$
\begin{aligned}
R'_P &= R_P/2 \\
&= 27\,\text{k}\Omega/2 \\
&= 13.5\,\text{k}\Omega\ (\text{use 12 k}\Omega\ \text{or 15 k}\Omega\ )
\end{aligned}
\tag{6.24}
$$

We will choose $R'_P = 12$ kΩ.

As explained above, we assign the original circuit cathode resistance value R_K to phase splitter resistor R_{K1}.

$$
R_{K1} = R_K = 820\ \Omega
\tag{6.25}
$$

Cathode resistor R_{K2} is chosen such that $R_{K1} + R_{K2} \cong R'_P$, so we have

$$
\begin{aligned}
R_{K2} &= R'_P - R_{K1} \\
&= 12\ \text{k}\Omega - 820\ \Omega \\
&= 11.18\ \text{k}\Omega\ (\text{use 12 k}\Omega\ )
\end{aligned}
\tag{6.26}
$$

The closest standard value is still 12 kΩ, so we simply go with $R_{K2} = 12$ kΩ.

Table 6.4 Q-point comparisons for cathodyne phase splitter

	Theoretical values	Phase splitter		
		PSpice simulation	GE (NOS) 5963	RCA (NOS) 12AU7
I_{PQ}	5 mA	4.95 mA	5.1 mA	5.0 mA
V_{PKQ}	125 V	127 V	122 V	124 V
V_{GKQ}	−4.0 V	−4.0 V	−4.16 V	−4.0 V

The phase splitter circuit was simulated using PSpice and built using an RCA 12AU7 and a GE 5963, which is an equivalent tube. The resulting currents and voltages are presented in Table 6.4. The agreement with theory is excellent.

AC Characteristics

Accurate voltage gain equations for the phase splitter are quite long and messy. The following approximate gain equation can be used if we assume that $R_{L1}, R_{L2} >> R_P$, R_{K2}. The plus sign applies to the cathode follower (V_{o2}), while the minus sign applies to the common cathode section (V_{o1}).

$$A_V = \pm \frac{R_{K1}}{R_{K1} + \dfrac{1}{g_m}}$$
$$= \pm \frac{820\ \Omega}{820\ \Omega + \dfrac{1}{2\ \text{mS}}} \tag{6.27}$$
$$= \pm 0.71$$

This is a typical gain for a phase splitter, which we will refer to several times in later applications. Because we had to use a standard resistor value for R_{K2}, there is not exact matching of the gain between the two halves of the phase splitter (because $R_{K1} + R_{K2} > R'_P$). The common cathode gain will be slightly lower than the gain of the cathode follower. When used as a driver in a push-pull power amp, this will introduce a small amount of distortion (not necessarily a bad thing).

The input resistance of the phase splitter is approximately equal to the value of the grid resistor, which as usual is normally very large.

$$R_{in} = R_G$$
$$= 470\ \text{k}\Omega$$

The output resistance of the cathodyne differs significantly between the two output terminals. The common cathode side output resistance looking into V_{o1} is

$$R_{o1} \cong R_P \parallel r_P$$
$$= 12 \text{ k}\Omega \parallel 7.2 \text{ k}\Omega \tag{6.28}$$
$$= 4.5 \text{ }\Omega$$

The output resistance of the cathode follower side, looking into V_{o2}, is approximately

$$R_{o2} \cong \frac{1}{g_m}$$
$$= \frac{1}{3 \text{ mS}} \tag{6.29}$$
$$333 \text{ }\Omega$$

Because the output resistance of the cathode follower section is fairly low, it is more effective at driving low-resistance loads. If the load resistances driven by the phase splitter are large (as they usually will be), the differences in output resistance will not have a significant effect on circuit operation.

The phase splitter of Fig. 6.19 is capable of producing output signals of about 60 V_{P-P}. The total differential output voltage is $V_{o2} - V_{o1} = 120$ V_{P-P}, which is slightly lower than the output voltage compliance of the original circuit. This is more than sufficient to serve as an output stage driver in the power amplifiers we will be looking at in the next chapter.

A Cathodyne Variation

A slightly different version of the cathodyne phase splitter is shown in Fig. 6.20. The only difference between this circuit and that of Fig. 6.19 is that the cathode follower output is taken from the bottom of R_{K1} instead of from the cathode itself. The gain matching between the two outputs of this version is slightly worse because the voltage divider formed between R_{K1} and R_{K2} reduces the cathode follower gain slightly. Either version of the cathodyne may be used in any of the push-pull amps presented in the next chapter.

Differential Pair Phase Splitter

There are several alternatives to the cathodyne phase splitter. The circuit shown in Fig. 6.21 is a differential amplifier, which can serve as a high-performance phase splitter. Compared to the cathodyne, the differential pair has the advantage of providing high voltage gain, and it has equal output resistance at both output terminals. Just like its bipolar transistor counterpart, the vacuum tube differential amplifier works best with matched devices.

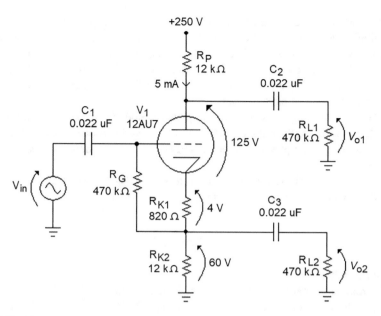

Fig. 6.20 Alternate version of the cathodyne phase splitter

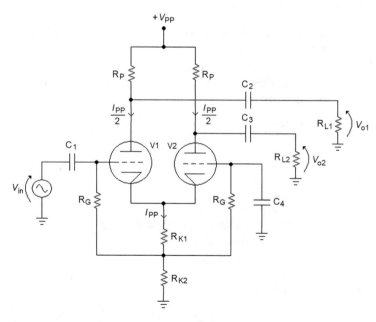

Fig. 6.21 Vacuum tube differential amplifier phase splitter

If we assume that V1 and V2 are closely matched, the total cathode current I_{PP} will divide evenly between the tubes. The split cathode resistance serves the same function as for the cathodyne. We could use either grid as the input terminal, but as configured here, output v_{o1} will be inverted, and v_{o2} will have the same phase as v_{in}. Capacitor C_4 is a bypass cap, shorting the grid of V2 to ground for AC signals, equalizing the gain of the two tubes.

When an input signal is present, the plate of tube V1 drives v_{o1} as a common cathode amplifier. At the same time, V1 acts as a cathode follower, driving V2 which is a common grid amplifier. The end result is that the two outputs have the same voltage gain.

Ideally, we would use a current source to supply I_{PP}. Since we are using a split cathode resistance to establish I_{PP}, the approximate voltage gain for v_{o2} and v_{o1} is

$$A_V \cong \pm \frac{g_m R'_P}{3} \qquad (6.30)$$

where $R'_P = r_P \parallel R_P \parallel R_L$.

The main advantage over the cathodyne is that the differential pair can have high voltage gain, while the cathodyne always has $A_V < 1$. The disadvantage is increased complexity.

A Practical Differential Phase Splitter

A basic differential phase splitter, derived from the 12AX7 amplifier of Fig. 6.12, is shown in Fig. 6.22. In this circuit, both triodes of a single 12AX7 are used. The dashed portion of the tube symbol indicates that this is a dual triode tube. It's very simple to convert the standard class A stage into an equivalent differential amplifier. We use the following steps:

1. Keep the original plate resistor value R_P (here, $R_P = 120$ kΩ).
2. Set R_{K1} the same as the original cathode resistor ($R_{K1} = R_K$).
3. Set $R_{K2} = R_P/2$ (in this case, $R_{K2} = 58$ kΩ).

The AC equivalent plate resistance for each side is

$$\begin{aligned} R'_P &= r_P \parallel R_P \parallel R_L \\ &= 65 \text{ k}\Omega \parallel 120 \text{ k}\Omega \parallel 470 \text{ k}\Omega \\ &= 38.7 \text{ k}\Omega \end{aligned}$$

The voltage gain for each side is

$$\begin{aligned} A_V &\cong \pm \frac{g_m R'_P}{3} \\ &= \pm \frac{1.5 \text{ mS} \times 38.7 \text{ k}\Omega}{3} \\ &= \pm 19.4 \end{aligned}$$

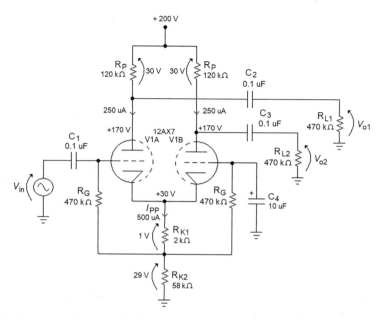

Fig. 6.22 12AX7 differential pair-based phase splitter

I built the circuit using several different 12AX7s and also ran PSpice simulations. In all cases, the experimental values were within 5% of the theoretical voltage, current, and voltage gain values.

Either grid may be used as the input terminal, but the unused input (the grid of V1B in this example) must be bypassed to ground. It is also possible to use both inputs simultaneously as well. This would form an amplifier with a differential input, similar to an op amp. In a practical application, we could use these two inputs to linearly mix two signal sources such as a reverb signal and a dry signal, for example.

Transformer-Coupled Phase Splitter

A transformer-coupled phase splitter is shown in Fig. 6.23. Like the cathodyne, the transformer-coupled phase splitter has the advantage of being less complex than the differential amplifier. Like the differential pair, the transformer-coupled phase splitter has equal impedance at both output terminals and may also provide significant voltage gain. Transformer coupling also allows for very easy impedance matching. The main disadvantages of the transformer-coupled phase splitter are that interstage coupling transformers are often hard to find and they are often relatively expensive.

The dots at the ends of the primary and secondary windings are called *phasing dots*. These dots indicate the relative phase of signals on the respective windings. The dotted ends of the windings will have the same relative phase.

Fig. 6.23 Transformer-
coupled phase splitter

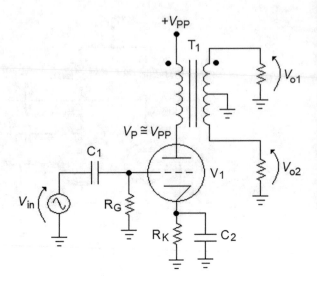

Resistance-coupled amplifiers are well suited for use in low-power, high-voltage-gain, and phase splitter applications, where high current and high power output are not a requirement. Efficient coupling of a tube amp to a low-impedance, high-current load, such as an 8 Ω loudspeaker, is most easily accomplished using transformer coupling.

Determination of Tube Parameters: g_m, r_P, and μ

This section is completely optional, but I wanted to provide some insight into how manufacturers determine the dynamic parameters that we use to analyze and design amplifiers. Now, we're not talking about deriving tube parameters from first principles of physics here. That is *way* beyond the scope of this book. We are going to cover the derivation of tube parameters using their characteristic curves.

Basic Definitions of g_m, r_P, and μ

The three parameters we are most interested in are transconductance (g_m), dynamic plate resistance (r_P), and amplification factor (μ). These parameters describe the dynamic characteristics of a given tube.

Remember that g_m tells us how sensitive the plate current is to changes in grid voltage. As a practical matter, the simplest way to define transconductance is by the equation

$$g_m \cong \frac{\Delta I_P}{\Delta V_G} \tag{6.31}$$

Mu (μ) tells us the maximum possible voltage gain that the tube can produce in the common cathode configuration. The actual voltage gain of an amplifier will always be less than μ. The equation we would use to determine μ experimentally is

$$\mu \cong \frac{\Delta V_P}{\Delta V_G} \tag{6.32}$$

The final parameter that is important to us is the dynamic plate resistance r_P. In a similar manner to g_m and μ, we can determine dynamic plate resistance using

$$r_P \cong \frac{\Delta V_P}{\Delta I_P} \tag{6.33}$$

If we divide (6.32) by (6.31), something very interesting happens. Take a look:

$$\begin{aligned}
\frac{\mu}{g_m} &= \frac{\Delta V_P}{\Delta V_G} \times \frac{\Delta V_G}{\Delta I_P} \\
&= \frac{\Delta V_P}{\Delta I_P} \\
&= r_P
\end{aligned}$$

We can now define any of these parameters in terms of the other two. This gives us the following very useful equations:

$$\mu = g_m r_P \tag{6.34}$$

$$r_P = \frac{\mu}{g_m} \tag{6.35}$$

$$g_m = \frac{\mu}{r_P} \tag{6.36}$$

Sometimes a data sheet will not list all parameters for a given tube. An example is the 3S4, a power pentode designed mainly for use in battery-powered applications. The first page of the 3S4 data sheet is shown in Fig. 6.24. Most of the typical parameters for use as a class A amplifier are listed, but we see that μ is missing.

Let's assume that we will operate the tube at $V_P = 90$ V, using the parallel filament connection. Examining the data sheet, we see that $g_m = 1575$ μS and $r_P = 0.1$ M$\Omega = 100$ kΩ. Mu is found using (6.34).

3S4

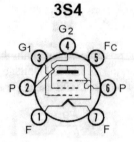

Maximum Average DC Ratings   (Design Center Values)

	Series Filament	Parallel[1] Filament
Plate Voltage	90	90 Volts Max.
Screen Voltage	67.5	67.5 Volts Max.
Cathode Current[2] (Zero Signal)	6	12 Ma Max.

CHARACTERISTICS AND TYPICAL OPERATION

Class A₁ Amplifier	Series Filament		Parallel[1] Filament		
Plate Voltage	67.5	90	67.5	90	Volts
Screen Voltage	67.5	67.5	67.5	67.5	Volts
Negative Grid Voltage . . .	−7	−7	−7	−7	Volts
Peak Signal Voltage	7	7	7	7	Volts
Zero Signal Plate Current . .	6.0	6.1	7.2	7.4	Ma
Zero Signal Screen Current . .	1.2	1.1	1.5	1.4	Ma
Transconductance	1400	1425	1550	1575	μmhos
Load Resistance	5000	8000	5000	8000	Ohms
Plate Resistance (approx.) . .	0.1	0.1	0.1	0.1	Megohms
Total Harmonic Distortion . .	12	13	10	10	Percent
Maximum Signal Power Output	160	235	180	270	Milliwatts

Fig. 6.24   3S4 pentode parameter summary

$$\mu = g_m r_P$$
$$= 1575 \ \mu S \times 100 \ k\Omega$$
$$= 157.5$$

Although it's not stated explicitly, the class A1 parameters listed in Fig. 6.24 apply to pentode-mode operation. The big hint that gives this away is that plate resistance $r_P = 100 \ k\Omega$, which is typical of a pentode. This is one of the exasperating aspects of working with these old data sheets. Modern transistor and IC data sheets are much more consistent and technically accurate.

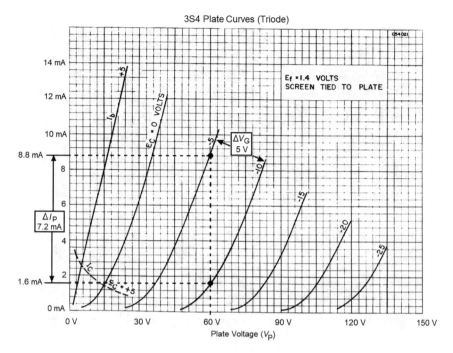

Fig. 6.25 Determining transconductance from triode plate curves

Determination of Triode-Mode Transconductance

We will now determine the triode-mode parameters for the 3S4. Refer to the triode plate curves of Fig. 6.25. Let's assume we plan to operate the tube with $V_{PQ} = 60$ V.

1. Draw a vertical line up from $V_P = 60$ V.
2. Determine I_P values where the vertical line intersects two V_G curves.
3. Determine the change in plate current and grid voltage: $\Delta I_P = 7.2$ mA and $\Delta V_G = 5$ V.
4. $g_m \cong \frac{\Delta I_P}{\Delta V_G} = \frac{7.2 \text{ mA}}{5 \text{ V}} = 1.44$ mS (or 1440 μS).

The g_m value derived from the curves is about the same as for the pentode mode given in Fig. 6.24.

Determination of Triode-Mode Mu

Again, let's assume we are operating near $V_{PQ} = 60$ V. Let's also assume $I_{PQ} \cong 3.6$ mA. Refer to the plate curves of Fig. 6.26.

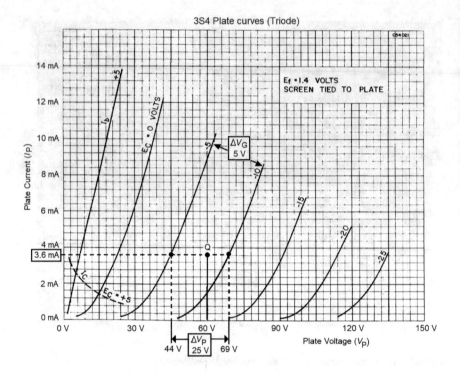

Fig. 6.26 Determination of μ from triode plate curves

1. Draw a horizontal line across at the desired operating current ($I_{PQ} \cong 3.6$ mA here).
2. Mark points on two plate curves surrounding the quiescent plate voltage ($V_{PQ} = 60$ V here).
3. Determine ΔV_P and ΔV_G for the points marked on the plate curves. Here, $\Delta V_P = 25$ V and $\Delta V_G = 5$ V.

$$\mu \cong \frac{\Delta V_P}{\Delta V_G} = \frac{25 \text{ V}}{5 \text{ V}} = 5$$

This tube has very low voltage gain when operating as a triode, but this is not unexpected. For example, from Table 6.1, we find that the 6 L6 has μ = 172 in the pentode mode and μ = 8 in the triode mode.

Determination of Triode Dynamic Plate Resistance

The dynamic plate resistance of the 3S4 is found as follows:

1. Choose a grid voltage curve that is close to the operating point of the tube. In this case, the $V_G = -5$ V curve is reasonable.

2. Choose two convenient plate current values that enclose the quiescent plate current. In this case, $I_{PQ} = 3.6$ mA so I chose $I_P = 2$ mA and 8 mA. This gives us

$$\Delta I_P = 6 \text{ mA}$$

Mark these points on the $V_G = -5$ V curve.

3. Draw lines down from the points marked on the plate curve and note these voltages. Here we find $V_P = 40$ V and 60 V. This gives us

$$\Delta V_P = 20 \text{ V}$$

4. Determine the dynamic plate resistance.

$$r_P \cong \frac{\Delta V_P}{\Delta I_P} = \frac{20 \text{ V}}{6 \text{ mA}} = 3.33 \text{ k}\Omega$$

This is a reasonable value for a low-μ triode (Fig. 6.27).

We could have determined the dynamic plate resistance of the 3S4 triode by applying (6.35).

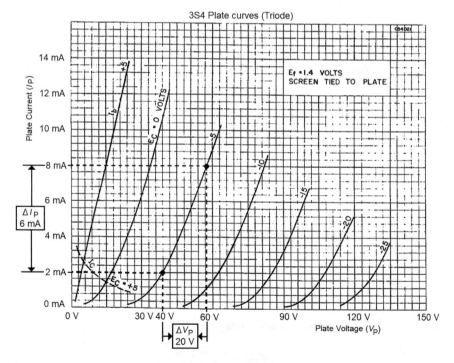

Fig. 6.27 Determination of r_P from triode plate curves

$$r_P = \frac{\mu}{g_m} = \frac{5}{1440\ \mu s} = 3.47\ k\Omega$$

This value differs somewhat from the value determined using the plate curves, but the difference is not significant. Choosing a different curve or slightly different points on the curve will yield a slightly different value as well.

Final Comments

The examples presented in this chapter should give you a good start on understanding the operation of existing amplifiers that you may already own and in designing your own amplifiers. Although many simplifying assumptions were made in developing the design and analysis procedures, their accuracy in predicting actual tube performance is quite good. The circuits studied here will be used along with the power amplifier stages presented in the next chapter to create complete guitar amplifiers of various power output levels.

Summary of Equations

Basic Cathode-Biased Tube Terminal Voltages

$$V_K = I_P R_K \tag{6.1}$$

$$V_{GK} = -I_P R_K \tag{6.2}$$

$$V_{PK} = V_{PP} - I_P(R_P + R_K) \tag{6.3}$$

$$V_{PK} \cong V_P = V_{PP} - I_P R_P(\text{ If } R_P \gg R_K) \tag{6.4}$$

Grid Biasing or Fixed Biasing

$$V_G = -V_{GG} \tag{6.5}$$

Typical Q-Point Plate Voltage Design Goal

$$V_{PQ} \cong \frac{V_{PP}}{2} \tag{6.6}$$

Quiescent (DC) Plate Power Dissipation

$$P_{DQ} = I_{PQ} V_{PQ} \tag{6.7}$$

Determining Required RP from DC Load Line

$$R_P = \frac{\Delta V_P}{\Delta I_P} \qquad (6.8)$$

$$R_P = \frac{V_{PM}}{I_{PM}} \qquad (6.9)$$

Equation to Determine Required R_K from (6.2)

$$R_K = \frac{-V_G}{I_{PQ}} \qquad (6.10)$$

AC Equivalent Plate Resistance

$$R_P' = r_P \parallel R_P \parallel R_L \qquad (6.11)$$

Bypassed Common Cathode Amp Voltage Gain

$$A_V = -g_m R_P' \qquad (6.12)$$

$$A_V = \mu \frac{-R_L'}{r_P + R_L'} \qquad (6.13)$$

Unbypassed Common Cathode Amp Voltage Gain

$$A_V = \frac{-g_m R_L'}{1 + g_m R_K} \qquad (6.14)$$

Input Resistance of Common Cathode Amplifier

$$R_{in} = R_G \qquad (6.15)$$

Formulas to Determine $V_{o(max)}$ (DC Load Line)

$$\Delta V_{PL} = V_{PQ} - V_{P(min)} \qquad (6.16)$$

$$\Delta V_{PR} = V_{PM} - V_{PQ} \qquad (6.17)$$

Maximum Allowable Input Voltage

$$V_{in\,(max)} = I_{PQ} R_K \qquad (6.18a)$$

$$V_{in(max)} = \frac{V_{o(max)}}{|A_V|}$$ (6.18b)

AC Equivalent Load Resistance

$$R'_L = R_P \parallel R_L$$ (6.19)

AC Load Line Endpoints

$$v_{PM} = V_{PQ} + I_{PQ}R'_L$$ (6.20)

$$i_{PM} = I_{PQ} + \frac{V_{PQ}}{R'_L}$$ (6.21)

Formulas to Determine $V_{o(max)}$ (AC Load Line)

$$\Delta V_{PL} = V_{PQ} - v_{P(min)}$$ (6.22)

$$\Delta V_{PR} = V_{PM} - V_{PQ}$$ (6.23)

Cathodyne Phase Splitter Resistor Values

$$R'_P = R_P/2$$ (6.24)

$$R_{K1} = R_K$$ (6.25)

$$R_{K2} = R'_P - R_{K1}$$ (6.26)

Cathodyne Voltage Gain and Output Resistance

$$A_V = \frac{R_{K1}}{R_{K1} + \frac{1}{g_m}}$$ (6.27)

$$R_{o1} \cong R_P \parallel r_P$$ (6.28)

$$R_{o2} \cong 1/g_m$$ (6.29)

Differential Phase Splitter Voltage Gain

$$A_V \cong \pm \frac{g_m R'_P}{3}$$ (6.30)

Basic Tube Parameter Relationships

$$g_m \cong \frac{\Delta I_P}{\Delta V_G}$$ (6.31)

$$\mu \cong \frac{\Delta V_P}{\Delta V_G} \qquad (6.32)$$

$$r_P \cong \frac{\Delta V_P}{\Delta I_P} \qquad (6.33)$$

$$\mu = g_m r_P \qquad (6.34)$$

$$r_P = \frac{\mu}{g_m} \qquad (6.35)$$

$$g_m = \frac{\mu}{r_P} \qquad (6.36)$$

Chapter 7
Vacuum Tube Power Amplifiers

Introduction

The main focus of this chapter is the basic design of guitar amplifier power output stages. Here, we will cover the important concepts behind transformer coupling and the determination of maximum amplifier output power. We will concentrate primarily on cathode-biased, class A output stages, using both single-ended and push-pull amplifier configurations. The operation of parallel-connected tubes and additional aspects of power supply design are also covered here.

A number of complete guitar amplifiers of varying output power ratings and complexity are presented. These amplifiers combine many of the circuits and concepts covered in earlier chapters. We will use the various low-power stages and phase splitters of Chap. 6, the tone controls of Chap. 2, and the reverb circuits of Chap. 5 in the design of these guitar amps.

Maximum Power Transfer

The maximum power transfer theorem states that maximum power will be transferred to a given load when the load resistance equals the output resistance of the amplifier. As an equation, this is simply

$$R_L = R_o \tag{7.1}$$

To illustrate the concept of maximum power transfer, consider an amplifier that has $R_o = 10\ \Omega$, as shown in Fig. 7.1. Assume that the amplifier produces an output voltage with peak amplitude $V_o = 20\ V_{pk}$. A graph of peak power delivered to the load P_L versus load resistance R_L is shown to the right, where we see that maximum output power occurs when $R_L = R_o$.

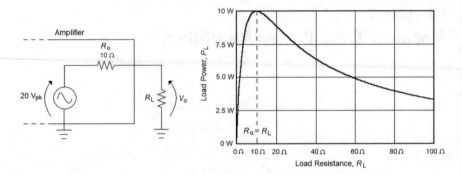

Fig. 7.1 Maximum power transfer occurs when $R_L = R_o$

In general, it is desirable to have maximum power transfer, so it would seem that a reasonable goal when designing a power amplifier is to have $R_L \cong R_o$. However, this often conflicts with other goals, such as a centered Q-point on the AC load line, which produces maximum output voltage compliance. We also have to work with the output transformers that are available as well. As we will see, we usually have a fair amount of latitude when choosing an output transformer, and good performance is usually achieved by using the manufacturers' recommended load resistance values.

Basic Transformer Operation

We are going to be using transformer coupling extensively in this chapter, so before we get into the amplifier design stuff, let's review the basics of transformer operation. Figure 7.2 shows a transformer with phasing dots placed by the windings. Recall that these dots indicate the relative phase of the primary and secondary windings. The turns ratio of the transformer is defined as

$$\text{Turns ratio} = \frac{n_P}{n_S} \tag{7.2}$$

The secondary voltage and current are given by

$$V_{\text{sec}} = V_{\text{pri}}\left(\frac{n_S}{n_P}\right) \tag{7.3}$$

$$I_{\text{sec}} = I_{\text{pri}}\left(\frac{n_P}{n_S}\right) \tag{7.4}$$

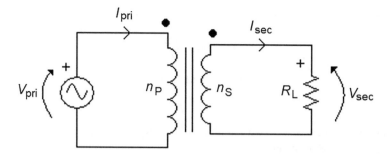

Fig. 7.2 Basic transformer voltage and current polarity relationships

The primary input power and secondary output power (load power) are given by

$$P_{in} = V_{pri}I_{pri} \tag{7.5}$$

$$P_o = V_{sec}I_{sec} \tag{7.6}$$

Reflected Load Resistance

Assuming 100% efficiency, the power delivered to the load equals the power input to the primary winding, $P_o = P_{in}$. Doing a little algebra involving (7.3) through (7.6), we come up with the following very important relationship:

$$R'_L = R_L\left(\frac{n_P}{n_S}\right)^2 \tag{7.7}$$

In (7.7), R'_L is called the *reflected load resistance*. The transformer is capable of making a given load resistance appear larger or smaller on the primary side. This allows transformers to be used for impedance matching.

We will tie all of these equations together with an example. Assume $n_P/n_S = 10$, $V_{pri} = 100$ V_{P-P}, and $R_L = 8$ Ω. The secondary voltage is

$$V_{sec} = V_{pri}\left(\frac{n_S}{n_P}\right)$$

$$= 100\ V_{P-P}\left(\frac{1}{10}\right)$$

$$= 10\ V_{P-P}$$

The reflected load resistance is

$$R'_L = R_L \left(\frac{n_P}{n_S}\right)^2$$
$$= 8\,\Omega \times 10^2$$
$$= 800\,\Omega$$

This transformer steps down the input voltage by a factor of 10 and steps up the effective load resistance by a factor of 100.

Manufacturers of audio output transformers don't usually list transformer turns ratios on their data sheets. Instead, the reflected load resistance as seen looking into the primary side is listed. The turns ratio of the transformer is found by rearranging (7.7), which produces

$$\frac{n_P}{n_S} = \sqrt{\frac{R'_L}{R_L}} \tag{7.8}$$

Let's apply this equation to a real transformer. The Hammond 1750J output transformer has the following specifications:

$$R'_L = 8\,k\Omega, \, R_L = 8\,\Omega$$

This transformer is designed to match an amplifier with $R_o \cong 8\,k\Omega$ to an $8\,\Omega$ speaker. The turns ratio is

$$\frac{n_P}{n_S} = \sqrt{\frac{8\,k\Omega}{8\,\Omega}}$$
$$= 31.6$$

Using this transformer in Fig. 7.2, a primary voltage of 100 $V_{P\text{-}P}$ would be stepped down to 3.16 $V_{P\text{-}P}$.

Magnetic Saturation

Ideally, the magnetic flux produced by current flow in the primary winding is confined to the iron core. The core flux density is proportional to the winding current. The greater the winding current, the greater the resulting magnetic flux will be. However, there is a limit to the amount of flux that the transformer core can carry. When this limit is reached, a further increase in current does not cause an increase in magnetic flux. The core is saturated and the transformer stops working properly.

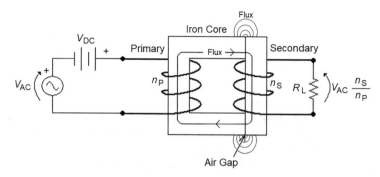

Fig. 7.3 Magnetic flux in the core of a transformer. An air gap is used in single-ended amplifiers

Single-ended, class A power amplifiers operate at relatively high DC plate current levels. In Fig. 7.3, the DC primary current is caused by the voltage source V_{DC}, which generates magnetic flux in the core. This static flux does not contribute to the output signal, but biases the core toward saturation. To prevent the core from saturating too soon on signal peaks due to this DC bias current, *single-ended output transformers* are designed with a small air gap which reduces the tendency for DC bias current to cause saturation.

The air gap also reduces the efficiency of the transformer and the inductance of the windings. For this reason, single-ended output transformers are much larger and heavier than normal output transformers of the same power rating. For example, the Hammond 125ESE and 125D transformers are both rated for $P_o = 10$ W, but the single-ended 125ESE weighs 3 lb. The equivalent push-pull 125D transformer weighs only 1 lb.

Transformer Coupling

The load on a guitar amplifier consists of one or more loudspeakers, typically having an equivalent load resistance ranging from 4 to 16 Ω. Capacitive coupling to such low-resistance loads is unsuitable in high-power audio output stage designs because tubes have relatively high output resistance. This is especially true in the case of the common cathode configuration.

Basically, tubes are high-impedance, high-voltage, low-current devices, while typical loudspeakers are low-impedance, low-voltage, high-current loads. The maximum power transfer theorem tells us that this is not a good match. As we learned in the last section, transformer coupling overcomes this problem by allowing easy and efficient impedance matching.

Advantages and Disadvantages

In addition to allowing impedance matching, transformer coupling also increases amplifier efficiency. You may recall that the capacitively coupled, class A, BJT amplifier has a maximum theoretical efficiency of $\eta = 25\%$. This is true of class A, vacuum tube, resistance-coupled amplifiers as well, except we also have the added inefficiency of the tube heater to consider. Transformer-coupled class A amplifiers have a maximum theoretical efficiency of 50%. In high-power applications, this is a significant benefit of transformer coupling.

On the downside, high-quality audio power transformers are large, heavy, and *very expensive*. I stress the expensive part here, especially in the case of class A power amplifiers. A typical 10 W rated, push-pull, output transformer may cost around $35.00 or so. A 10 W rated, single-ended power output transformer will cost something like $50.00 or more. I am talking about non-high-fidelity transformers, suitable for use in guitar amplifiers here. High-fidelity audio output transformers are usually about twice as expensive.

Recall that high-fidelity audio amplifiers require a frequency response of at least 20 Hz–20 kHz. Realistically, the frequency response of a guitar amplifier output transformer needs only to extend from about 80 Hz to perhaps 10 kHz. This means we can get away with a smaller, lighter, and less expensive transformer than we would require for a hi-fi amp design.

The size and cost associated with transformers have eliminated their use in nearly all transistor circuit designs, except for radiofrequency and switching power supply applications. Also, solid-state class B and AB push-pull emitter followers have very low output resistance which is ideal for driving low-impedance loads, so transformer coupling is not required for impedance matching. Another important consideration is that amplifier designs that don't use transformer coupling (or capacitive coupling) naturally have frequency response down to DC. And even if DC response is not needed, it is generally much less expensive to use capacitive coupling if possible.

You may be asking yourself, if transformer coupling has so much going against it, why is it used so extensively in vacuum tube circuit design? Aside from ease of impedance matching, another big reason is that while transistor circuit designers have the advantage of complementary NPN and PNP devices, there is no equivalent of the PNP transistor in the world of vacuum tubes. Fundamentally, the tube circuit designer has fewer circuit topologies to choose from. Because of this, and because most loads are low impedances, the use of transformers in tube circuit design is almost unavoidable.

Table 7.1 Selected audio output transformers

Hammond part number	$P_{o(rms)}$ (W)	Z_{pri} (Ω)	Z_{sec} (Ω)	$I_{DC(max)}$ (mA)	Weight (lb)	Comments
125ASE	3	2.5 k–10 k	4–32	25	0.51	Single-ended output
125BSE	5	2.5 k–10 k	4–32	45	1.1	Single-ended output
125DSE	10	2.5 k–10 k	4–32	70	2	Single-ended output
125ESE	15	2.5 k–10 k	4–32	80	3.0	Single-ended output
125GSE	25	2.5 k–10 k	4–32	100	6.0	Single-ended output
125A	3	1.2 k–25 k	1.5–15	20 × 2	0.19	Push-pull, CT output
125B	5	1.2 k–25 k	1.5–15	35 × 2	0.3	Push-pull, CT output
125D	10	1.2 k–25 k	1.5–15	55 × 2	1.0	Push-pull, CT output
125E	15	1.2 k–25 k	1.5–15	60 × 2	1.5	Push-pull, CT output
1750J	40	4 k–8 k	4–8	—	4	Push-pull, CT output

A Sampling of Audio Output Transformers

Because audio output transformers are so crucial to the remaining material in this chapter, I compiled a brief list of readily available audio output transformers in Table 7.1. This short list of transformers won't cover all of our needs, but they cover a reasonable variety of reflected load resistance and power output requirements. This is just a small sample of the transformers made by Hammond Manufacturing, Ltd. (www.hammondmfg.com). See their website or the Antique Electronic Supply website (www.tubesandmore.com) for more information and prices. Replacement transformers for specific amplifiers made by Marshall, Vox, Fender, and others are also available.

The speaker/primary impedance connections for the standard 125x series transformers are shown in Table 7.2. The speaker/primary impedance hookup options for the 125xSE single-ended output transformers are listed in Table 7.3. Wiring identification diagrams for both standard and single-ended output transformers are shown in Fig. 7.4.

The transformers with part numbers ending in SE are designed for single-ended, class A operation, which requires the output transformer to operate with a large DC primary winding current. Push-pull output transformers also operate with large DC currents in class A designs. However, DC core saturation is not a problem in the push-pull amplifier topology.

Class A, Single-Ended Amplifiers

We will start with a walk through the design (with some analysis mixed in as well) of a single-ended, class A, transformer-coupled output stage. As in the design of any power amplifier stage, some of the important factors we have to consider are the desired power output, what kind of tube(s) we wish to use, and cost.

Table 7.2 125 series push-pull output transformer connections

Secondary lugs connected to speaker	Speaker impedance (Ω)							
	1.5	2	3.2	4	6	8	12	15
	Reflected plate-to-plate load resistance, R'_{PP} (kΩ) (across entire primary winding)							
1 and 2	27	–	–	–	–	–	–	–
2 and 3	18	24	–	–	–	–	–	–
3 and 4	16.5	22	–	–	–	–	–	–
4 and 5	10	13.5	21.6	27	–	–	–	–
5 and 6	8.5	11	18	22	–	–	–	–
1 and 3	5.4	7.2	11.5	15	21.6	–	–	–
2 and 4	4.2	5.6	9	11.2	16.8	22.5	–	–
3 and 5	3.3	4.4	7	8.8	13.2	17.6	26.4	–
4 and 6	2.4	3.2	5.1	6.4	9.6	12.8	19.2	24
1 and 4	2.15	2.9	4.6	5.8	8.7	11.6	17.4	21.5
2 and 5	1.55	2.05	3.3	4.1	6.15	8.2	12.3	15.5
3 and 6	1.3	1.7	2.7	3.4	5.1	6.8	10.2	12.8
1 and 5	–	1.4	2.2	2.8	4.2	5.6	8.4	10.2
2 and 6	–	–	1.7	2.1	3.15	4.2	6.3	8
1 and 6	–	–	1.2	1.5	2.3	3	4.5	5.6

Table 7.3 125SE single-ended output transformer connections

	Speaker impedance (Ω)			
	4	8	16	32
Secondary wires connected to speaker	Reflected load resistance, R'_L (Ω)			
Black-orange	10 k	–	–	–
Black-green	5 k	10 k	–	–
Black-yellow	2.5 k	5 k	10 k	–
Black-white	–	2.5 k	5 k	10 k

In general, class A power amplifiers have a few significant disadvantages; they are large, heavy, expensive, and inefficient and produce relatively low output power. However, it is my opinion (and this is totally subjective on my part) that single-ended class A power amplifiers produce a very nice, warm distortion when overdriven. If you are not after ear-shattering sound levels and you like a nice, bluesy overdrive tone, then single-ended class A is a good choice.

6L6GC Triode-Mode, SE Amplifier Design Example

The circuit we will work with in this example is shown in Fig. 7.5. We will assume that the DC winding resistances of the primary and secondary are negligible ($R_{DC} \cong 0\ \Omega$). Except for output transformer coupling, this circuit is very similar to the class

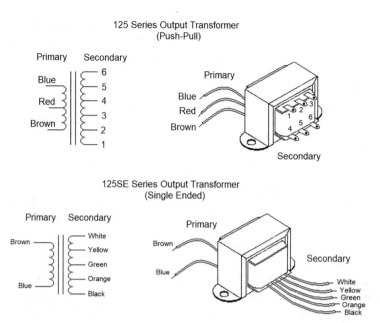

Fig. 7.4 Primary and secondary lead identification for 125 and 125SE series transformers

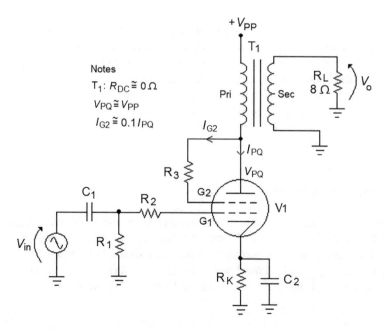

Fig. 7.5 Triode-connected pentode, class A, SE output stage

A amplifiers covered in Chap. 6. A pentode is shown in the schematic, but with the screen grid G2 connected to the plate through R_3, the tube operates as a triode.

Unlike control grid G1, recall that the screen grid G2 will carry some current, usually around 10% of the plate current ($I_{G2} \cong 0.1 I_{PQ}$). Resistor R_3 keeps G2 at a slightly lower voltage than the plate. This limits G2 power dissipation and current, but does not cause a major deviation from triode operation. Like before, control grid resistor R_1 is typically set to a high value, say 100 to 470 kΩ, depending on the maximum allowable value listed in the tube data sheet, in order to give the amplifier high input resistance.

Power triodes are inherently low-mu tubes, so the voltage gain of this stage will not be very high. Typical gain from plate to grid will be on the order of $A'_V \leq 10$. In order to effectively match the 8 ohm load to the higher dynamic resistance r_P of the tube, transformer T_1 will step down the plate signal voltage significantly. This will result in a reduction of the overall voltage gain of the stage to perhaps 1 or even less. In order to maximize voltage gain, cathode bypass capacitors are nearly always used in this application.

Resistor R_2, which is often called a *snubber resistor* or *grid stopper*, is used to limit the high frequency response of the amplifier which helps to prevent parasitic oscillation. The stopper resistor, along with the stray capacitance at the grid, forms a low-pass filter at the input of the amplifier. This resistor can also limit grid current, should the grid be overdriven by a large input signal. Typical values for the grid stopper resistors are

$$1\,k\Omega \leq R_2 \leq 10\,k\Omega$$

The effective grid capacitance (accounting for the Miller effect) of the typical power pentode such as the 6L6 or EL34 is on the order of 100 pF. A 10 kΩ stopper sets the LP input corner frequency to about 160 kHz, which allows audio frequencies to be amplified normally. Because the input resistance of the control grid G_1 is so high, R_2 has no effect on the gain of the amplifier at audio frequencies.

Resistor R_1 and stopper resistor R_2 form what is sometimes called a *gamma network*. Though it sounds impressive, this name simply derives from the fact that the on the schematic, the positions of R_1 and R_2 resemble the Greek letter gamma, Γ.

Choosing an Output Tube

One of the first things we need to do is select an output tube. In this example, we are using the 6L6GC, but in general, tube selection depends on many factors such as desired output power, cost, parts availability, and so on. For single-ended triode-mode power output up to about 5 watts, the EL34 or 6L6GC is a good choice. For output power levels up to about 3 W, the EL84 or 6V6GT are suitable choices.

Using the data from Table 6.1 (or any convenient 6L6 data sheet), the 6L6GC class A parameters are

6L6GC parameter summary (class A, $V_{PP} = 350$ V)

	Pentode mode	Triode mode	
$P_{D(max)}$	30 W	30 W	Max DC plate power dissipation
$I_{P(max)}$	80 mA	54 mA	Max DC plate current
$V_{P(max)}$	500 V	450 V	Max DC plate voltage
g_m	5.2 mS	4.7 mS	Transconductance
r_P	33 kΩ	1.7 kΩ	Internal dynamic plate resistance
$R'_{L(typ)}$	4.2 kΩ	5 kΩ	Typical reflected load resistance
μ	172	8	Amplification factor
Heater	6.3 V, 900 mA		

The values listed were taken from the GE 6L6GC data sheet (ET-T1515A). There is quite a large difference between $R'_{L(typ)}$ (the recommended reflected load resistance at the output transformer primary) and the dynamic plate resistance r_P, especially for pentodes. This is very common and serves to show that maximum power transfer is not necessarily the most important consideration in audio output stage design.

In general, a higher R'_L will reduce peak plate current but increase peak plate voltage. Lower R'_L increases peak plate current but reduces peak plate voltage. We will address these ideas in more detail when we construct the AC load line.

Depending on the supply voltage available, it's reasonable to expect a single-ended, maximum output power of around 3–5 watts using the 6L6GC in the triode mode, driving an 8 Ω load. Operation in the pentode mode may yield a maximum output power of about 5–10 watts.

Choosing Power Supply Voltage

A reasonable assumption for the power supply voltage would be 200 V $\leq V_{PP} \leq$ 400 V. The higher the supply voltage, the higher the potential output power of the amplifier will be. We will use

$$V_{PP} = 350 \text{ V}$$

Choosing Quiescent Plate Current I_{PQ}

Using a high quiescent plate current will help to ensure good power output from the amp. We will be a bit conservative in this example and set plate current for

$$I_{PQ} = 45 \text{ mA}$$

This is well below the maximum DC value of 54 mA.

Output Transformer Selection

We must now choose an output transformer. Since this is a single-ended amplifier, we must use a single-ended output transformer that is designed to operate with a high DC primary winding current. Referring to Table 7.1, we find that the 125BSE, rated for a DC bias current of 45 mA, just meets our requirements. The 125DSE, which will operate with a DC bias current of up to 70 mA, would work as well, but it is larger and more expensive.

The DC Load Line

The slope of the DC load line is determined by the total resistance in series with the plate and cathode of the tube. Applying Ohm's law, the maximum DC plate current is

$$I_{PM} = V_{PP}/(R_P + R_K)$$

The lower the total resistance, the steeper the load line will be. When transformer coupling is used, the DC load line will be almost vertical when plotted on the scale of the plate curve graph. This happens because the primary winding of the typical output transformer will have a low DC resistance $R_{DC} \leq 100 \ \Omega$ and the cathode resistance is typically $R_K \leq 1000 \ \Omega$. Because of this, it is impractical to plot the DC load line.

Two Useful Approximations

Since we are making the approximation $R_{DC} \cong 0 \ \Omega$, we may write

$$V_{PQ} \cong V_{PP} \tag{7.9}$$

Equation (7.10) is another approximation that simplifies the design process. Here, we are disregarding the voltage drop $I_{PQ}R_K$. This will introduce some error into our calculations, but it will also result in a more conservative design which will increase tube lifetime. Recall that we used this approximation when designing amps back in Chap. 6 (see Fig. 6.2).

$$V_{PKQ} \cong V_{PQ} = V_{PP} \tag{7.10}$$

The AC Load Line

1. Determine Q-point.

Our Q-point coordinates are already set as

$$V_{PQ} \cong 350 \text{ V}, \ I_{PQ} = 45 \text{ mA}$$

Mark the Q-point on the 6L6GC plate characteristic curves, as shown in Fig. 7.6.

The approximate quiescent power dissipation of the tube is given by

$$
\begin{aligned}
P_{DQ} &\cong V_{PQ}I_{PQ} \\
&= 350 \text{ V} \times 45 \text{ mA} \\
&= 15.75 \text{ W}
\end{aligned}
\tag{7.11}
$$

This is well below the maximum DC power dissipation rating of 30 W for the 6L6GC. Recall that for a class A amplifier, maximum power dissipation occurs under no-signal conditions and average plate power dissipation actually decreases as output voltage swing increases.

2. Determine required cathode resistance R_K.

Since the Q-point doesn't fall on any of the 6L6 curves, we can estimate the grid bias voltage graphically as being

$$V_G = -32 \text{ V}$$

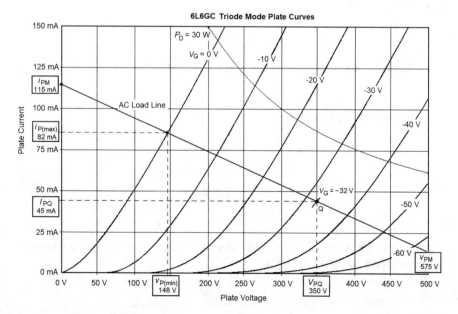

Fig. 7.6 AC load line for the 6L6, class A, single-ended amplifier

The cathode resistance is given by

$$R_K = \frac{-V_G}{I_{PQ}}$$
$$= \frac{32\,\text{V}}{45\,\text{mA}} \tag{7.12}$$
$$= 711\,\Omega \ (\text{use } 680\,\Omega \text{ or } 820\,\Omega)$$

Using $R_K = 680\,\Omega$ will operate the tube hotter at a slightly higher plate current, producing a bit higher maximum power output. Using $820\,\Omega$ would run the tube cooler, which will slightly reduce power output and extend tube life somewhat. Either value will work well here.

We will use $R_K = 680\,\Omega$. The power dissipation of the resistor is

$$P_{RK} = I_{PQ}^2 R_K$$
$$= 45\,\text{mA}^2 \times 680\,\Omega$$
$$= 1.4\,\text{W}$$

We need at least a 2 W rated resistor, though I'd probably go with 5 W just to be safe.

3. Select reflected load resistance R'_L.

   We have already chosen the 125BSE output transformer. Checking Table 7.3, we find that for $R_L = 8\,\Omega$, we can choose $R'_L = 10\,\text{k}\Omega$, $5\,\text{k}\Omega$, or $2.5\,\text{k}\Omega$. The data sheet for the 6L6GC lists a typical value $R'_L = 5\,\text{k}\Omega$, so we will use the recommended value.

$$R'_L = 5\text{k}\Omega$$

The black and yellow secondary wires of the 125BSE are connected to the $8\,\Omega$ speaker.

4. Determine maximum plate voltage v_{PM}.

   We use (6.20), which is renumbered for convenience.

$$v_{PM} = V_{PQ} + I_{PQ}R'_L$$
$$= 350\,\text{V} + (45\,\text{mA} \times 5\,k\Omega) \tag{7.13}$$
$$= 350\,\text{V} + 225\,\text{V}$$
$$= 575\,\text{V}$$

The 6L6GC can easily withstand this peak voltage, but the point is off the scale of the graph, so we can't plot it.

5. Determine maximum plate current i_{PM}.

$$
\begin{aligned}
i_{PM} &= I_{PQ} + \frac{V_{PQ}}{R'_L} \\
&= 45\text{mA} + \frac{350\text{V}}{5000\ \Omega} \\
&= 45\text{mA} + 70\text{mA} \\
&= 115\text{mA}
\end{aligned}
\tag{7.14}
$$

Plot the i_{PM} point on the plate characteristic curves as shown in Fig. 7.6.
6. Plot AC load line.

Draw a line from i_{PM} through the Q-point. We find that if projected to the right, the AC load line would have an x-intercept at $v_{PM} \cong 575$ V.

Q-Point Voltage Limits

The Q-point can move to the left until it intersects the $V_G = 0$ V curve. The coordinates of this point are

$$
v_{P(\min)} = 148\ \text{V}, i_{P(\max)} = 82\ \text{mA}
$$

The Q-point left- and right-going limits are

$$
\begin{aligned}
\Delta V_{PL} &= V_{PQ} --v_{P(\min)} \\
&= 350\text{V}--148\text{V} \\
&= 202\text{V}
\end{aligned}
\tag{7.15}
$$

$$
\begin{aligned}
\Delta V_{PR} &= v_{PM}--V_{PQ} \\
&= 575\text{V}--350\text{V} \\
&= 225\text{V}
\end{aligned}
\tag{7.16}
$$

We will denote the smaller of ΔV_{PL} and ΔV_{PR} as ΔV_P. Here, we have

$$
\Delta V_P = \Delta V_{PL} = 202\ \text{V}
$$

AC Characteristics of the Amplifier

We will now determine the voltage gain, the output voltage compliance, and the maximum output power of the amplifier. First, we determine the turns ratio of the output transformer using (7.8).

$$\frac{n_P}{n_S} = \sqrt{\frac{R'_L}{R_L}}$$
$$= \sqrt{\frac{5000\ \Omega}{8\ \Omega}}$$
$$= 25$$

Calculate the output voltage compliance using ΔV_P and the reciprocal of the turns ratio.

$$V_{o(max)} = \Delta V_P \left(\frac{n_S}{n_P} \right)$$
$$= 202\ \text{V} \left(\frac{1}{25} \right) \qquad (7.17)$$
$$= 8.1\ \text{V}_{pk} = 16.2\ \text{V}_{P-P}$$

The maximum output power is

$$P_{o(max)} = \frac{V^2_{o(max)}}{R_I}$$
$$= \frac{8.1\ \text{V}^2}{8\ \Omega} \qquad (7.18)$$
$$= 7.8\ \text{W}_{pk}$$

For a sinusoidal output signal, the RMS output power is half the peak power output

$$P_{rms} = P_{o(max)}/2$$
$$= 7.8\ \text{W}_{pk}/2 \qquad (7.19)$$
$$= 3.9\ \text{W}_{rms}$$

The voltage gain at the 6L6 plate terminal with respect to the grid A'_V is found using (7.20), where g_m and r_P are found in Table 6.1 or any convenient 6L6 data sheet.

$$A'_V = g_m(R'_L \parallel r_P)$$
$$= 4.7\text{mS}(5\text{k}\Omega \parallel 1.7\text{k}\Omega)$$
$$= 4.7\text{mS} \times 1.27\text{k}\Omega \qquad (7.20)$$
$$= 6$$

The tube voltage gain A'_V is stepped down by the output transformer. To find the overall voltage gain of the amplifier A_V, we multiply by the reciprocal of the turns ratio.

$$A_V = A_V' \left(\frac{n_S}{n_P}\right)$$

$$= 6\left(\frac{1}{25}\right) \tag{7.21}$$

$$= 0.24$$

We need to know the input voltage V_{in} required to drive the amplifier to maximum output. The maximum unclipped plate voltage was found to be $\Delta V_P = 202$ V. We use this in the following equation:

$$V_{in(max)} = \frac{\Delta V_P}{A_V'}$$

$$= \frac{202 \text{ V}}{6} \tag{7.22}$$

$$= 33.7 \text{ V}_{pk} = 67.4 \text{ V}_{P-P}$$

Note that $\Delta V_P/A'_V = 33.7$ V$_{pk}$ is close to the magnitude of the grid voltage $|V_G| = 32$ V, determined on the plate curves of Fig. 7.6. The difference is not significant in this example, but if the difference is large, the smaller of these voltages will determine when the amplifier is overdriven. Driving the control grid more positive than the cathode will overdrive the amp, but if we include a grid stopper resistor (gamma network), then the grid current will be held to a safe level.

Grid Resistors and the Gamma Network

We finish this design by choosing gamma network resistor values. The data sheet for the 6L6GC specifies a maximum G1 resistance of 500 kΩ using cathode feedback biasing. We will use the following resistor values:

$$R_1 = 270 \text{ k}\Omega, R_2 = 10 \text{ k}\Omega$$

Screen grid G_2 will draw around 4 mA or so. A good value for the screen grid resistor is

$$R_3 = 1 \text{ k}\Omega, 1 \text{ W}$$

The finished amplifier is shown in Fig. 7.7.

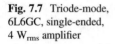

Fig. 7.7 Triode-mode, 6L6GC, single-ended, 4 W_{rms} amplifier

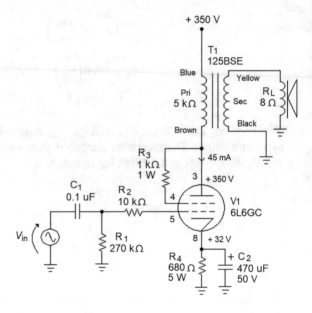

Substituting an EL34

Technically, the 6L6 is not a true pentode. It is a *beam power tetrode*. Beam power tetrodes don't have a suppressor grid. Instead they have beam-forming electrodes that keep the electron stream tightly bunched as it travels from the cathode to the plate, which helps to suppress secondary electrons like a true pentode. Because they behave essentially like pentodes, beam power tetrodes are often called *beam power pentodes* or simply pentodes.

The EL34 is a true pentode that is quite similar to the 6L6GC, except that it can carry slightly more plate current and has about half the dynamic plate resistance and the heater draws 1.5 A (versus 900 mA for the 6L6GC). Switching to an EL34 will result in a different sound and different distortion/overdrive characteristics. Here is a summary of primary EL34 parameters.

EL34 parameter summary (class A)

	Pentode mode	Triode mode	
$P_{D(max)}$	25 W	25 W	Max DC power dissipation
$I_{P(max)}$	100 mA	70 mA	Max DC plate current
$V_{P(max)}$	500 V	450 V	Max DC plate voltage
g_m	10.6 mS	4.7 mS	Transconductance
r_P	15 kΩ	1.7 kΩ	Internal dynamic plate resistance
R'_L	3.5 kΩ	3.5 kΩ	Typical reflected load resistance
μ	160	8	Amplification factor
Heater	6.3 V, 1.5 A		

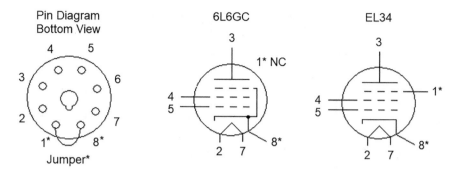

Fig. 7.8 Pin diagrams for the 6L6GC and EL34. Shorting pin 1 to pin 8 of the socket allows use with either the 6L6 or the EL34

The 6L6 and EL34 have the same pin designations, except that the suppressor grid of the EL34 (pin 1) must be connected to the cathode (pin 8) externally. The pin diagrams for both tubes are shown in Fig. 7.8.

Since pin 1 of the 6L6 is not connected to anything internally, we can wire it to the cathode (jumper socket pin 1 to pin 8) which allows us to substitute the EL34 directly in its place. From now on, this connection will be made in all circuits using the 6L6GC, as well as the EL34.

With the jumper in place, we can substitute an EL34 for the 6L6GC used in the last example with no problems. This will be verified in the upcoming analysis, but in general, it's not a good idea to make this substitution without studying the amplifier circuit of interest to make sure it's ok.

Q-Point Analysis

Rather than designing the amplifier, we are now analyzing the circuit. I think design is more fun, but analysis still cool. We use the EL34 transconductance curves to determine I_{PQ} as follows.

To find the Q-point location, we use (7.12), which is $R_K = -V_G/I_{PQ}$, to plot the bias line on the EL34 triode-mode transconductance curve graph as shown in Fig. 7.9.

1. Choose a convenient grid voltage on the x-axis (in this case, I chose $V_G = -60$ V), and mark Point 1 on the plate current axis using

$$I_1 = \frac{-V_G}{R_K} = \frac{60 \text{ V}}{680 \text{ }\Omega} \cong 90 \text{ mA}$$

2. Draw a line from Point 1 to the origin of the graph. The intersection of this line with the curve for $V_P = 350$ V is the Q-point.

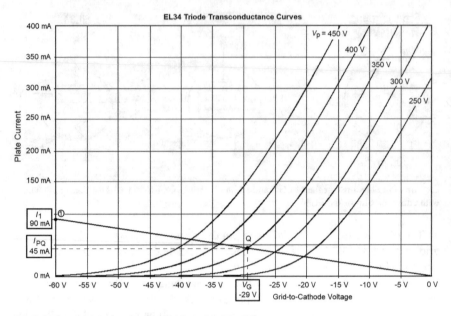

Fig. 7.9 Q-point location for the EL34 used in Fig. 7.7

3. Read the coordinates of the Q-point, which are approximately

$$V_P = 350 \text{ V}, I_{PQ} = 45 \text{ mA} \quad (\text{and } P_{DQ} = 15.75 \text{ W})$$

This is identical to the Q-point of the 6L6GC. You might want to plot an AC load line for the EL34 in this circuit to verify that the tube operates correctly.

It's worth repeating that it's normally not a good idea to arbitrarily substitute different tubes in power output stages. It's possible for a tube that is different than the original specification to bias incorrectly, resulting in too much plate current or excessive plate power dissipation. This can result in destruction of the tube or worse yet destruction of the output transformer. Also, different tubes may have vastly different heater current requirements, which can cause overheating of the main power transformer. In this case, the circuit will work with either tube, and if you happened to have one handy, you could even try a KT88 in this circuit. The KT88 is capable of much higher power operation and would be biased quite cold here.

6L6GC Pentode-Mode, SE Amplifier Design Example

In this example, we will redesign the previous 6L6GC triode amplifier for operation in the pentode mode. The schematic is shown in Fig. 7.11.

Choosing Power Supply Voltage

As in the previous design, we will use

$$V_{PP} = 350 \text{ V}$$

Choosing Quiescent Plate Current I_{PQ}

In the pentode mode, the 6L6GC can operate with a DC plate current as high as 80 mA. We'll be a bit conservative and go with 60 mA for this design.

$$I_{PQ} = 60 \text{ mA}$$

Output Transformer Selection

Referring again to Table 7.1, we find that the 125DSE will operate with a DC bias current of up to 70 mA. This is sufficient for our amplifier.

The AC Load Line

1. Determine Q-point.
   The Q-point coordinates are $V_{PQ} \cong V_{PP} = 350$ V and $I_{PQ} = 60$ mA. Mark this as the Q-point on the 6L6GC plate characteristic curves, as shown in Fig. 7.10.
   The approximate quiescent power dissipation of the tube is

$$P_{DQ} \cong V_{PP}I_{PQ}$$
$$= 350 \text{ V} \times 60 \text{ mA}$$
$$= 21 \text{ W}$$

This is well below the maximum power dissipation rating of 30 W for the 6L6GC. The Q-point parameters are summarized below.

$$V_{PQ} = 350 \text{ V}, I_{PQ} = 60 \text{ mA}, P_{DQ} = 21 \text{ W}$$

2. Determine cathode resistance R_K.
   The Q-point is slightly above the $V_G = -20$ V curve, so we can graphically estimate the grid voltage to be

$$V_G = -19 \text{ V}$$

The cathode resistance is given by (7.12).

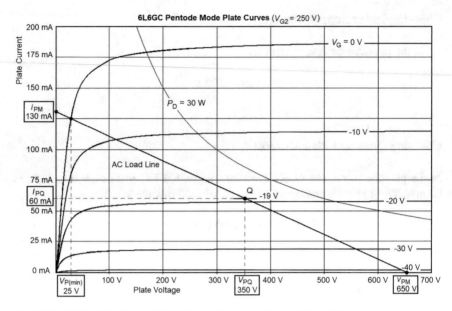

Fig. 7.10   AC Load line for the 6L6GC pentode-mode, class A, SE amplifier

$$R_K = \frac{-V_G}{I_{PQ}}$$

$$= \frac{19\ V}{60\ mA}$$

$$= 317\ \Omega \quad (\text{use } 330\ \Omega)$$

Using 330 Ω will operate the tube a bit cooler which will reduce output power slightly, but will also extend tube life somewhat. There will be no noticeable difference in performance.

The power dissipation of the cathode resistor is

$$P_{RK} = I_{PQ}^2 R_K$$

$$= 60mA^2 \times 330\ \Omega$$

$$= 1.2\ W$$

This resistor is labeled R_4 in the schematic. We can use a 2 W resistor or larger.

3. Select reflected load resistance, R'_L.

As before, we use Table 7.3, which tells us that for $R_L = 8\ \Omega$, we can choose $R'_L = 2.5, 5,$ or 10 kΩ. The data sheet for the 6L6GC lists a typical value $R'_L = 4.2$ kΩ. We will go with the choice that is closest to General Electric's recommendation and use the 5 kΩ connections of the 125DSE (the black and yellow secondary wires).

$$R'_L = 5k\Omega$$

4. Determine maximum plate voltage v_{PM} using (7.13).

$$\begin{aligned} v_{PM} &= V_{PQ} + I_{PQ}R'_L \\ &= 350\,V + (60\,mA \times 5\,k\Omega) \\ &= 350\,V + 300\,V \\ &= 650\,V \end{aligned}$$

Plot this point on the voltage axis of the 6L6GC plate curves as in Fig. 7.10.
5. Determine maximum plate current i_{PM} using (7.14).

$$\begin{aligned} i_{PM} &= I_{PQ} + \frac{V_{PQ}}{R'_L} \\ &= 60\ mA + \frac{350\ V}{5000\,\Omega} \\ &= 60\,mA + 70\,mA \\ &= 130\,mA \end{aligned}$$

6. Plot AC load line.
    Draw a line from i_{PM} to v_{PM}. This line should go right through the Q-point.

Q-Point Voltage Limits

The Q-point can move to the left until it intersects the $V_G = 0$ V curve. The coordinates of this point are

$$v_{P(min)} = 25\ V,\ i_{P(max)} = 125\ mA$$

Notice that $v_{P(min)}$ is much lower for the pentode than for the triode (25 V vs. 148 V). This means that the plate voltage can swing in the negative direction much further than for a triode. All things being equal, the pentode is typically capable of significantly higher output voltage swing and output power.
    The Q-point limits are

$$\Delta V_{\text{PL}} = V_{\text{PQ}} -- v_{\text{P(min)}}$$
$$= 350 \text{ V} --25 \text{ V}$$
$$= 325 \text{ V}$$
$$\Delta V_{\text{PR}} = v_{\text{PM}} -- V_{\text{PQ}}$$
$$= 650 \text{ V} -- 350 \text{ V}$$
$$= 300 \text{ V}$$

We will denote the smaller of ΔV_{PL} and ΔV_{PR} as ΔV_{P}. Here, we have

$$\Delta V_{\text{P}} = \Delta V_{\text{PR}} = 300 \text{ V}$$

AC Characteristics of the Amplifier

We will now determine the voltage gain, the output voltage compliance, and the maximum output power of the amplifier. First, we determine the turns ratio of the output transformer using (7.8).

$$\frac{n_{\text{P}}}{n_{\text{S}}} = \sqrt{\frac{R_{\text{L}}'}{R_{\text{L}}}}$$
$$= \sqrt{\frac{5000 \ \Omega}{8 \ \Omega}}$$
$$= 25$$

We now calculate the maximum output voltage using ΔV_{P} and the reciprocal of the turns ratio.

$$V_{\text{o(max)}} = \Delta V_{\text{P}} \left(\frac{n_{\text{S}}}{n_{\text{P}}} \right)$$
$$= 300 \text{ V} \left(\frac{1}{25} \right)$$
$$= 12 \text{ V}_{\text{pk}} = 24 \text{ V}_{\text{P--P}}$$

The maximum output power is

$$P_{\text{o(max)}} = \frac{V_{\text{o(max)}}^2}{R_{\text{L}}}$$
$$= \frac{12 \text{ V}^2}{8 \ \Omega}$$
$$= 18 \text{ W}_{\text{pk}}$$

The RMS output power is

$$P_{rms} = P_{o(max)}/2$$
$$= 18 \text{ W}/2$$
$$= 9 \text{ W}_{rms}$$

This is over twice the power output of the triode-mode 6L6GC amplifier, using the same supply voltage. This is why pentodes are commonly used in power amp stages.

The voltage gain magnitude at the 6L6 plate, referred to the grid, is found using (7.20) and data taken from Table 6.1.

$$A'_V = g_m(R'_L \parallel r_P)$$
$$= 5.2\text{mS} \ (5\text{k}\Omega \parallel 33\text{k}\Omega)$$
$$= 5.2\text{mS} \times 4.3\text{k}\Omega$$
$$= 22.4$$

Using (7.21), the overall voltage gain of the amplifier is

$$A_V = A'_V \left(\frac{n_S}{n_P}\right)$$
$$= 22.4 \left(\frac{1}{25}\right)$$
$$= 0.9$$

The maximum unclipped plate voltage was found to be $\Delta V_P = 300$ V. The input voltage that will just produce clipping is

$$V_{in(max)} = \frac{\Delta V_P}{A'_V}$$
$$= \frac{300 \text{ V}}{22.4}$$
$$= 13.4 \text{ V}_{pk} = 26.8 \text{ V}_{P-P}$$

Note in this example that output clipping will occur before the control grid is overdriven because $\Delta V_P/A'_V < |V_G|$ (i.e., 13.4 V < 19 V).

Grid Resistor and Gamma Network

We now need to choose the grid resistor. The maximum allowable G1 resistance is 500 kΩ using cathode feedback biasing, so as before, we use

$$R_1 = 270 \text{ k}\Omega, R_2 = 10 \text{ k}\Omega$$

The Screen Grid Resistor, R_3

The plate curves of Fig. 7.10 are most accurate for $V_{G2} = 250$ V, but the actual value is not critical. We can usually provide an appropriate screen supply voltage by placing a dropping resistor in series with the $+V_{PP}$ rail that supplies earlier stages with power. Recall that the screen grid resistor serves to limit current should the tube be driven into clipping. In power amplifiers, this is usually a 1 or 2 W resistor ranging from 470 Ω to 2.2 kΩ. We will use

$$R_3 \; = 1 \text{ k}\Omega, 1\text{W}$$

The complete output stage is shown in Fig. 7.11. We make the usual approximation that the screen grid runs at about 10% of the plate current I_{PQ}. The 20 mA current shown at the top of the schematic represents current delivered to preceding stages that are not shown. This is where we derive the screen grid supply voltage.

EL84 Pentode-Mode, SE Amplifier Design Example

The EL84 (or equivalent 6BQ5) is a very popular power pentode that has very impressive performance considering its relatively small physical size. The EL84 uses a miniature 9-pin (E9-1, or *Noval base*) socket, just like the popular 12AX7 and similar tubes. The EL84 has the following basic specs.

EL84 (6BQ5) parameter summary

	Pentode mode	Triode mode	
$P_{D(max)}$	12 W	12 W	Max DC power dissipation
$I_{P(max)}$	60 mA	40 mA	Max DC plate current
$V_{P(max)}$	300 V	250 V	Max DC plate voltage
g_m	11 mS	5.4 mS	Transconductance
r_P	38 kΩ	3.5 kΩ	Internal dynamic plate resistance
$R'_{L(typ)}$	5.2 kΩ	3.5 kΩ	Typical reflected load resistance
μ	418	19	Amplification factor
Heater	6.3 V, 760 mA		

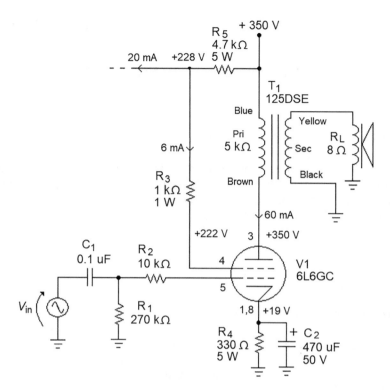

Fig. 7.11 6L6GC pentode-mode, SE, 9 W amplifier

Choosing Power Supply Voltage

For this amplifier, we will use $V_{PP} = 250$ V.

Q-Point Selection

Since there is more than one way to design a given circuit, we will begin this design by choosing the quiescent power dissipation of the tube. To leave a reasonable safety margin, the EL84 will be operated at $P_{DQ} = 10$ W, which is about 17% less than the maximum DC power dissipation rating of 12 W.

The next step is to determine the EL84 plate current. We solve (7.11) for I_{PQ} which gives us

$$I_{PQ} = \frac{P_{DQ}}{V_{PP}}$$
$$= 10 \text{ W}/250 \text{ V}$$
$$= 40 \text{ mA}$$

(7.23)

The Q-point is plotted on the EL84 pentode plate curves of Fig. 7.12. The Q-point summary is

$$V_{PQ} = 250 \text{ V}, I_{PQ} = 40 \text{ mA}, P_{DQ} = 10 \text{ W}$$

Output Transformer and R'_L Selection

Use the 125BSE, which has a maximum DC current rating of 45 mA. The pentode-mode dynamic plate resistance is $r_P = 38$ kΩ, and the data sheet lists typical reflected load resistance as $R'_L = 5.2$ kΩ. Again, for an 8 Ω load, it looks like the best choice is to use the 5 kΩ leads (yellow and black).

$$R'_L = 5 \text{ k}\Omega$$

Determine Cathode Resistance R_4

The grid biasing voltage falls right on the curve

$$V_G = -8 \text{ V}$$

The cathode resistance is

$$R_4 = \frac{-V_G}{I_{PQ}}$$
$$= \frac{8V}{40mA}$$
$$= 200\Omega \ (\text{use } 200 \ \Omega)$$

$$P_{R4} = I_{PQ}^2 R_4$$
$$= 40mA^2 \times 220 \ \Omega$$
$$= 0.35 \text{ W}$$

Use a 0.5 W or higher rated resistor.

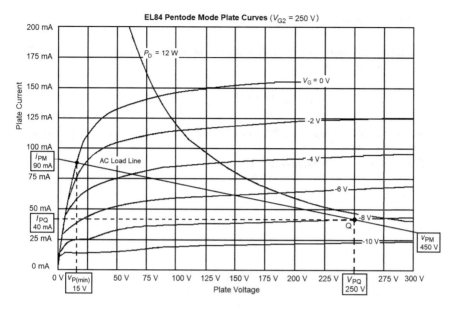

Fig. 7.12 AC load line for the EL84 pentode-mode, SE amplifier

The AC Load Line

Determine Maximum Plate Voltage v_{PM} Using (7.13).

$$v_{PM} = V_{PQ} + I_{PQ}R'_L$$
$$= 250 \text{ V} + (40 \text{ mA} \times 5 \text{ } k\Omega)$$
$$= 250 \text{ V} + 200 \text{ V}$$
$$= 450 \text{ V}$$

This point is off the scale of the graph. Now we find i_{PM}.

$$i_{PM} = I_{PQ} + \frac{V_{PQ}}{R'_L}$$
$$= 40 \text{ mA} + \frac{250 \text{ V}}{5000 \, \Omega}$$
$$= 40 \text{ mA} + 50 \text{ mA}$$
$$= 90 \text{ mA}$$

Plotting the AC load line, as seen in Fig. 7.12, it seems very plausible that the right endpoint would intersect the voltage axis at $v_{PM} = 450$ V.

Q-Point Voltage Limits

The Q-point can move to the left until it intersects the $V_G = 0$ V curve. The coordinates of this point are

$$v_{P(min)} \cong 15 \text{ V}, \, i_{P(max)} \cong 85 \text{ mA}$$

As we expected, $v_{P(min)}$ is nice and low which helps give the amp good output voltage compliance.

The Q-point voltage excursion limits are

$$\begin{aligned} \Delta V_{PL} &= V_{PQ} -- v_{P(min)} \\ &= 250\text{V} -- 15\text{V} \\ &= 235\text{V} \end{aligned}$$

$$\begin{aligned} \Delta V_{PR} &= v_{PM} -- V_{PQ} \\ &= 450\text{V} -- 250\text{V} \\ &= 200\text{V} \end{aligned}$$

$$\Delta V_P = \Delta V_{PR} = 200 \text{ V}$$

AC Characteristics of the Amplifier

We will now determine the voltage gain, the output voltage compliance, and the maximum output power of the amplifier. Like before, we determine the turns ratio of the output transformer using (7.8).

$$\begin{aligned} \frac{n_P}{n_S} &= \sqrt{\frac{R_L'}{R_L}} \\ &= \sqrt{\frac{5000 \, \Omega}{8 \, \Omega}} \\ &= 25 \end{aligned}$$

We now calculate the maximum output voltage using ΔV_P and the reciprocal of the turns ratio.

$$\begin{aligned} V_{o(max)} &= \Delta V_P \left(\frac{n_S}{n_P} \right) \\ &= 200 \text{ V} \left(\frac{1}{25} \right) \\ &= 8 \text{ V}_{pk} = 16 \text{ V}_{P-P} \end{aligned}$$

The maximum and rms output power are

$$P_{o(max)} = \frac{V^2_{o(max)}}{R_I}$$

$$= \frac{8^2}{8}$$

$$= 8 \, W_{pk} = 4 \, W_{rms}$$

Using a supply voltage of 250 V, this amplifier has the same maximum output power as the 6L6GC operating in the triode mode, with a 350 V supply. Once again, we see the advantage of the pentode over the triode when we are after high power.

The EL84 plate voltage gain magnitude referred to the grid is

$$A'_V = g_m(R'_L \parallel r_P)$$
$$= 11mS \ (5k\Omega \parallel 38k\Omega)$$
$$= 11mS \times 4.4k\Omega$$
$$= 48.4$$

The net voltage gain of the amplifier is

$$A_V = A'_V \left(\frac{n_S}{n_P}\right)$$

$$= 48.4\left(\frac{1}{25}\right)$$

$$= 2$$

The maximum unclipped plate voltage was found to be $\Delta V_P = 200$ V. The maximum input voltage is

$$V_{in(max)} = \frac{\Delta V_P}{A'_V}$$

$$= \frac{200V}{48.4}$$

$$= 4.1V_{pk} = 8.2V_{P-P}$$

The grid bias voltage is $V_G = -8$ V, so clipping will occur before the grid is overdriven.

Grid Resistor and Gamma Network

The maximum allowable G1 resistance for the EL84 is 1 MΩ using cathode feedback biasing, so we choose

$$R_1 = 470 \text{ k}\Omega, R_2 = 10 \text{ k}\Omega$$

The Screen Grid Resistor

The plate curves of Fig. 7.10 are accurate for $V_{G2} = 250$ V, which happens to be our plate supply voltage V_{PP}. We need to run the screen grid at a lower voltage than the plate in the pentode mode. As before, we will assume that preceding stages are present that allow us to drop some voltage across R_5 to provide the screen grid bias voltage.

The screen grid resistor is set to $R_3 = 1$ kΩ, 1 W.

The final amplifier is shown in Fig. 7.13.

EL34 Pentode-Mode, SE Amplifier Design Example

We will finish this section with the design of an output stage using an EL34 pentode. We work out this design using the following starting point:

$$V_{PP} = 300 \text{ V}, I_{PQ} = 65 \text{ mA}, R_L = 8\Omega$$

By now, the basic design process for this type of amplifier has become very familiar to you, so we will fly through this design pretty quickly.

Plot Q-Point

The Q-point coordinates are

$$V_{PQ} = 300 \text{ V}, I_{PQ} = 65 \text{ mA}, P_{DQ} = 19.5 \text{ W}$$

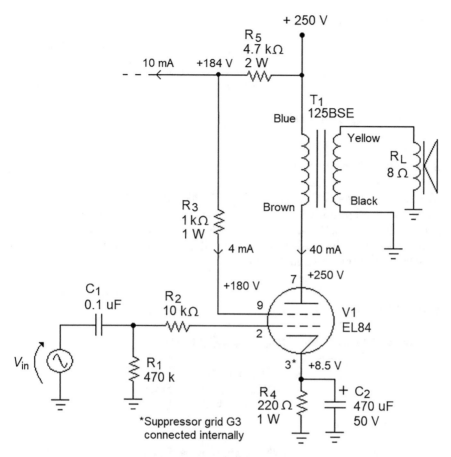

Fig. 7.13 EL84 pentode-mode, SE, 4 W amplifier

Determine Cathode Resistance R_K

The Q-point is close enough to the $V_G = -15$ V curve, so let's go with it in our calculations.

$$V_G \cong -15 \text{ V}$$

The cathode resistance is

$$
\begin{aligned}
R_K &= \frac{-V_G}{I_{PQ}} \\
&= \frac{15 \text{ V}}{65 \text{ mA}} \\
&= 231 \ \Omega \quad (\text{use } 220\Omega \text{ or } 270\Omega)
\end{aligned}
$$

We will choose $R_K = 270\ \Omega$.

$$
\begin{aligned}
P_{RK} &= I_{PQ}^2 R_K \\
&= 65\ \text{mA}^2 \times 270\Omega \\
&= 1.14\text{W}
\end{aligned}
$$

Use a 2 W or higher resistor.

Select Output Transformer

Since $I_{PQ} = 65$ mA, we select the 125DSE which is rated for $I_{DC} = 70$ mA (see Table 7.1). In the pentode mode, the EL34 has a typical recommended load $R'_L = 3.5$ kΩ. Use the yellow and black secondary leads, which give a reflected load resistance of 5 kΩ when driving an 8 Ω load. We will soon verify that this reflected load resistance works very well in this example.

Plot AC Load Line

Determine Maximum Plate Voltage v_{PM} Using (7.13).

$$
\begin{aligned}
v_{PM} &= V_{PQ} + I_{PQ} R'_L \\
&= 300\text{V} + (65\text{mA} \times 5\text{k}\Omega) \\
&= 300\text{V} + 325\text{V} \\
&= 625\text{V}
\end{aligned}
$$

This point is off the scale of the graph. Now we find i_{PM}.

$$
\begin{aligned}
i_{PM} &= I_{PQ} + \frac{V_{PQ}}{R'_L} \\
&= 65\text{mA} + \frac{300\text{V}}{5000\ \Omega} \\
&= 65\text{mA} + 60\text{mA} \\
&= 125\text{mA}
\end{aligned}
$$

The AC load line is shown in Fig. 7.14. The minimum plate voltage and maximum plate current are found to be

$$
v_{P(\min)} = 20\ \text{V},\ i_{P(\max)} = 120\ \text{mA}
$$

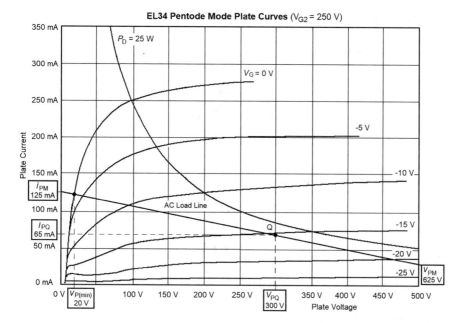

Fig. 7.14 AC load line for pentode-mode EL34, SE amplifier

Q-Point Voltage Compliance

$$\Delta V_{PL} = V_{PQ} -- v_{P(\min)}$$
$$= 300V -- 20V$$
$$= 280V$$

$$\Delta V_{PR} = v_{PM} -- V_{PQ}$$
$$= 625V --300\ V$$
$$= 325V$$

$$\Delta V_P = \Delta V_{PL} = 280\ V$$

AC Characteristics of the Amplifier

The transformer turns ratio is

$$\frac{n_P}{n_S} = \sqrt{\frac{R'_L}{R_L}}$$
$$= \sqrt{\frac{5000\ \Omega}{8\ \Omega}}$$
$$= 25$$

The maximum output voltage is

$$V_{o(max)} = \Delta V_P \left(\frac{n_S}{n_P}\right)$$
$$= 280\ V \left(\frac{1}{25}\right)$$
$$= 11.2\ V_{pk} = 22.4\ V_{P-P}$$

The peak and rms output power are

$$P_{o(max)} = \frac{V^2_{o(max)}}{R_I}$$
$$= \frac{11.2^2}{8}$$
$$= 15.7\ W_{pk} = 7.9\ W_{rms}$$

The voltage gain magnitude A'_V, from grid to plate, is

$$A'_V = g_m (R'_L \parallel r_P)$$
$$= 10.6mS\ \ (5k\Omega \parallel 15k\Omega)$$
$$= 10.6mS \times 3.75k\Omega$$
$$= 39.8$$

The overall voltage gain of the amplifier is

$$A_V = A'_V \left(\frac{n_S}{n_P}\right)$$
$$= 39.8 \left(\frac{1}{25}\right)$$
$$= 1.6$$

The maximum unclipped plate voltage was found to be $\Delta V_P = 280$ V. The input voltage that will drive the output into clipping is

$$V_{in(max)} = \frac{\Delta V_P}{A_V'}$$
$$= \frac{280V}{39.8}$$
$$= 7V_{pk} = 14V_{P-P}$$

Output clipping will occur before the grid is overdriven because $I_{PQ}R_K \cong 15$ V.

Grid Resistors and Gamma Network

The grid resistor/gamma network resistor values are

$$R_1 = 270 \text{ k}\Omega, R_2 = 10 \text{ k}\Omega$$

Assume that $V_{G2} = 250$ V. The screen grid resistor is

$$R_3 = 1 \text{ k}\Omega, 1W$$

The final amplifier is shown in Fig. 7.15.

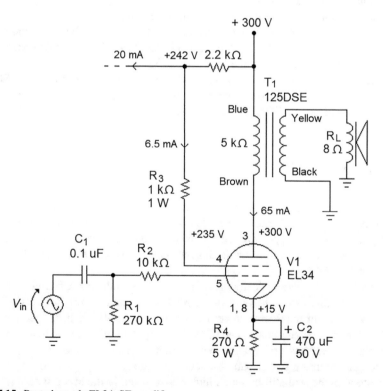

Fig. 7.15 Pentode-mode EL34, SE amplifier

Selectable Triode/Pentode Operation

We have seen that for a given tube, operation in the pentode mode will result in approximately a doubling of maximum power output compared to triode-mode operation. In addition, an amplifier built using a 6L6GC, EL34, EL84, or similar pentode will have very different sound qualities in overdrive conditions when operated as a triode versus as a pentode. Because of this, many guitarists like to be able to select between these two modes of operation to tailor their sound to a given situation. There are several approaches that can be taken to switch an amplifier between triode and pentode operation.

Switching R_K and Suppressor Grid G2

Since we previously designed both triode- and pentode-mode 6L6GC amps that operate from the same supply voltage, we will use these circuits as the basis for our first example of a dual-mode amplifier. The idea here is very simple; simultaneously switch the cathode resistor and the suppressor grid connections using a DPDT switch, as shown in Fig. 7.16.

The advantage of this switching method is that the amplifier will operate at a proper Q-point for the given mode selection. In Fig. 7.16, in the triode mode, the tube will operate as described by the AC load line of Fig. 7.6. In the pentode mode, tube operation is described by the load line in Fig. 7.10.

Note that the design of the original triode-mode amplifier used the 125BSE for T_1. This transformer is not designed to handle the higher $I_{PQ} = 60$ mA of pentode-mode operation, so the 125DSE must be used here.

Switching Only G2

It is possible to switch between triode and pentode modes of operation without switching the cathode resistor. When this is done, the cathode resistor must be the appropriate value for operation in the triode mode. In the case of the 6L6GC amplifier we've been working with, this means we must use $R_K = 680 \ \Omega$, 5 W, as shown in Fig. 7.17.

The tube will be biased ideally for operation in the triode mode, but will run quite cold in the pentode mode because of the high-value cathode resistor. This will not harm the tube, but maximum power output will be perhaps half of that normally available in the pentode mode. The 6L6GC in Fig. 7.17 will operate at roughly the same plate current and grid voltage in both triode and pentode modes.

Though maximum power output is reduced, the sensitivity (gain) of the amp in the pentode mode will still be much higher than when in the triode mode. Just for

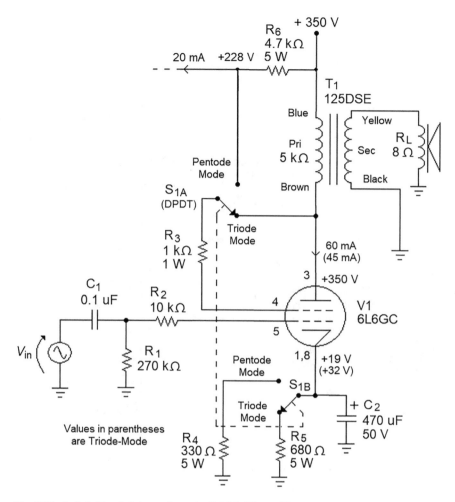

Fig. 7.16 Switchable triode/pentode-mode 6L6GC, SE amplifier

fun, you might like to perform an analysis of the circuit in pentode mode to see how the 680 Ω cathode resistor affects operation.

Parallel-Connected Tubes

Just like when using transistors, effective tube current and power handling capability can be doubled by connecting two tubes in parallel. The two parallel-connected tubes may be treated as a single equivalent tube with twice the transconductance and

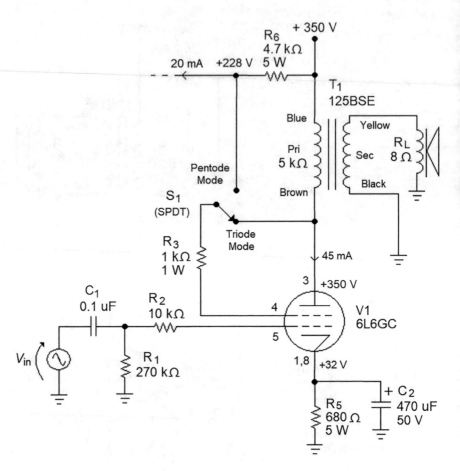

Fig. 7.17 Alternate dual-mode, 6L6GC, SE amp

half the dynamic plate resistance of a single tube. In equation form, we can write this as

$$g_{m(eq)} = 2g_m \tag{7.24}$$

$$r_{P(eq)} = r_P/2 \tag{7.25}$$

We don't need new plate curves in order to design an amplifier using parallel-connected tubes. Instead, we design the amp as usual for a single tube and then connect the tubes in parallel, using half the usual cathode resistance.

An example based on the EL84 amplifier of Fig. 7.13 is shown in Fig. 7.18. In the original circuit, the cathode resistor was $R_K = 220\ \Omega$. Here, we use

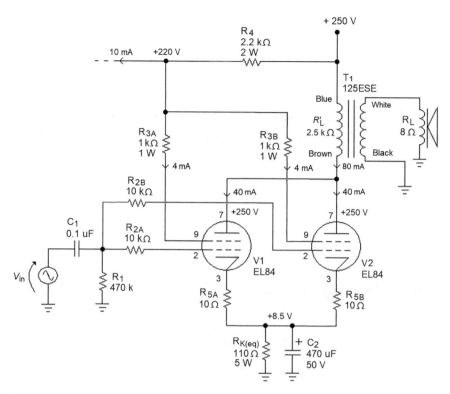

Fig. 7.18 Parallel-connected EL84s

$$R_{K(eq)} = R_K/2$$
$$= 220\,\Omega/2 \tag{7.26}$$
$$= 110\,\Omega$$

The 10 Ω cathode swamping resistors are not absolutely necessary, but they help to equalize the division of plate current between the tubes should they be slightly mismatched.

There are a few other changes to the original schematic that should be noted. Because the total plate current has doubled, we are now using the 125ESE output transformer which can handle the increased current ($I_{DC} = 80$ mA). Also, supply dropping resistor R_4 was reduced to compensate for doubling of the screen grid current.

In the original design the $R'_L = 5$ kΩ, yellow and black secondary leads were used. Because the parallel tubes have $r_{P(eq)} = r_P/2$, the 2.5 kΩ reflected load connections are now used (white and black secondary leads).

Connecting the tubes in parallel has allowed us to cut the reflected load resistance in half. The effect on the output voltage swing and output power level is dramatic. Using the 2.5 kΩ connection for T1, the turns ratio is

$$\frac{n_P}{n_S} = \sqrt{\frac{R'_L}{R_L}}$$

$$= \sqrt{\frac{2500\ \Omega}{8\ \Omega}}$$

$$= 17.7$$

Connecting the tubes in parallel does not change the maximum primary (plate side) voltage limit, so taking values from the original design and using the new turns ratio, we have

$$V_{o(max)} = \Delta V_P \left(\frac{n_S}{n_P}\right)$$

$$= 200\ V \left(\frac{1}{17.7}\right)$$

$$= 11.3\ V_{pk} = 22.6\ V_{P-P}$$

The maximum output power is

$$P_{o(max)} = \frac{V^2_{o(max)}}{R_I}$$

$$= \frac{11.3^2}{8}$$

$$= 16\ W_{pk} = 8\ W_{rms}$$

As you probably expected, this is double the output power of the original amplifier design, but it's nice to see that the math confirms our suspicions.

Complete SE Amplifier Examples

We've spent a lot of time looking at single-ended power amplifiers in isolation, so now it's time we put it all together and look at some complete amplifiers, with associated power supplies. A spring reverb option will also be presented.

6AQ5 Low-Power, SE Practice Amp

In the first two editions of the book, I started off with a pretty complex amp design at this point. I thought I'd start with a simpler, reasonably inexpensive low-power amplifier this time around. The schematic for the amplifier is shown in Fig. 7.19. I designed and built this amplifier just for fun in early 2020, and I liked its

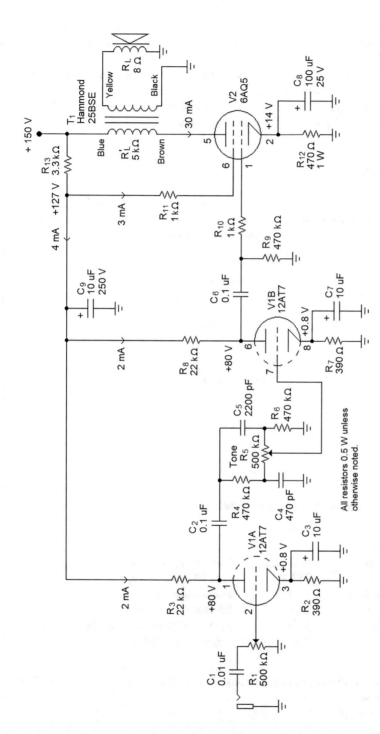

Fig. 7.19 Low-power, single-ended practice amplifier

performance so much I made a nice Fender-style, tweed cabinet for it. I probably have ten amplifiers of varying power output sitting around, but this is my favorite practice amp.

The first two stages are built around the 12AT7 dual triode using a simple, single-pot tone control. The output stage uses a 6AQ5 pentode (in the pentode mode). I chose this tube because they are readily available and inexpensive (I bought 15 6AQ5s for $20.00 on eBay) and have decent power output capability.

Amplifier Analysis

The load lines for the 12AT7 sections of the amplifier are not presented here, but the main parameters for these stages are given below. You should be able to verify these values fairly easily if you worked through previous examples.

Stage 1 and 2 Analysis Summary

$$V_G = -0.8V$$

$$V_{PQ} = 80 \text{ V}$$

$$I_{PQ} = 2 \text{ mA}$$

$$V_{o(\text{max})} = 40 \text{ V}_{pk} = 80 \text{ V}_{P-P}$$

$$A_V = 40$$

The 6AQ5 has the following characteristics:

	Pentode mode	Triode mode	
$P_{D(max)}$	12 W	9 W	Max DC power dissipation
$I_{P(max)}$	45 mA	49.5 mA	Max DC plate current
$V_{P(max)}$	250 V	250 V	Max DC plate voltage
g_m	3.7 mS	4.8 mS	Transconductance
r_P	58 kΩ	2 kΩ	Internal dynamic plate resistance
$R'_{L(typ)}$	5.5 kΩ	3.5 kΩ	Typical reflected load resistance
μ	215	9.5	Amplification factor
Heater	6.3 V, 450 mA		

In the design of the output stage, I chose $I_{PQ} = 30$ mA with $V_{PP} = 150$ V. The Q-point values are

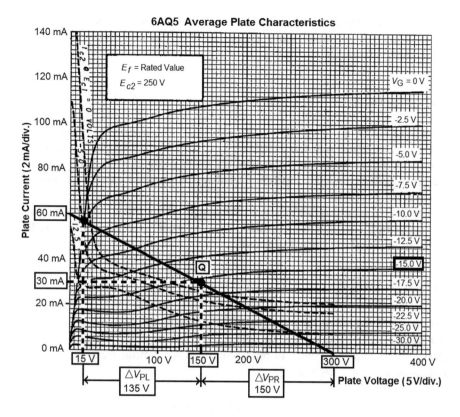

Fig. 7.20 AC load line for the 6AQ5 output stage

$$V_{PQ} = 150 \text{ V}, I_{PQ} = 30 \text{ mA}, P_{DQ} = 4.5 \text{ W}$$

A Hammond 125BSE output transformer was used which provides $R'_L = 5 \text{ k}\Omega$ with an 8 Ω load and a turns ratio of $n_P/n_S = 25$. The AC load line for the output stage is shown in Fig. 7.20. The maximum unclipped plate voltage swing is $\Delta V_{PL} = 135 \text{ V}$. The Q-point is very close to the center of the AC load line, so we will get almost symmetrical clipping and very good power output, considering the relatively low plate supply voltage.

Based on the transformer step-down factor of $n_S/n_P = 0.04$, the voltage gain of the output stage is $A_{V3} \cong 0.68$. The output voltage compliance and output power are

$$V_{o(\text{max})} = 5.4 \text{ V}_{\text{pk}} = 10.8 \text{ V}_{\text{P-P}}$$

$$P_{o(\text{max})} = 3.6 \text{ W}_{\text{pk}} = 1.8 \text{ W}_{\text{rms}}$$

The overall gain of the amplifier is $A_V \cong 272$, which allows the amp to be driven into clipping with an input signal $V_{in(max)} = 20$ mV$_{pk}$. Any guitar could easily overdrive this amp.

The Power Supply

The power supply is shown in Fig. 7.21. A silicon diode rectifier was used to keep cost low. The 269EX can supply 72 mA from the HV winding and 2 A from the 6.3 V winding, which makes it a good match for this amp. A neon power indicator was used in this circuit, but the 6.3 V winding has plenty of capacity to drive an incandescent lamp instead.

This power supply is fairly simple. Resistor R_2 limits initial charging current, protecting the diodes at turn-on. Resistor R_4 is a bleeder resistor that discharges the filter capacitors when power is turned off.

Efficiency

We've mentioned efficiency a number of times throughout the book, so let's take a moment and actually determine the efficiency of this amplifier. Efficiency η is the ratio of useful output power to total power delivered from the power supply. Expressed as a percentage, this is

$$\eta = \frac{P_{out}}{P_{in}} 100\%$$

The total power supplied by the high-voltage (V_{PP}) section of the power supply is

$$P_{PP} = 37 \text{ mA} \times 270 \text{ V} \cong 10 \text{ W}$$

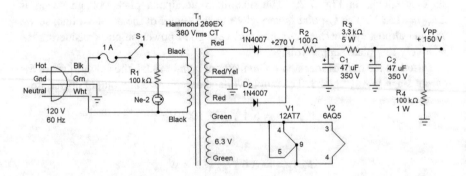

Fig. 7.21 Power supply for the 6AQ5 practice amp

The power dissipation of the tube heaters is

$$P_{\text{heater}} = 6.3 \text{ V}(300 \text{ mA} + 450 \text{ mA}) = 4.7 \text{ W}$$

The approximate total DC power input is

$$P_{\text{in}} = P_{\text{PP}} + P_{\text{heater}} = 14.7 \text{ W}$$

Maximum efficiency will occur when the amplifier is producing maximum unclipped output, which is $P_{\text{o(max)}} = 1.8 \text{ W}_{\text{rms}}$. This gives us an overall efficiency of

$$\eta = \frac{P_{\text{out}}}{P_{\text{in}}} 100\%$$
$$= \frac{1.8 \text{ W}}{14.7 \text{ W}} 100\%$$
$$\cong 12.2\%$$

As you can see, class A amplifiers are not very efficient, especially when implemented with vacuum tubes. Even if we could eliminate the heater power dissipation (e.g., by using transistors), the efficiency only increases to 18%.

The Finished Amplifier

A photo of the partially completed aluminum chassis and the finished amp in the cabinet I built is shown in Fig. 7.22. Terminal strips were used for the point-to-point

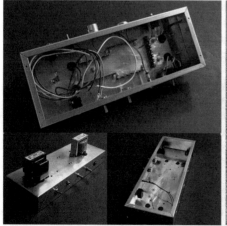

Fig. 7.22 Practice amp chassis and complete amplifier

wiring, but turret board construction could be used as well. The speaker used in this build was a Celestion G10 Greenback, a 10-inch, 30 W speaker which sounds very good with this amp, either clean or distorted.

I originally designed this amp to use either the 6AQ5 or a 6BF5 output tube, using a switched cathode resistor. The 6BF5 is a similar pentode, with the same pinout as the 6AQ5. This option was not included here to simplify the circuit. If you would like to see the alternative version, as well as scope waveforms showing clipping levels and other details, check out the Facebook page for the book.

6L6/EL34 Dual-Mode, SE Amplifier

A more complex amplifier is shown in Fig. 7.23. The voltage gain stages of this amp are implemented using the 12AU7 amplifier designed previously (Fig. 6.7). The output stage is the 6L6GC triode/pentode amp of Fig. 7.16. An EL34 tube may be substituted for the 6L6 with no problems, if you would like to experiment a bit. The amplifier was designed for operation with a power supply voltage $V_{PP} = +350$ V but will operate well with a supply voltage of 300 V or even lower. A suitable power supply is presented in the next section.

As discussed previously, the output stage is switchable for either triode or pentode operation. Power output in the triode mode is about 4 W_{rms}. Pentode-mode output power is 9 W_{rms}. The voltages listed in the schematic were obtained using PSpice to simulate the circuit, with a 6L6GC output tube, and $V_{PP} = +350$ V. These voltages are meant to be used as a general guide for testing and trouble-shooting purposes. Actual circuit voltages may vary by about $\pm 10\%$ using either a 6L6GC or EL34.

The simple single-pot tone control circuit is used again in this amp, but the overall gain of the amplifier is more than sufficient to easily drive the output into clipping using any of the other tone control circuits covered. If greater sensitivity is desired, R_{12} may be reduced to 47 kΩ.

Two volume controls are used on this amplifier. Adjusting R_1 for maximum input signal and adjusting R_{13} for low output drive allow the preamp section to be overdriven, while the output operates at low power. Turning up both R_1 and R_{13} allows the output stage to be overdriven, which produces distinctively different overdrive characteristics. The second volume control, Vol 2, is often called the "Master Volume" control.

The Power Supply

The schematic for the power supply used with this amplifier is shown in Fig. 7.24. The value of the supply voltage V_{PP} depends on which rectifier tube is used in the supply. The 5AR4 produces $V_{PP} \cong 345$ V. A slightly lower supply voltage of about

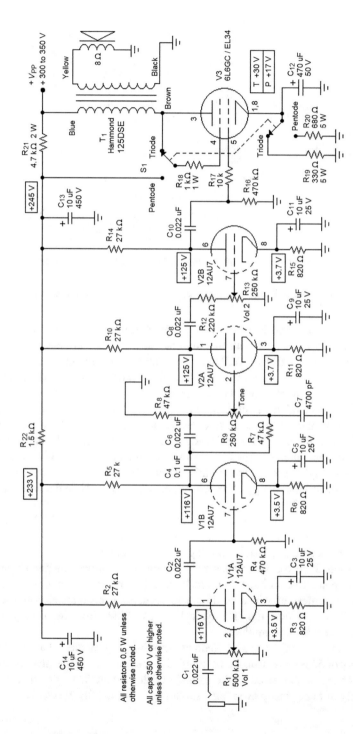

Fig. 7.23 Complete dual-mode (triode/pentode), SE amplifier

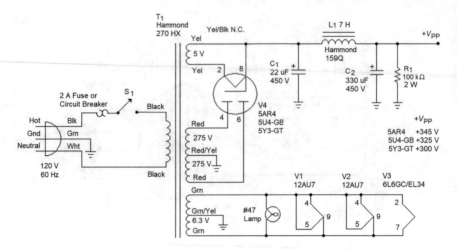

Fig. 7.24 Power supply for the 4 W/9 W single-ended, 6L6-GC/EL34 amplifier

$V_{PP} \cong 325$ V can be obtained using a 5U4-GB. Using a 5Y3-GT results in $V_{PP} \cong$ 300 V.

The 5AR4, 5U4, and 5Y3 have the same pin designations, and any of them can be used in this power supply with the amplifier in Fig. 6.23. In general, you should never substitute a 5AR4 for a 5U4 or 5Y3 unless you are sure the amplifier can tolerate the higher supply voltage. These three tubes also have different heater current requirements that must be considered.

The 5AR4 is a relatively efficient, indirect-heated dual diode. The 5U4 and 5Y3 are both filamentary-cathode, dual diodes. Each diode of the 5U4 can carry 300 mA load current, while the 5Y3 is designed to carry 125 mA per diode. Effectively, the 5Y3 has higher internal resistance (lower perveance) than the 5U4, which is the reason for its higher voltage drop in this circuit. All three rectifier tubes operate with a heater voltage $V_H = 5$ V_{rms}. The 5U4 requires $I_H = 3$ A, while the 5AR4 and 5Y3 both require $I_H = 2$ A.

The 5AR4 can be mounted in any position, while the 5U4 and 5Y3 are designed to be mounted vertically. Horizontal mounting of the 5U4 or 5Y3 can cause the heater filament to sag, possibly shorting the tube internally.

Operation at lower V_{PP} values reduces maximum power output slightly and reduces Q-point power dissipation. I really couldn't tell any difference in maximum loudness regardless of which rectifier was used, so I recommend going with the 5Y3GT for longer tube life. The typical current drawn from the supply is about 100 mA or less for this amplifier.

The center tap of the 6.3 V filament winding is grounded. This reduces the possibility of 60 Hz hum from being coupled to the cathodes of the tubes. Sometimes the center tap is connected to the cathode of the output tube, which reduces the possibility of hum pickup even further, but this is not usually necessary.

Fig. 7.25 Partially constructed 6L6-GC/EL34 SE amplifier

Eliminating the Choke

The 7 H choke (Hammond 159Q) is large, heavy, and fairly expensive (about $35 USD at the time of writing). In order to save some expense and weight, the 159Q may be replaced with a 100 Ω, 5 W resistor with very little effect on performance.

Amplifier Performance

Maximum output power is about 4 W_{rms} in the triode mode and about 9 W_{rms} in pentode mode. This isn't a lot of power, but the amplifier really sounds great when driving a moderately efficient speaker. On a cost per watt basis, this amplifier is very expensive, and if you are looking for lots of power at a lower cost, then you should look at the push-pull amplifiers that are covered in the next section.

Photos of the chassis I used for this amplifier, with tube sockets, transformers, filter capacitors, and the AC line cord installed, are shown in Fig. 7.25. An extra 9-pin tube socket was included so that I could easily add another amplifier section to incorporate a reverb unit.

Adding A Spring Reverb

Reverb is a basic effect that can really enhance the sound of an amplifier. All of the amplifiers presented in this chapter sound good (in my opinion, anyway), but the addition of a reverb tank really brings out the best in these amps.

There are three basic options available for adding reverb to an amp. The most convenient method is to use an external effect stomp box. The other two alternatives are to add a dedicated reverb section to the amplifier using either a spring reverb tank or a solid-state digital reverb module.

From a strictly aesthetic standpoint, I prefer to use reverb springs in old-school tube amp designs. Actually, I prefer not to use any transistors or op amps in tube amp designs either, but these are subjective preferences. This being said, there are definite

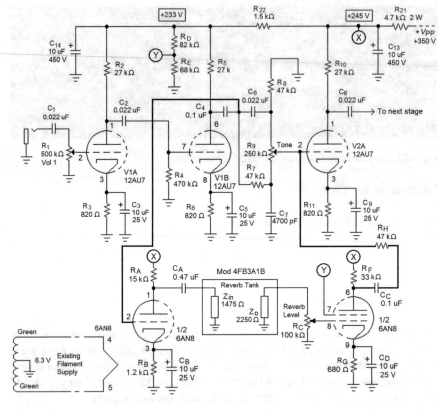

Fig. 7.26 Reverb circuit added to the amplifier of Fig. 7.23

advantages to using a digital reverb module. Digital reverb modules are smaller and easier to fit onto the amp chassis. Digital reverbs are also less likely to pick up hum. When placing a spring reverb tank in an amplifier, you should move the tank around to find a location that produces minimum hum pickup. The main inconvenience of using a digital reverb unit is that a regulated low-voltage DC power supply must be added to the amp.

Take a look at Fig. 7.26. The reverb section is grafted into the existing 6L6GC SE amplifier from Fig. 7.23, in parallel with the tone control section. Because reverb springs usually have rather low input coil impedance, it is common to use transformer coupling to the driver (the triode section of the 6AN8). Because the reverb tank has a reasonably high Z_{in} of 1475 Ω, simple capacitive coupling is used here, avoiding the need for a transformer. The output of the reverb unit is amplified by the pentode section of the 6AN8.

Because of the relatively long distance between the reverb tank and the amp, it is imperative that shielded cables be used to prevent hum. Most reverb tanks have RCA jacks for input and output connectors. You can adapt inexpensive standard shielded audio or video cables to work in this application.

The 6AN8 dual triode/pentode we used in Chap. 6 is well suited for the reverb driver/amp application. The triode section is a medium-mu device, very similar to the 12AU7 with $\mu = 21$, $g_m = 4.5$ mS, and $r_P = 4.7$ kΩ. The relatively low plate resistance of the 6AN8 triode allows this tube to drive the low-impedance reverb tank reasonably well. The 4FB3A1B reverb spring used here only requires a few milliamps of drive current, which is easily handled by the 6AN8 triode.

The pentode section of the 6AN8 provides high voltage gain ($\mu = 1241$), which is important because the output of the reverb tank is typically not much greater than 15 or 20 mV$_{P-P}$. In this circuit, the voltage gain of the 6AN8 pentode should be over 100. The output of the pentode is capacitively coupled to the grid of V2A. Resistor R_H is used to help prevent the 6AN8 pentode from loading down the tone control circuit.

The Reverb Tank

In the previous edition of this book, the reverb tank I used in this design was a Belton BS3EB2C1B. I used this tank because it was what I happened to have on hand and it worked pretty well. The BS3EB2C1B has $Z_{in} = 800$ Ω and $Z_{out} = 2575$ Ω. This Z_{in} is a bit low, and the 6AN8 has to work pretty hard to drive it. In order to improve the performance of the reverb and reduce the load on the 6AN8 triode driver, a Mod 4FB3A1B reverb tank was substituted. The 4FB3A1B has $Z_{in} = 1475$ Ω which the 6AN8 was able to drive more efficiently. The output impedance of the 4BF3A1B is 2250 Ω, which is essentially equivalent to the previous Belton reverb tank (2575 Ω). Reverb performance was noticeably improved with this change.

Reverb Design Considerations

When I designed the reverb circuit for these amps, I was a bit worried about the possibility of a positive feedback path being created through the tone control network, which could result in oscillation. In terms of phase, the output of the reverb spring is impossible to predict; there are just too many unknown variables that affect this signal, including spring position and nonlinearities, drive level, tube gain, tone control settings, the notes being played, and so on. Because of this, I wasn't sure if stability would be an issue.

The possibility of instability exists because the tone control network is not *unilateral*. A unilateral circuit allows the flow of signal energy in one direction only. Tubes (and transistors) are essentially unilateral devices. Driving the grid of a tube produces a large signal at the plate terminal, but applying a signal to the plate will not produce an appreciable output at the grid.

I thought to myself what if the addition of the reverb had made the amp unstable? What would I do to correct the problem? The solution is fairly simple; include a unilateral device in the signal path. If the input to the reverb driver is taken from the wiper of volume pot R_4, then triode V1B makes the tone control network unilateral, and the output of the reverb tank circuit will not feedback around to its input. The problem with this idea is that the signal level at this point is much smaller than at the output of V1B, which would not drive to reverb spring sufficiently. Fortunately, the reverb works with no problem as it is, but I thought this was a good place to provide some additional insight into the design process and to introduce the concept of the unilateral circuit.

EL84, 4 Watt, SE Amplifier

A complete 4 W, SE amplifier based on the EL84 and 12AX7 designs used previously is shown in Fig. 7.27. The 12AX7 provides high voltage gain, which allows a single preamp/driver tube to be used even with a high loss tone control.

The total gain of the 12AX7 stages, including the loss incurred by the Baxandall tone control circuit, is about 170. The EL84 stage requires about 10 V_{P-P} input for full output, so an input signal of about 60 mV_{P-P} will drive the amp into clipping, with both volume controls set to maximum. The dual volume controls allow overdrive of the input stage, the output stage, or a combination of both.

Adding a Reverb

The 4W, EL84 amplifier with reverb added is shown in Fig. 7.28. This is similar to the modification that was done to the 6L6 amplifier. Notice that capacitors C_2 and C_8 have been increased to 0.1 μF. This was done to compensate for the loading effect caused by adding the reverb.

The Power Supply

A power supply that works well with the EL84 amplifier is shown in Fig. 7.29. Solid-state diodes are used in the rectifier to save the cost of a larger power transformer and tube diode. This power supply is nearly identical to the supply in Fig. 7.21. As before, the function of R_2 is to limit the initial charging current of C_1 when power is first applied. This resistor is not needed when a tube rectifier is used because the filter cap would charge gradually as the diode tube heated up.

Resistors R_5 and R_6 are used to cause the ends of the 6.3 V winding of the transformer to be symmetrical about ground. This is called *hum balancing*. Hum balancing reduces coupling of the AC filament voltage to the cathodes of the tubes, which in turn reduces the likelihood of filament supply-induced hum. These resistors are not usually necessary, and they may be omitted if desired.

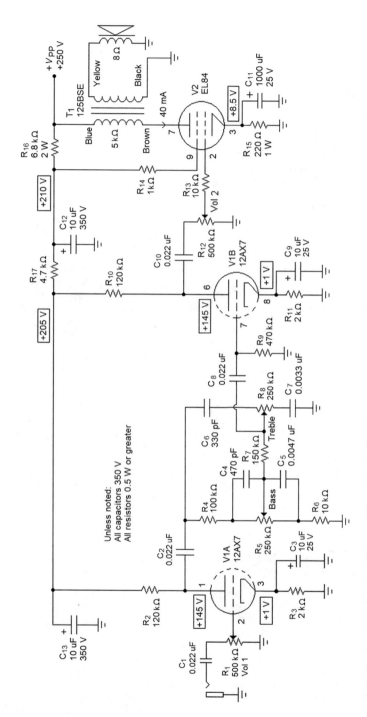

Fig. 7.27 Single-ended, class A, 4 Watt amplifier using the EL84

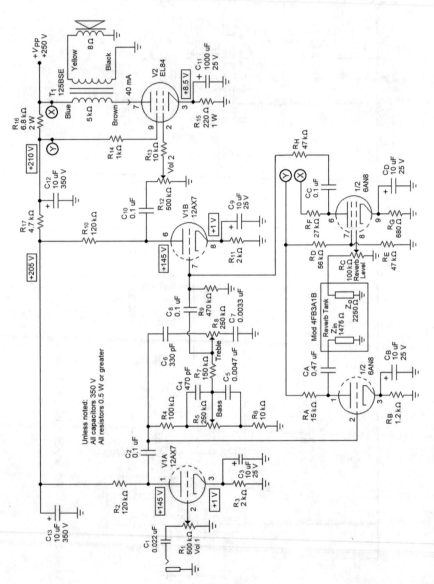

Fig. 7.28 EL84 pentode, 4 Watt, SE amplifier with optional reverb

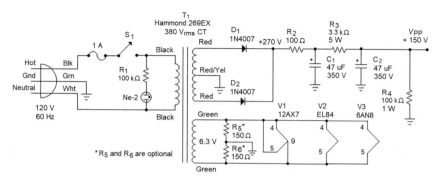

Fig. 7.29 Power supply for the EL84 amplifiers of Figs. 7.27 and 7.28

Class A, Single-Ended Amp Distortion

Like NPN transistors, class A biased tubes have greater sensitivity to positive-going input signals than to negative-going signals. This asymmetry means that the transfer function of a single-ended amplifier (tube or transistor) will have neither even nor odd symmetry. Both even and odd harmonics will be created in the output signal, which is usually perceived as a warm sounding distortion.

Push-Pull Amplifiers

If you are after real power, the push-pull output stage is the way to go. In this section, we examine the basic operation of the cathode-biased, class A, push-pull stage. A basic push-pull stage, driven by a cathodyne phase splitter, is shown in Fig. 7.30.

Basic DC Operation

The output stage of Fig. 7.30 is a cathode-biased, class A, push-pull amplifier. Class A biasing is rarely, if ever used in solid-state push-pull stages but it is quite common in vacuum tube amplifier designs. In this circuit, both output tubes are operated at the same, relatively high quiescent plate current ($I_{PQ1} = I_{PQ2} = I_{PQ}$). The primary winding of the output transformer carries a large DC current which is

$$I_{DC} = 2I_{PQ} \tag{7.27}$$

In the push-pull circuit, the primary current I_{DC} splits evenly between opposite halves of the winding. Because the plate currents are flowing in opposite directions through the respective halves of the primary, there is cancellation of the magnetic

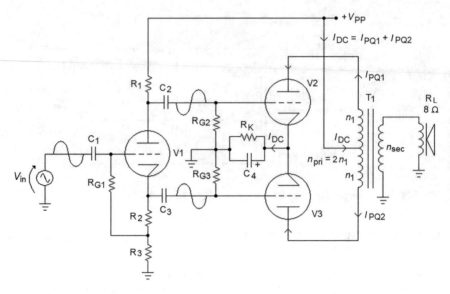

Fig. 7.30 Cathode-biased, class A, push-pull output stage with phase splitter drive

flux, and the core does not bias toward saturation. This allows us to use a smaller, lighter, more efficient, and less expensive output transformer than in an equivalent SE amplifier (see Table 6.3).

Basic AC Operation

When a signal is applied to the input of the phase splitter, it produces output signals of equal amplitude but opposite phase which are applied to the tubes of the push-pull stage. As the grid of tube V2 is driven positive, its plate current increases. At the same time, the grid of tube V3 is driven negative, decreasing its plate current. This creates an unbalanced current flow through the two halves of the primary winding, which couples to the secondary winding producing an output signal. When the polarity of the phase splitter output reverses, the grid of V2 goes negative which decreases its plate current, while the grid of V3 goes positive increasing its plate current.

If the push-pull tubes are matched and they behave linearly, the circuit will have perfect symmetry, and since this is a class A stage, the total DC primary current will remain constant (if I_{P1} increases by say 10%, I_{P2} will decrease by 10% and vice versa). If the circuit behaved ideally, the output would be a scaled, exact replica of the input signal, as long as the amp was not overdriven. In practice, the tubes won't be perfectly matched, and they are not close to being linear, especially when driven to their limits, which is likely to occur when this circuit is in normal use.

Because the push-pull stage is biased for class A operation, there will be no crossover distortion. Class A biasing also means that the efficiency of this stage will not be as high as for a class B or class AB design, because the output tubes have high P_{DQ}.

AC Analysis of the Push-Pull Output Transformer

The analysis of the push-pull stage is a bit tricky. We will assume that the tubes are matched and behave linearly. Using the simplified equivalent circuit of Fig. 7.31, the plate-to-plate reflected load resistance across the entire primary winding R'_{PP} is given by (7.7), which is repeated here. This is the primary impedance that manufacturers list on push-pull transformer data sheets (see Table 7.2).

$$R'_{PP} = R_L \left(\frac{n_P}{n_S} \right)^2 \tag{7.7}$$

Since this stage is biased class A1 and the tubes are operating symmetrically in series, there should be no AC signal current flow out the center tap. The effective reflected load resistance R'_L "seen" by each plate is half of the rated plate-to-plate primary resistance. That is,

Fig. 7.31 Simplified equivalent push-pull output stage

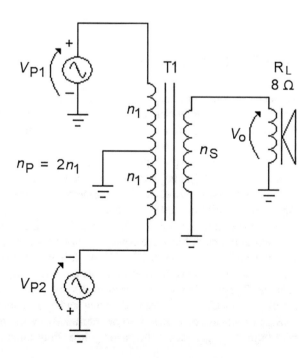

$$R'_L = \frac{R'_{PP}}{2} \tag{7.28}$$

This equivalent resistance is factored into the construction of the AC load line for a given tube in the push-pull stage.

Interestingly, if the output stage was biased for class B operation, the reflected load resistance seen by each plate would be $R'_{PP}/4$. This happens because the center tap would carry the full AC signal current, effectively halving the turns ratio of the transformer.

The output voltage is given by

$$V_o = (V_{P1} + V_{P2})\left(\frac{n_S}{n_P}\right)$$

Assuming that the tubes are matched, $V_{P1} = V_{P2} = V_P$, and we can write

$$V_o = 2V_P\left(\frac{n_S}{n_P}\right) \tag{7.29}$$

And finally, the effective turns ratio for each half of the primary is

$$\frac{n_1}{n_S} = \frac{1}{2}\left(\frac{n_P}{n_S}\right) \tag{7.30}$$

Push-Pull Class A Distortion

The class A, push-pull stage does not exhibit crossover distortion. However, the nonlinear transconductance of the output tubes will still cause distortion (the squashing-stretching effect). But because of the symmetry of the circuit, the even-order harmonics tend to cancel. Most of the distortion that does occur consists of odd harmonics because the push-pull amplifier transfer function has odd symmetry. The more closely the tubes are matched, the more perfect the symmetry of the transfer function of the push-pull stage. This is generally preferred in high-fidelity amplifiers, but not necessarily in guitar amplifiers.

Sometimes different tubes will be used in each half of the push-pull stage, or different numbers of tubes are connected in parallel in each half of the circuit. For example, an EL34 may be used for V1, while a 6L6GC may be used for V2. Another example would be to operate one tube as a triode, while the other is operated as a pentode. These techniques are used to disrupt the odd symmetry of the transfer function, producing more even harmonics and a nicer sounding distortion.

Significant distortion may also be introduced by the phase splitter as well. It is very difficult to get exactly the same gain at both outputs of the cathodyne phase splitter, which causes asymmetrical operation of the output tubes. The unequal output resistances of the common cathode and cathode follower outputs of the phase splitter can also contribute to asymmetry.

EL34 Triode-Mode, Push-Pull Design Example

Since we used the 6L6GC first in the single-ended amp section, let's start with the EL34 this time. We will base this design on the circuit of Fig. 7.30, with $V_{PP} = 350$ V and $R_L = 8\ \Omega$.

Q-Point Location and Determination of R_K

Let's begin by choosing the Q-point power dissipation for the output tubes. Since the maximum DC power dissipation rating of the EL34 is 25 W, setting $P_{DQ} = 20$ W is a reasonable goal. The quiescent plate current for each tube is found using (6.23).

$$I_{PQ} = \frac{P_{DQ}}{V_{PP}}$$
$$= 20\ \text{W}/350\ \text{V}$$
$$\cong 57\ \text{mA}$$

So, our Q-point values for each tube are

$$V_{PQ} = 350\ \text{V},\ I_{PQ} = 57\ \text{mA},\ P_{DQ} = 20\ \text{W}$$

Plot the Q-point on the EL34 triode-mode plate curves as shown in Fig. 7.32. There is no grid voltage curve at the Q-point location, but we can graphically estimate the grid voltage to be

$$V_G = -27\ \text{V}$$

The total current carried by the transformer primary winding is

$$I_{DC} = 2I_{PQ}$$
$$= 2 \times 57\ \text{mA}$$
$$= 114\ \text{mA}$$

The required cathode resistance is

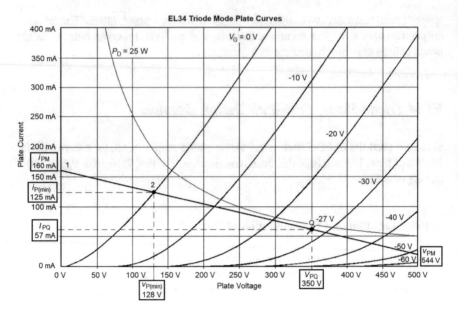

Fig. 7.32 AC load line for the EL34 triode-mode, push-pull amplifier stage

$$R_K = \frac{-V_G}{I_{DC}}$$
$$= \frac{27 \text{ V}}{114 \text{ mA}} \qquad (7.31)$$
$$= 237 \, \Omega \ (\text{use } 270 \, \Omega)$$

We will use the next higher standard value $R_K = 270 \, \Omega$ which will run the tubes a bit cooler. The power dissipation of the cathode resistor is

$$P_{RK} = I_{DC}^2 R_K$$
$$= 114 \text{ mA}^2 \times 270 \, \Omega$$
$$= 3.5 \text{ W (use 5 W)}$$

Output Transformer Selection

We have determined $I_{PQ} = 57$ mA for each of the tubes in the amplifier ($I_{DC} = 114$ mA). Looking at Table 7.1, the 125D is rated for 55 mA per side, which is slightly underrated for this application. Further analysis will show that this amplifier is capable of producing maximum output power of about 11 W$_{rms}$ into an 8 Ω load. The 125E, which is rated at 60 mA/side and 15 W$_{rms}$, is the best choice for this amplifier.

From Table 6.1, the typical recommended reflected load resistance for the EL34 operating as a triode is $R'_L = 3.5$ kΩ. Table 7.2 lists reflected load resistances across the entire primary winding, but taking into account the relationship we derived in (7.28), the desired plate-to-plate resistance is

$$
\begin{aligned}
R'_{PP} &= 2R'_L \\
&= 2 \times 3.5\text{k}\Omega \\
&= 7\text{k}\Omega
\end{aligned}
\tag{7.32}
$$

Consulting Table 7.2, we find that connecting 125E lugs 3 and 6 to an 8 Ω speaker will give us $R'_{PP} = 6.8$ kΩ, which is close enough. The actual reflected load resistance seen by the plate of each EL34 will be

$$
\begin{aligned}
R'_L &= R'_{PP}/2 \\
&= 6.8\text{k}\Omega/2 \\
&= 3.4\text{k}\Omega
\end{aligned}
\tag{7.33}
$$

The AC Load Line

We will now plot the AC load line for a single output tube using the same basic procedure that was used when we designed the SE amp stages. The load line data obtained here is used to tell us the overall behavior of the push-pull stage.

Determine the maximum plate voltage using (7.13).

$$
\begin{aligned}
v_{PM} &= V_{PQ} + I_{PQ}R'_L \\
&= 350\text{V} + (57\text{mA} \times 3.4\text{k}\Omega) \\
&= 350\text{V} + 194\text{V} \\
&= 544\text{V}
\end{aligned}
$$

Determine the maximum plate current using (7.14).

$$
\begin{aligned}
i_{PM} &= I_{PQ} + \frac{V_{PQ}}{R'_L} \\
&= 57 \text{ mA} + \frac{350 \text{ V}}{3.4k\Omega} \\
&= 57 \text{ mA} + 103 \text{ mA} \\
&= 160 \text{ mA}
\end{aligned}
$$

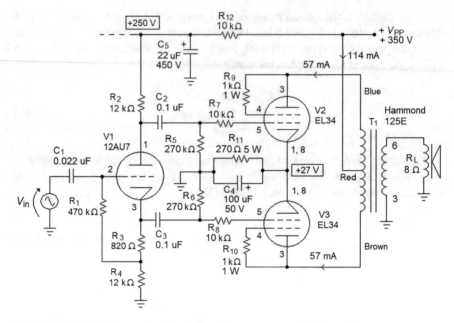

Fig. 7.33 EL34, 11 Watt, push-pull, triode-mode output stage with phase splitter

As usual, the AC load line extends from i_{PM} to v_{PM}, passing through the Q-point. In this example, v_{PM} is off the scale, but the maximum plate current is on the scale. Place a mark at $i_{PM} = 160$ mA (Point 1 in Fig. 7.32).

Draw the load line from $I_{PM} = 160$ mA through the Q-point. Visualizing the projection of the load line to the V_P axis, it seems reasonable that $v_{PM} \cong 544$ V.

The complete power output stage is shown in Fig. 7.33. The 12AU7 phase splitter of Fig. 6.19 is used in this application. Note the use of gamma network resistors R_7 and R_8 (the grid stoppers) that help prevent parasitic oscillation. It is also assumed that this will be part of a larger, complete amplifier, so the V_{PP} supply rail is extended to the left. Resistor R_{12} may have a different value, depending on how much current is drawn by preceding stages, which are not shown here.

AC Characteristics of the Amplifier

Let's start by determining the turns ratio of the 125E output transformer.

$$\frac{n_P}{n_S} = \sqrt{\frac{R'_{PP}}{R_L}}$$

$$= \sqrt{\frac{6.8\,k\Omega}{8\,\Omega}}$$

$$= 29.2$$

The maximum unclipped AC plate voltage for a single tube is the smaller of ΔV_{PL} and ΔV_{PR}. Moving along the load line to the left until $V_G = 0$ V (Point 2), we find

$$\begin{aligned}
\Delta V_{\text{PL}} &= V_{\text{PQ}} -- v_{\text{P(min)}} \\
&= 350\text{V} -- 128\text{V} \\
&= 222\text{V}
\end{aligned}$$

Graphically, moving along the AC load line from the Q-point to the right (or by using $\Delta V_{\text{PR}} = I_{\text{PQ}} R'_{\text{L}}$), the maximum plate voltage swing ΔV_{PR} is

$$\begin{aligned}
\Delta V_{\text{PR}} &= v_{\text{PM}} -- V_{\text{PQ}} \\
&= 544\text{V} -- 350\text{V} \\
&= 194\text{V}
\end{aligned}$$

The smaller of ΔV_{PL} and ΔV_{PR} still serves as the limit of the peak plate voltage swing for each tube, which in this case is

$$\Delta V_{\text{P}} = \Delta V_{\text{PR}} = 194 \text{ V}$$

Because the tubes are effectively working series-aiding, the total primary winding voltage swing is twice the voltage of a single tube. Slightly modifying (7.29), we have

$$\begin{aligned}
V_{\text{o(max)}} &= 2\Delta V_{\text{P}} \left(\frac{n_S}{n_P} \right) \\
&= 2 \times 194 \text{ V} \times \frac{1}{29.2} \\
&= 13.3 \text{ V}_{\text{pk}}
\end{aligned} \tag{7.34}$$

The maximum output power is found using (7.18).

$$\begin{aligned}
P_{\text{o(max)}} &= \frac{V_{\text{o(max)}}^2}{R_{\text{L}}} \\
&= \frac{13.3\text{V}^2}{8\Omega} \\
&= 22\text{W}_{\text{pk}}
\end{aligned} \tag{7.18}$$

Assuming a sinusoidal output signal, the RMS output power is

$$\begin{aligned}
P_{\text{rms}} &= P_{\text{o(pk)}}/2 \\
&= 22 \text{ W}/2 \\
&= 11 \text{ W}_{\text{rms}}
\end{aligned}$$

The voltage gain from grid to plate of each tube is found using the familiar equation

$$
\begin{aligned}
A'_V &= g_m(R'_L \parallel r_P) \\
&= 4.7\text{mS}(3.4\text{k}\Omega \parallel 1.7\text{k}\Omega) \\
&= 4.7\text{mS} \times 1.13\text{k}\Omega \\
&= 5.3
\end{aligned}
\tag{7.35}
$$

For each EL34, the required AC grid input voltage for full output is

$$
\begin{aligned}
V_{G(pk)} &= \frac{\Delta V_P}{A'_V} \\
&= \frac{194\text{V}}{5.3} \\
&= 36.6\text{V}_{pk}
\end{aligned}
\tag{7.36}
$$

Remember, the voltages at the output terminals of the phase splitter move equally in opposite directions. At maximum output, the plate voltage of the 12AU7 will change by ± 36.6 V, while the cathode voltage changes by ∓ 36.6 V.

Recall from back in Chap. 6 that the gain of the typical cathodyne phase splitter is

$$
A_{V(PS)} \cong 0.7 \quad \text{(typical)}
$$

For maximum unclipped output power, the peak voltage required at the input of the phase splitter is given by (7.37)

$$
\begin{aligned}
V_{in} &= V_{G(pk)}/A_{V(PS)} \\
&= 36.6\text{ V}/0.7 \\
&= 52.3\text{ V}_{pk} = 104.6\text{ V}_{P-P}
\end{aligned}
\tag{7.37}
$$

This is the required output voltage compliance of the stage driving the phase splitter.

We will now determine the overall voltage gain of the cathodyne and push-pull stage. The turns ratio of the output transformer was determined to be

$$
n_p/n_s = 29.2
$$

We have already determined the grid-to-plate voltage gain of the EL34s, which was $A'_V = 5.3$. This is stepped down by the transformer to an effective push-pull voltage gain $A_{V(PP)}$ of

$$A_{V(PP)} = A'_V \left(\frac{n_S}{n_P}\right)$$
$$= 5.3 \left(\frac{1}{29.2}\right)$$
$$= 0.182$$

The overall gain of this stage is

$$A_{V(overall)} = A_{V(PP)} A_{V(PS)}$$
$$= 0.182 \times 0.7$$
$$= 0.127$$

The Cathode Bypass Capacitor

The ideal class A, push-pull output stage has perfectly matched tubes with linear dynamic characteristics (constant g_m, μ, and r_P). We assume the transformer is perfectly linear as well. If this is the case, the output tubes operate with perfect symmetry, and the cathode resistor (R_{11} in this circuit) only carries the DC bias current. There should be no AC signal current flow through R_{11}, and so ideally the bypass cap serves no purpose.

In a real circuit, we don't have perfect matching, and the tubes are definitely not linear. The output transformer will not be perfect either, but is usually not as nonlinear as the tubes. These factors cause unbalanced operation of the tubes, which in turn causes an AC signal current to flow through the bypass capacitor to ground. If the bypass cap was not used, some negative feedback would be produced, slightly reducing the gain of the stage and producing slightly more linear operation.

So, we see that for an ideal class A, push-pull stage, the cathode bypass capacitor serves no function, but in a real circuit, the bypass cap maximizes gain at the expense of slightly greater distortion. This begs the question of whether the bypass cap is even necessary. I have listened to amplifiers with and without the bypass cap, and I couldn't detect any audible difference in performance. Although the push-pull stages I've used here include cathode bypassing, this feature can be omitted if desired.

This could be an interesting aspect of amplifier design to explore further. For example, if you were to design an amplifier using two different output tubes, say a 6L6 and an EL34, then the presence of the bypass capacitor might have a significant effect on distortion and overdrive characteristics because of the extreme asymmetry of the stage. Just something to think about.

A Complete EL34 Triode-Mode Amplifier

Any of the common cathode amplifiers designed in Chap. 6 could easily drive this output stage into clipping. An example of a complete amplifier using this phase splitter/push-pull section is shown in Fig. 7.34. This circuit uses the Marshall-type tone control, but any tone control you prefer could be substituted. The sensitivity of this amplifier is very high, so cathode bypass capacitors are not used on triodes V1B and V2A.

If you would like to experiment, this amplifier can be modified in many different ways. For example, you could bypass the cathodes of V1B and V2A, or you might consider using 12AT7-based voltage gain stages in place of the 12AX7. You might also consider redesigning the phase splitter to use a differential pair.

A suitable power supply for this amplifier is shown in Fig. 7.35. Either the 5AR4 or the 5U4GB dual diodes may be used. The 5AR4 will deliver close to +350 V at the V_{PP} supply rail, which matches our design assumptions. Use of the 5U4GT results in a supply rail voltage of about +325 V which will work just fine, producing slightly lower maximum output power and running the tubes just a bit cooler.

6L6GC Pentode-Mode, Push-Pull Design Example

As was the case for the single-ended amp designs, operation in the pentode mode results in much greater output power from a given push-pull amplifier. Most of this design example will be identical to the previous example, so things will move more quickly this time. Let's assume the power supply voltage and load are

$$V_{PP} = 350 \text{ V}, \ R_L = 8 \ \Omega$$

Q-Point Location and Determination of R_K

The maximum power dissipation rating of the 6L6GC is 30 W. Setting $P_{DQ} = 25$ W, the quiescent plate current for each tube is found using (7.23).

$$\begin{aligned} I_{PQ} &= \frac{P_{DQ}}{V_{PP}} \\ &= 25 \text{ W}/350 \text{ V} \\ &\cong 71 \text{ mA} \end{aligned}$$

In summary, the Q-point values for each tube are

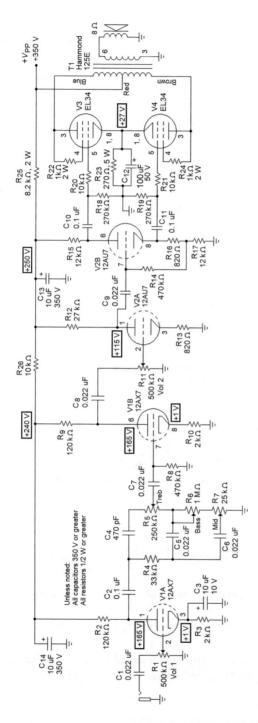

Fig. 7.34 Complete 11 Watt, 6L6GC triode-mode, push-pull amp with tone control

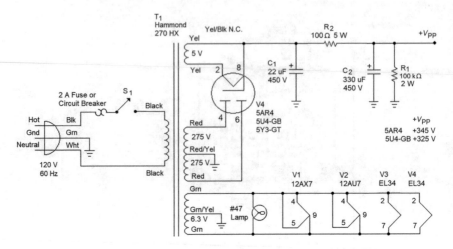

Fig. 7.35 Suggested power supply for the amplifier of Fig. 7.34

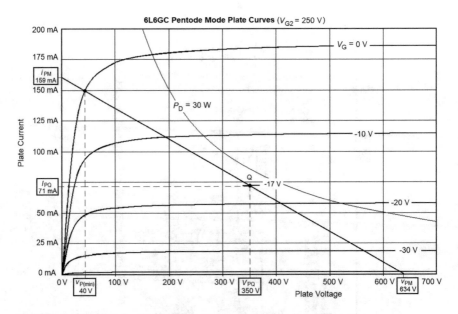

Fig. 7.36 AC load line for the 6L6GC pentode-mode, push-pull amplifier stage

$$V_{PQ} = 350 \text{ V}, I_{PQ} = 71 \text{ mA}, P_{DQ} = 25 \text{ W}$$

Plot the Q-point on the 6L6GC pentode-mode plate curves, as shown in Fig. 7.36. The Q-point is within the SOA, but it didn't fall on a grid voltage curve, so we estimate

$$V_G = -17 \text{ V}$$

The total cathode current, flowing in the primary winding, is

$$I_{DC} = 2I_{PQ}$$
$$= 2 \times 71 \text{ mA}$$
$$= 142 \text{ mA}$$

The required cathode resistance is

$$R_K = \frac{-V_G}{I_{DC}}$$
$$= \frac{17 \text{ V}}{142 \text{ mA}}$$
$$= 120 \text{ } \Omega$$

The power dissipation of the cathode resistor is

$$P_{RK} = I_{DC}^2 R_K$$
$$= 142 \text{ mA}^2 \times 120 \text{ } \Omega$$
$$= 2.42 \text{ W (use 5 W)}$$

Output Transformer Selection

From Table 6.1, the typical recommended reflected load resistance for the 6L6GC operating as a pentode is $R'_L = 4.2$ kΩ. Using (7.32), the optimal plate-to-plate reflected resistance is

$$R'_{PP} = 2R'_L$$
$$= 2 \times 4.2\text{k}\Omega$$
$$= 8.4\text{k}\Omega \text{ (Recommended value from data sheet)}$$

We have determined $I_{PQ} = 71$ mA for each of the tubes in the amplifier ($I_{DC} = 142$ mA). Consulting Table 7.2, we find that the 1750J (40 W, $R'_{PP} = 8$ kΩ for an 8 Ω load) is the suitable choice for this amplifier. Using this transformer, the actual reflected load resistance seen by the plate of each 6L6 is given by (7.33).

$$R'_L = R'_{PP}/2$$
$$= 8\text{k}\Omega/2$$
$$= 4\text{k}\Omega \text{ (Actual 1750J transformer value)}$$

The AC Load Line

We will now plot the AC load line for a single output tube. Determine the maximum plate voltage using (7.13).

$$v_{PM} = V_{PQ} + I_{PQ}R'_L$$
$$= 350 \text{ V} + (71 \text{ mA} \times 4 \text{ k}\Omega)$$
$$= 350 \text{ V} + 284 \text{ V}$$
$$= 634 \text{ V}$$

Determine the maximum plate current using (7.14).

$$i_{PM} = I_{PQ} + \frac{V_{PQ}}{R'_L}$$
$$= 57 \text{ mA} + \frac{350 \text{ V}}{4k\Omega}$$
$$= 71 \text{ mA} + 88 \text{ mA}$$
$$= 159 \text{ mA}$$

In Fig. 7.36, the AC load line is plotted from $v_{PM} = 634$ V to $i_{PM} = 159$ mA, and we find that it intersects the Q-point as required.

The complete output stage is shown in Fig. 7.37. As in the previous amplifier, the 12AU7 phase splitter of Fig. 6.19 is used here.

AC Characteristics of the Amplifier

As before, we begin by determining the turns ratio of the output transformer. Using (7.8), we obtain

$$\frac{n_P}{n_S} = \sqrt{\frac{R'_{PP}}{R_L}}$$
$$= \sqrt{\frac{8 k\Omega}{8\Omega}}$$
$$\cong 32$$

The maximum unclipped AC plate voltage limits are

$$\Delta V_{PL} = V_{PQ} -- v_{P(\min)}$$
$$= 350V -- 40V$$
$$= 310V$$

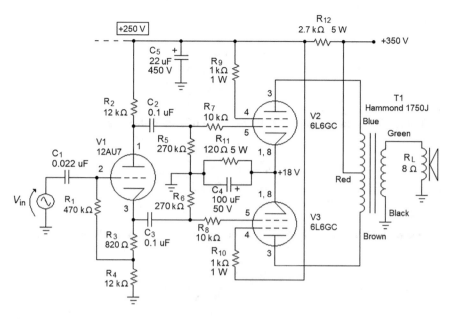

Fig. 7.37 6L6GC, 20 Watt, push-pull pentode-mode amp with phase splitter

$$\Delta V_{PR} = v_{PM} -- V_{PQ}$$
$$= 634V -- 350V$$
$$= 284V$$

The maximum peak, unclipped plate voltage swing is

$$\Delta V_P = \Delta V_{PR} = 284 \text{ V}$$

Using (7.34), we determine the maximum unclipped output voltage to be

$$V_{o(max)} = 2\Delta V_P \left(\frac{n_S}{n_P}\right)$$

$$= 2 \times 291 \text{ V} \times \frac{1}{32}$$

$$= 18 \text{ V}_{pk}$$

The peak output power is

$$P_{o(max)} = \frac{V_{o(max)}^2}{R_L}$$

$$= \frac{18^2}{8}$$

$$\cong 40W_{pk}$$

The RMS output power is

$$P_{\text{rms}} = P_{\text{o(pk)}}/2$$
$$= 40 \text{ W}/2$$
$$= 20 \text{ W}_{\text{rms}}$$

Using (7.35), the grid-to-plate voltage gain of each 6L6GC is

$$A'_V = g_m(R'_L \parallel r_P)$$
$$= 5.2\text{mS} \times (4\text{k}\Omega \parallel 33\text{k}\Omega)$$
$$= 5.4\text{mS} \times 3.6\text{k}\Omega$$
$$\cong 19.4$$

For each 6L6, the required AC grid voltage for full output is given by (7.36).

$$V_{G(\text{pk})} = \frac{\Delta V_P}{A'_V}$$
$$= \frac{284 \text{ V}}{19.4}$$
$$= 14.6 \text{ V}_{\text{pk}}$$

Assuming that $A_{V(\text{PS})} \cong 0.7$, the peak voltage required at the input of the phase splitter for full output is

$$V_{\text{in}} = V_{G(\text{pk})}/A_{V(\text{PS})}$$
$$= 14.6 \text{ V}/0.7$$
$$\cong 21V_{\text{pl}} = 42V_{\text{P-p}}$$

The grid-to-plate gain of each 6L6 is $A'_V = 19.4$, which is stepped down by the transformer to

$$A_{V(\text{PP})} = A'_V \left(\frac{n_s}{n_p}\right)$$
$$= 19.4\left(\frac{1}{32}\right)$$
$$= 0.606$$

The overall gain of this stage is

$$A_{V \text{ (overall)}} = A_{V(PP)}A_{V(PS)}$$
$$= 0.606 \times 0.7$$
$$= 0.424$$

A Complete 6L6GC Amplifier

The amplifier in Fig. 7.38 incorporates the 6L6GC/12AU7 pentode push-pull stage with phase splitter, along with the 12AX7 voltage gain stages we used previously. The tone control used is the Baxandall type, which as usual may be replaced with an alternative circuit of your own choosing. The reverb may be eliminated to simplify the circuit if desired.

The power supply of Fig. 7.35 may be used with this amplifier (don't forget to add heater connections for the 6AN8, if used). Alternatively, the power supply of Fig. 7.39 may be used. This power supply uses silicon diodes in place of the tube rectifier to lower costs.

Standby Operation

The power supply of Fig. 7.39 also shows an optional standby switch for the high-voltage supply rail. Standby switching, which is common in many guitar amps, is a somewhat controversial topic. The idea is that allowing the tube filaments to heat up before turning on the high-voltage supply will help to prevent *cathode stripping* from occurring in the tubes. Cathode stripping is the erosion of cathode emission-enhancing coating caused by a high anode voltage being present when filament heating occurs. In reality, cathode stripping is only potentially significant at DC plate voltages of well over 1 kV. The main use of the standby switch is to quiet the amplifier for short periods of time while allowing the tube heaters to remain in operation.

Switching the V_{PP} rail of an amplifier that uses a choke filter in the power supply can cause very high voltage transients to occur when the standby mode is switched on. What happens is that when the standby switch is opened, the relatively large supply current is suddenly reduced to zero which causes the magnetic field around the choke winding to collapse. The collapsing field induces a high-voltage spike across the choke winding that adds in series with the output of the rectifier. This phenomenon is often called "inductive kick." Transients of several thousand volts across the choke can be produced which can cause arcing across the standby switch or even break down the insulation around the choke winding. Because of this, I would not recommend using standby switching on supplies that use a choke in the filter. The inductive kick effect is more likely to cause high-voltage spikes in

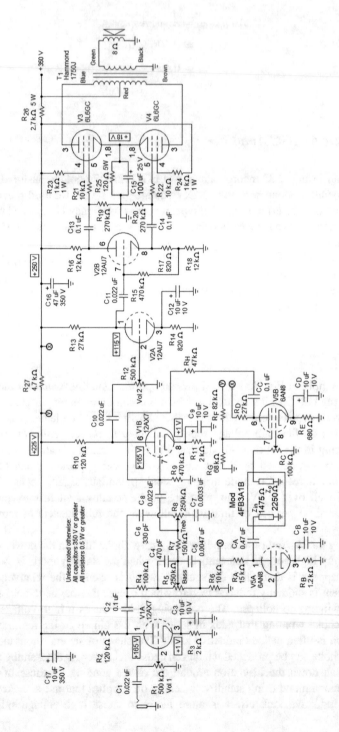

Fig. 7.38 Complete 20 Watt, push-pull amp with reverb

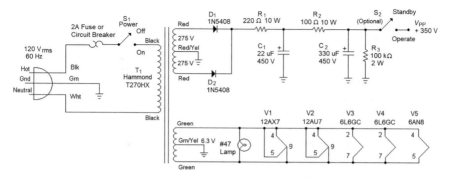

Fig. 7.39 Power supply for the amplifier of Fig. 7.38

amplifiers that use class A biased power output stages because they have very high quiescent current draw, which creates high flux density in the choke.

6V6 Amplifier with Effects Loop

I wanted to design a new amplifier, and I happened to have a P-TF41316 power transformer in a box of tube-related stuff I had collected. This transformer is used in the Fender 5E3 chassis, Deluxe Reverb amp. This transformer has three secondaries, with ratings of 660 V_{rms}, CT at 120 mA, 6.3 V_{rms} at 3 A, and 5 V_{rms} at 3 A. Using this as a starting point, I started thinking through the design.

I didn't want to do all the metal working and cabinet making this time, so I ordered a premade 5E3 chassis that fit the transformer and a tweed cabinet. The cabinet is set up for a 12-inch speaker and mounting of the 5E3 chassis. The chassis is nickel plated and pre-drilled/punched for all components. Also, since the chassis is pre-drilled for a turret board, that's the wiring technique used here.

The Effects Loop

The 5E3 chassis has holes for four input jacks, which is more than necessary, but convenient for the inclusion of an effects loop. The effects loop is a common feature that is present on many newer amplifiers. An effects loop allows the insertion of stomp boxes and other effects in between the early stages of an amplifier. The idea behind the effects loop is to operate external effects at signal levels somewhat higher than the instrument output level so that noise is reduced.

Some effects loops output *consumer line level* signals. Others may produce *pro line level output*. Still others may simply output instrument or *stomp level* signals. The standard line level voltages are:

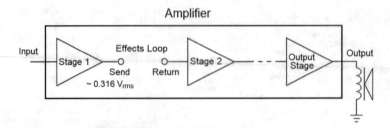

Fig. 7.40 Adding an effects loop to an amplifier

$$\text{Consumer line level} \quad 316\text{mV}_{\text{rmss}}\left(0.447\text{V}_{\text{pk}}\right)$$

$$\text{Pro line level} \quad\quad\quad 1.228\ \text{V}_{\text{rms}}\left(1.736\ \text{V}_{\text{pk}}\right)$$

Aside from overdrive, most stomp-box effects have close to unity gain, so their output is considered instrument or stomp level. Many stomp-box effects are not able to handle inputs much higher than instrument level. These effects should be used as usual, right after the guitar.

The output of the amplifier that drives the effects loop is usually called s*end*. The input to the effects loop is usually called *return*. A block diagram representation for an amplifier with an effects loop is shown in Fig. 7.40. Peak guitar pickup output voltage varies over a range from around 100 to 500 mV$_{\text{pk}}$. For this amplifier, I decided to go with $A_V \cong 2$ prior to the send output, putting the signal close to consumer line level, which should allow operation with many stomp-box effects.

Most often the effects loop send driver is implemented using a cathode follower, which will also usually be a cathodyne-based design. This approach is used because the output resistance of a cathode follower can easily be made close to 600 Ω, which is the standard output resistance for a line driver. The downside here is that cathode followers have less than unity gain and we need to compensate with an additional gain stage. In order to keep the size and complexity of this amp design somewhat reasonable, I decided to sacrifice the low output impedance and go with a common cathode send stage.

Low-Power Sections

The schematic for the amplifier is shown in Fig. 7.41. The first stage is unbypassed, which gives a gain $A_{V(1)} \cong 2$. The level output to the effects loop Send jack is adjusted with R_4. You will have to experiment with your equipment to determine which effects are suitable for use here and what Vol 1 setting works best. If the effects loop is not used, a short patch cord should be jumped between the Send and Return jacks. You could also eliminate the effects loop completely if so desired.

The effects loop returns to the second stage ($A_{V(2)} \cong 15$), which in turn drives the tone control circuit. Just to change things up a bit, I decided to use a different tone

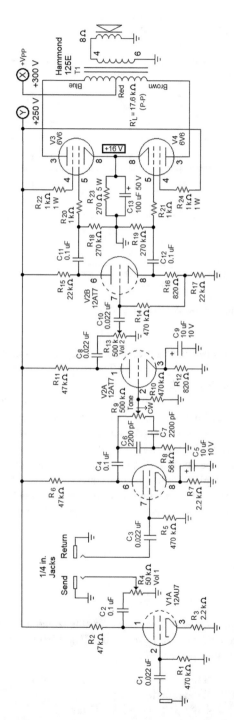

Fig. 7.41 6V6 amplifier with effects loop

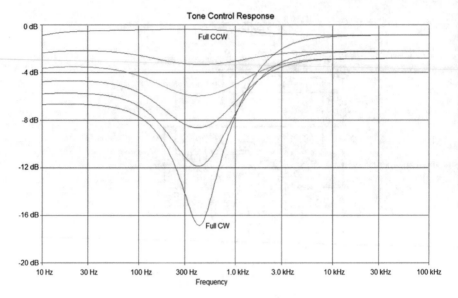

Fig. 7.42 Frequency response of the 6V6 tone control circuit

control circuit here. This tone control produces a scooped midrange response, with an average loss of about 6 dB ($A_{V(\text{tone})} = 0.5$). The frequency response curves for the tone control are shown in Fig. 7.42.

We need to make up for the tone circuit loss and the low gain of the first stage, so the third stage was designed using a 12AT7. This stage has gain $A_{V(3)} = 45$. Since the 5E3 chassis has a position for a second volume control, it was placed after this stage. This control could be eliminated if desired.

The fourth stage uses the second 12AT7 triode as a cathodyne phase splitter. Recall that the gain of the typical cathodyne is about $A_{V(\text{PS})} \cong 0.7$. It would be a good practice and a lot of fun to do load line analyses on the first four stages of the amplifier. Assume that the external effects loop resistance is about 100 kΩ and the tone control input resistance is 200 kΩ.

The Output Stage

The output stage is fairly typical of the earlier circuits covered in this chapter. The main parameters for the 6V6-GT are listed below. The 6V6 is used in the pentode mode to obtain higher output power.

6V6-GT parameters

	Pentode mode	Triode mode	
$P_{D(\text{max})}$	12 W	9 W	Max DC power
$I_{P(\text{max})}$	35 mA	35 mA	Max DC plate current

$V_{P(max)}$	315 V	315 V	Max DC plate voltage
g_m	3.7 mS	5 mS	Transconductance
r_P	80 kΩ	1.96 kΩ	Internal dynamic plate resistance
$R'_{L(typ)}$	8 kΩ	3.5 kΩ	Typical reflected load resistance
μ	296	9.8	Amplification factor
Heater	6.3 V, 450 mA		

Each 6V6-GT is biased for the following Q-point:

$$V_{PQ} = 300 \text{ V}, \, I_{PQ} = 30 \text{ mA}, \, P_{DQ} = 9 \text{ W}$$

A Hammond 125E, which fits perfectly on the 5E3 chassis, was chosen as the output transformer. With an 8 Ω load, connection to secondary lugs 4 and 6 produces a plate-to-plate reflected load resistance of $R'_{PP} = 17.6$ kΩ. Each 6V6 will see a reflected load of

$$R'_L = \frac{R'_{PP}}{2} = \frac{17.6 \text{ k}\Omega}{2} = 8.8 \text{k}\Omega$$

The suggested reflected load resistance of the 6V6 operating as a pentode is 8 kΩ, so this is a good match. The AC load line for a single 6V6 is shown in Fig. 7.43. The maximum unclipped plate voltage swing is

$$\Delta V_P = \Delta V_{PR} = 264 \text{ V}$$

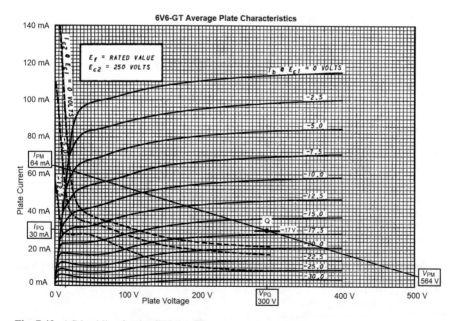

Fig. 7.43 AC load line for the 6V6 amplifier

The turns ratio of the transformer is

$$\frac{n_P}{n_S} = \sqrt{\frac{R'_{PP}}{R_L}}$$

$$= \sqrt{\frac{17.6 \text{ k}\Omega}{8 \ \Omega}}$$

$$\cong 47$$

The maximum output voltage swing and output power are

$$V_{o(max)} = 2\Delta V_P \left(\frac{n_S}{n_P}\right)$$

$$= 2 \times 264\text{V} \times \frac{1}{47}$$

$$= 11.2 \text{ V}_{pk} = 22.4 \text{ V}_{P-P}$$

The peak output power is

$$P_{o(max)} = \frac{V^2_{o(max)}}{R_L}$$

$$= \frac{(11.2\text{V})^2}{8\Omega}$$

$$\cong 15.7\text{W}_{pk} = 7.9\text{W}_{rms}$$

The grid-to-plate voltage gain of each 6V6 is

$$A'_V = g_m(R'_L \parallel r_P)$$
$$= 3.7\text{mS} \times (8.8\text{k}\Omega \parallel 80\text{k}\Omega)$$
$$= 3.7\text{mS} \times 8\text{k}\Omega$$
$$= 29.6$$

For each 6V6, the required peak AC grid voltage for full output is given by (7.36).

$$V_{G(pk)} = \frac{\Delta V_P}{A'_V}$$

$$= \frac{264 \text{ V}}{29.6}$$

$$\cong 9\text{V}_{pk}$$

Assuming that $A_{V(PS)} \cong 0.7$, the peak voltage required at the input of the phase splitter for full output is

$$V_{\text{in}} = \frac{V_{G(\text{pk})}}{A_{V(\text{PS})}}$$
$$= 9V/0.7$$
$$\cong 13V_{\text{pk}} = 26V_{\text{P-P}}.$$

These voltages are easily within the output voltage compliance of the phase splitter and the 12AT7 driver stage. The overall gain of the amplifier is the product of the individual stage gains, which is

$$A_V = A_{V(1)}A_{V(2)}A_{V(\text{tone})}A_{V(3)}A_{V(\text{PS})}A'_{V(4)}\left(\frac{n_S}{n_P}\right)$$
$$= 2 \times 15 \times 0.5 \times 45 \times 0.7 \times 29.6 \times 0.021$$
$$= 294$$

For $V_{o(\text{max})} = 11.2\ V_{\text{pk}}$, the input signal required for full, unclipped output is

$$V_{\text{in(max)}} = 30\ \text{mV}_{\text{pk}}$$

The AC load line for the 6L6 output pentodes is shown in Fig. 7.43.

The Power Supply

The power supply for the 6V6 amp is shown in Fig. 7.44. A tube rectifier (5Y3GT) and standby switch are used in this design because the 5E3 chassis has holes and silk-screened labeling present. If you are using a homemade chassis, you could easily eliminate the standby switch and go with a solid-state rectifier. If a solid-state

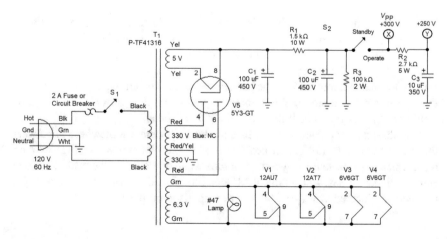

Fig. 7.44 Power supply for the 6V6 amplifier

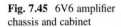

Fig. 7.45 6V6 amplifier chassis and cabinet

rectifier is used, a redesign with a different power transformer would also be recommended.

The chassis and tweed cabinet for the amp are shown in Fig. 7.45. I used a Celestion G12M-65, 12-inch Creamback speaker for this build. This speaker can handle up to 65 W, but sounds great with this amplifier.

Ultralinear Amplifiers

We will finish this chapter with a brief discussion of *ultralinear* (UL) amplifiers. An ultralinear push-pull output stage, based on the output stage of Fig. 7.30, is shown in Fig. 7.46. The ultralinear design approach connects the screen grids of the push-pull pentodes to special primary winding taps on the output transformer. These taps are usually located at about 40% of the primary winding. The UL taps typically operate the screen grids at about 10 V less than the plates, and they provide local negative feedback at the output which linearizes the gain of the amp.

The ultralinear tap connections cause the tube to operate somewhere in between the triode mode and full-blown pentode mode. The plate curves of Fig. 7.47 are representative of those for a hypothetical tube when operating in the pentode, triode, and ultralinear modes, with a typical AC load line plotted on each graph as well. The Q-points are located at $V_{PQ} = 250$ V for comparison purposes.

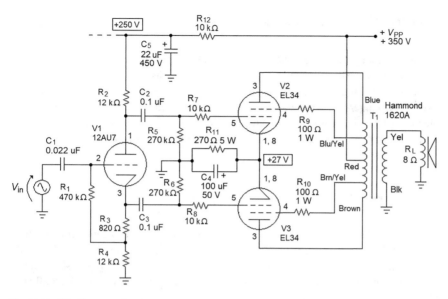

Fig. 7.46 Ultralinear push-pull output stage

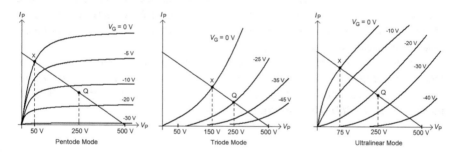

Fig. 7.47 Representative plate curves for pentode, triode, and ultralinear operation

Examining these curves, we see that in the pentode mode, the Q-point can move to the left to $V_{P(min)} \cong 50$ V. In the triode mode, the Q-point can move left to $V_{P(min)} \cong 150$ V. In the ultralinear mode, the Q-point can swing to $V_{P(min)} \cong 75$ V, which allows output voltage compliance close to operation in the pentode mode. The second important thing to notice is that the ultralinear mode curves are not as flat as the pentode curves. This means that the dynamic plate resistance of the ultralinear mode tube ($r_P = \Delta V_P / \Delta I_P$) will be similar to operation in the triode mode.

In summary, ultralinear operation results in output resistance and clipping characteristics similar to triode operation while achieving voltage gain and output compliance that is closer to pentode operation. Ultralinear stages are quite common in vacuum tube high-fidelity audio amplifiers, but not as commonly used in guitar amplifier designs.

Construction Techniques and Tips

We will wrap up this chapter with a few tips relating to the physical construction of tube amplifiers, but before moving on, I really have to stress once again that tube amplifiers are not good beginners' projects. The voltages present in these circuits are extremely dangerous. Until you have gained a few years' experience working with electronic circuits, or unless you have the assistance of an experienced mentor, it is best to stick with low-voltage, solid-state amplifiers and effects.

Chassis Materials

Probably the best chassis material to use in the construction of tube amps is steel. Steel provides excellent electromagnetic shielding and is strong but can be pretty difficult to work with. Aluminum does not provide the electromagnetic shielding of steel but is much easier to work with, so unless I buy a premade chassis, I usually use aluminum when I build tube amps.

Be prepared for a lot of drilling and hole-punching if you decide to build a tube amp. An assortment of chassis punches is invaluable for making holes for tube sockets, large can-type filter capacitors, and larger panel lamps. Chassis punches are relatively expensive but are preferred if you are working with a steel chassis. Chassis punch sets usually sell for $50.00 and up.

Step drills are also very handy for making large holes in the chassis and are a lot less expensive than chassis punches. Step drills work ok with aluminum, but they are not the best choice for drilling large holes in a steel chassis. An assortment of chassis punches and step drill bits is shown in Fig. 7.48

Fig. 7.48 Chassis punches and step drills

Fig. 7.49 Twisting filament wiring helps reduce hum

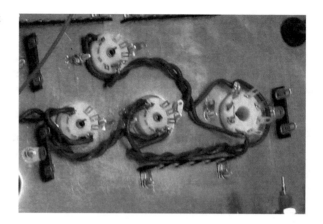

Wiring Tips

The concept of star grounding was mentioned previously in Chap. 4 (see Fig. 4.9). Star grounding helps to prevent circuit instability and reduce hum by eliminating ground loops. Ideally, all paths to ground should have zero resistance. Ground loops are ground paths that have relatively significant resistance, therefore producing a significant voltage drop. A single star ground helps to reduce this effect.

Tube filament wiring can be a source of 60 Hz hum in the output of the amplifier. When building tube amplifiers, it is a good idea to run twisted wires to the tube filaments and if possible to keep these wires away from signal lines. This is especially true for the input stage of the amplifier. Twisted filament wiring is shown in Fig. 7.49.

There are some very common-sense things that can be done to reduce hum when building an amp. First, it is a good idea to keep the input stage circuitry far from the power supply section. This reduces magnetic coupling between the power transformer/choke and the wiring that carries low-level signals.

The input jacks of the amp should be located close to the first stage, and potentiometers should be located close to the circuitry to which they are connected. It's also a good idea to use shielded cable for connections to input jacks and potentiometers as well.

Long wires tend to act like antennas so keeping wiring short helps to reduce this effect. Wires that are oriented parallel to one another act as if they were coupled together by a capacitor. Magnetic coupling between adjacent wires also occurs. The longer the wires, and the closer together they are, the greater the capacitive and magnetic coupling. Magnetic coupling is dominant in high-current wiring, such as in the filament supply. Capacitive coupling is more significant at higher frequencies.

To minimize this coupling effect, wires that carry power and those that carry signals should not be run parallel to one another and should cross at 90° angles if possible. This is called *orthogonal wiring*.

Testing Tips

It's important to be extremely careful when working in an amplifier that is powered up. It is a good idea to keep one hand in your pocket at all times while taking measurements in high voltage circuits in order to lessen the danger of accidental electric shock. If you are tired or distracted, stop working, and come back to it after you've had a chance to rest.

When you are building a tube amplifier (or any circuit for that matter), it helps to be very neat and methodical. A good way to keep track of your work and lessen the chances of making a wiring error is to mark the schematic with a highlighter as you solder in components and wires. Some people even like to create a pictorial wiring diagram before starting construction.

Build and Test the Amp in Stages

Generally, it is a good idea to construct and test an amplifier in stages. Construction and testing of the power supply is usually completed first. Keep in mind that initially, since the remaining sections of the amplifier are not present, the supply will be under no-load conditions and is likely to produce a much higher than expected output voltage. I do not recommend operating the power supply with no load, as components can sometimes be subjected to higher voltages than their ratings allow.

It is a good practice to connect a resistor from V_{PP} to ground to provide a load on the power supply when performing initial tests. A suitable load resistor for the power supplies presented in this book is $R_L = 10$ kΩ, 20 W. This resistor draws about 35 mA at 350 V while dissipating about 12 watts. Examples of several load resistors I have used during amplifier testing are shown in Fig. 7.50.

Fig. 7.50 Assorted load resistors used for power supply and amplifier testing

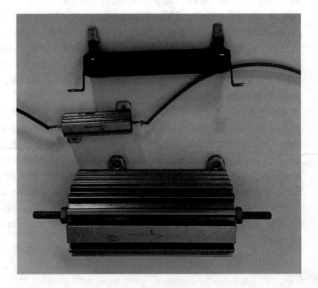

After initial power supply testing, I like to build the output stage first because it draws the most current and will load the power supply down close to its actual operating voltage. Once the output stage is wired and tested, I will usually work forward from the first stage, testing for proper operation as I go.

Should you see smoke, hear sizzling or frying sounds, or see any components glowing inappropriately, turn off power immediately. In almost all cases, such problems are the result of incorrect wiring. Unplug the circuit and allow all capacitors to discharge to 10 V or less before checking for construction errors.

When making voltage measurements at various points in the circuit, you can generally expect to find voltage variations of 10% or so, depending on supply loading conditions, alternate tube substitutions, component tolerances, and approximations made during the design process.

Never Operate the Amp Without a Load

It is imperative that a suitable load be connected to the output stage before power is applied. Under no-load conditions, the output transformer will have an extremely high primary side impedance. This results in a nearly horizontal AC load line for the amplifier and can easily subject the output tubes and output transformer primary windings to several thousand volts. This can cause destructive arcing in the transformer, the output tubes, or both. You may use a loudspeaker as a load, but this can become loud and irritating very quickly. In place of the loudspeaker, a power resistor of 8 ohms (or whatever your intended load may be), rated for 20 watts or higher, will work fine for testing purposes with the amplifiers presented in this book.

Final Comments

It is interesting that characteristic curves are used extensively when designing tube audio circuits, while device curves are rarely used in the design of transistor audio circuits. There are several reasons for this. First, tubes tend to have relatively uniform characteristics. The published tube curves are really quite accurate and useful. Important transistor parameters such as beta, transconductance, and small-signal dynamic collector resistance vary widely from one device to another. A second reason for the predominance of graphical techniques in tube circuit design and analysis is because that's the way it was done back in the glory days of tubes, before calculators and computers were readily available. The old graphical methods are easy to learn and yield excellent results, so they are still used today. I have to admit though that PSpice simulation is a great convenience as long you have good tube models.

This brings us to the end of the book. It's been a pretty long and sometimes intense journey, but I hope you found at least some of the circuits and concepts interesting and useful. Be sure to check the Facebook page for the book for additional circuit designs and other related information. Feel free to post your own designs, suggestions, and ideas for experimentation there as well.

Summary of Equations

Maximum Power Transfer

$$R_L = R_o \tag{7.1}$$

Transformer Turns Ratio

$$\frac{n_P}{n_S} \tag{7.2}$$

Transformer Secondary Voltage and Current

$$V_{sec} = V_{pri}\left(\frac{n_S}{n_P}\right) \tag{7.3}$$

$$I_{sec} = I_{pri}\left(\frac{n_P}{n_S}\right) \tag{7.4}$$

Transformer Input and Output Power

$$P_{in} = V_{pri}I_{pri} \tag{7.5}$$

$$P_o = V_{sec}I_{sec} \tag{7.6}$$

Transformer Reflected Load Resistance

$$R'_L = R_L\left(\frac{n_P}{n_S}\right)^2 \tag{7.7}$$

Determining Turns Ratio Given Reflected Load Resistance

$$\frac{n_P}{n_S} = \sqrt{\frac{R'_L}{R_L}} \tag{7.8}$$

Design Approximations

$$V_{PQ} \cong V_{PP} \qquad (7.9)$$

$$V_{PKQ} \cong V_{PQ} = V_{PP} \qquad (7.10)$$

Q-Point Power Dissipation

$$P_{DQ} = V_{PQ}I_{PQ} \qquad (7.11)$$

Calculating R_K from Q-Point Location

$$R_K = \frac{-V_G}{I_{PQ}} \qquad (7.12)$$

AC Load Line Limits

$$v_{PM} = V_{PQ} + I_{PQ}R'_L \qquad (7.13)$$

$$i_{PM} = I_{PQ} + \frac{V_{PQ}}{R'L} \qquad (7.14)$$

$$\Delta V_{PL} = V_{PQ} - -v_{P(min)} \qquad (7.15)$$

$$\Delta V_{PR} = v_{PM} - -V_{PQ} \qquad (7.16)$$

$$V_{o(max)} = \Delta V_P \left(\frac{n_S}{n_P}\right) \qquad (7.17)$$

Output Power Formulas

$$P_{o(max)} = \frac{V^2_{o(max)}}{R_L} \qquad (7.18)$$

$$P_{rms} = P_{o(max)}/2 \qquad (7.19)$$

Plate-to-Grid Voltage Gain (Transformer Coupling)

$$A'_V = g_m(R'_L \parallel r_P) \tag{7.20}$$

Actual Voltage Gain (Transformer Coupling)

$$A_V = A'_V \left(\frac{n_S}{n_P}\right) \tag{7.21}$$

Max Input Voltage for Unclipped Output

$$V_{in(\max)} = \frac{\Delta V_P}{A'_V} \tag{7.22}$$

Formula for I_{PQ} Given P_{DQ}

$$I_{PQ} = \frac{P_{DQ}}{V_{PP}} \tag{7.23}$$

Parameters for Two Parallel-Connected Tubes

$$g_{m(eq)} = 2g_m \tag{7.24}$$

$$r_{P(eq)} = r_P/2 \tag{7.25}$$

$$R_{K(eq)} = R_K/2 \tag{7.26}$$

Total DC Primary Current for Push-Pull Stage

$$I_{DC} = 2I_{PQ} \tag{7.27}$$

Push-Pull Plate Reflected Load Resistance for Each Plate (Class A)

$$R'_L = \frac{R_{PP}}{2} \tag{7.28}$$

Output Voltage for Push-Pull Stage (Class A)

$$V_o = 2V_P \left(\frac{n_S}{n_P}\right) \tag{7.29}$$

Effective Turns Ratio per Tube (Push-Pull, Class A)

$$\frac{n_1}{n_S} = \frac{1}{2}\left(\frac{n_P}{n_S}\right) \tag{7.30}$$

Cathode Resistor for Push-Pull Stage (Class A)

$$R_K = \frac{-V_G}{I_{DC}} \tag{7.31}$$

Plate-to-Plate Reflected Load Resistance (Push-Pull, Class A)

$$R_{PP} = 2R'_L \tag{7.32}$$

$$R'_L = R_{PP}/2 \tag{7.33}$$

Push-Pull Peak Output Voltage

$$V_{o(pk)} = 2\Delta V_P\left(\frac{n_S}{n_P}\right) \tag{7.34}$$

Push-Pull Plate-to-Grid Voltage Gain

$$A'_V = g_m(R'_L \parallel r_P) \tag{7.35}$$

Max Input Voltage for Unclipped Output (Push-Pull)

$$V_{G(pk)} = \frac{\Delta V_P}{A'_V} \tag{7.36}$$

Required Phase Splitter Input Voltage for Full Output

$$V_{in(PS)} = V_{G(pk)}/A_{V(PS)} \tag{7.37}$$

Appendices

Appendix A: Some Basic Circuit Theory

Voltage and Current Polarities and Conventions

Unless we are dealing specifically with the flow of electrons through a vacuum tube, we will assume current flows from positive to negative. This is called *conventional current flow*. Conventional current flow is used by virtually all manufacturers today on transistor and IC data sheets, application notes, etc.

As shown in Fig. A.1, voltage drops across components will be designated using curved arrows. Think of these arrows as voltmeters that sense the voltage across a given component. The point of the arrow is equivalent to the positive lead of a voltmeter. Currents will be represented with arrowheads drawn on the wires in which the current is flowing. For the resistor and the diode shown in Fig. A.1, if the current is flowing in the direction indicated, the current is positive. The voltage dropped across the components will also be positive for the arrow directions given here.

Linear Circuits

Ideally, resistors, capacitors, inductors, voltage sources, and current sources are linear circuit elements. If you graph the current versus voltage characteristic (transconductance) for a linear component, you get a straight line, as shown for the resistor in Fig. A.1.

One of the nice things about linear components is that we can use Ohm's law ($V = IR$) to relate I, V, and R. Because the plot of current vs. voltage is a straight line, we can use the coordinates of any point on the line and get the same result for $R = V/I$.

Fig. A.1 Linear (resistor) and nonlinear (diode) devices

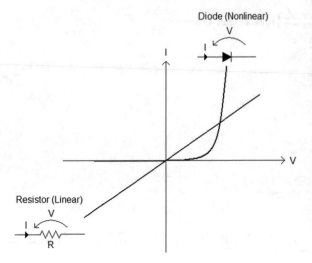

Diodes, transistors, and tubes are all nonlinear devices. An example is the graph of current versus voltage for a typical diode shown in Fig. A.1. We can't use Ohm's law to characterize nonlinear devices because the quotient V/I is not a constant. Even though circuits that contain diodes, transistors, and tubes are nonlinear, in many cases, we can treat them as if they are approximately linear and obtain useful analysis results. And, even though Ohm's law is not applicable to nonlinear devices, Kirchhoff's laws still apply.

Series Circuits

A typical series circuit is shown in Fig. A.2. All circuit elements (resistors and a battery in this case) carry the same current in a series circuit.

The various voltage drops across the circuit elements can be determined using Ohm's law and Kirchhoff's voltage law.

Note that I have oriented the voltage-sensing arrows pointing counterclockwise on the resistors so that the voltage drops across the resistors will be positive. I did this because I like to work with positive numbers, but the arrows could have been oriented the other direction, resulting in negative voltage values.

Fig. A.2 A series circuit

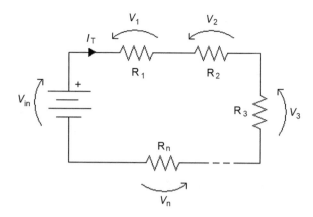

Ohm's Law

Ohm's law relates current, voltage, and resistance for linear circuit elements such as resistors, capacitors, and inductors. The basic Ohm's law relationships are

$$V = IR \qquad I = \frac{V}{R} \qquad R = \frac{V}{I}$$

Regardless of whether a component is linear or nonlinear, its power dissipation is given by

$$P = VI$$

For a resistance, power dissipation can be calculated in a number of different ways. Here are three of the most commonly used equations:

$$P = VI \qquad P = I^2 R \qquad P = \frac{V^2}{R}$$

Conductance G is the reciprocal of resistance, $G = 1/R = I/V$. The unit of conductance is the siemen S. Older references use the mho, an upside-down omega.

Kirchhoff's Voltage Law

Kirchhoff's voltage law (KVL) says that the algebraic sum of all voltages around a closed loop is zero. The series circuit of Fig. A.2 forms a closed loop. If we assume there are no other resistors than those shown, we sum the voltage drops moving clockwise from the battery to obtain the KVL equation

$$0 = V_{in} - V_1 - V_2 - V_3 - \ldots - V_n$$

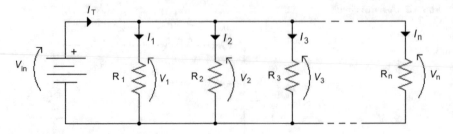

Fig. A.3 A parallel circuit

Subtracting V_{in} from both sides and multiplying by -1 gives us

$$V_{in} = V_1 + V_2 + V_3 + \ldots + V_n$$

The general rule is that we add voltages whose sensing arrows are pointed in the direction we travel around the loop. We subtract voltages whose sensing arrows point against the direction we have chosen.

Parallel Circuits

A typical parallel circuit is shown in Fig. A.3. In a parallel circuit, all elements have the same voltage across them.

The voltage drops across all elements in a parallel circuit are equal. That is,

$$V_1 = V_2 = V_3 = \ldots = V_n = V_{in}$$

The current flowing through a given element may be determined using Ohm's law, Kirchhoff's current law, or a combination of the two.

Nodes

A node is a point where two or more circuit elements are connected together. For example, in Fig. A.3, the entire top line of the schematic is a node—the top terminal of all of the resistors could be drawn connected at a single point or node. The circuit in Fig. A.3 has a total of two nodes. In Fig. A.2, each junction between resistors is a node, so this circuit has $n + 1$ nodes.

Branches

A branch is a path through which current can flow. The circuit in Fig. A.1 has one branch. That is, there is only one path through which current can flow.

The circuit of Fig. A.2 has $n + 1$ branches. However, we could have started our numbering with I_T renamed I_1, in which case there would be n branches.

Kirchhoff's Current Law

Kirchhoff's current law (KCL) states that the algebraic sum of all currents entering and leaving a node is zero. A node with five branches extending from it is shown in Fig. A.4.

You can arbitrarily choose the polarity of the current sensing arrows. For example, if we assume that arrows pointing to the node are positive, while those pointing out of the node are negative, the KCL equation for Fig. A.4 is

$$0 = I_1 - I_2 + I_3 + I_4 - I_x$$

Let's say that we know the values $I_1 = 10$ mA, $I_2 = 2$ mA, $I_3 = -4$ mA, and $I_4 = 1$ mA. We can find the value of I_x by adding it to both sides of the equation and substituting the given current values as follows:

$$I_x = 10 \text{ mA} - 2 \text{ mA} + (-4 \text{ mA}) + 1 \text{ mA}$$
$$= 5 \text{ mA}$$

This means that 5 mA is flowing out from the node via the path indicated by the I_x arrow, in the direction indicated.

Fig. A.4 Example of a node with five branches

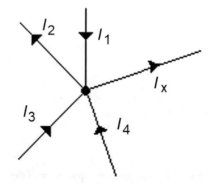

The Superposition Principle

The principle of superposition is used to analyze linear circuits that contain multiple voltage and/or current sources. The idea behind superposition is that we can determine the response for each source acting individually, and then we add these responses to get the complete response. For example, to analyze the circuit in Fig. A.5, we can start by "killing" current source I_1 and finding the output V'_o caused by voltage source V_1 acting alone.

An ideal current source has infinite internal resistance, so when we kill a current source, it is simply replaced with an open circuit. Redrawing the circuit with source I_1 killed as shown in Fig. A.5, we can find V'_o using the voltage divider relationship. Note that since R_3 is sticking out into space, it has no effect on this calculation.

$$V'_0 = V_1 \left(\frac{R_2}{R_1 + R_2} \right)$$

$$= 12\,\text{V} \left(\frac{2\,k\Omega}{1\,k\Omega + 2\,k\Omega} \right)$$

$$= 8\,\text{V}$$

Next, we kill voltage source V_1 and determine the output voltage caused by current source I_1 acting alone. The internal resistance of an ideal voltage source is zero, so when we kill a voltage source, we replace it with a short circuit. Redrawing the circuit with voltage source V_1 killed, we find $R_1 \parallel R_2 = 667\,\Omega$. As in the previous step, resistor R_3 has no effect on the calculation. The second component of the output voltage is

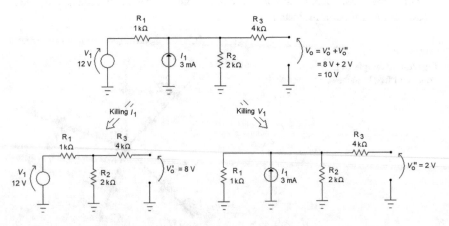

Fig. A.5 Example of the application of superposition in circuit analysis

$$V_o'' = I_1 \left(R_1 \middle\| R_2 \right)$$
$$= 3 \text{ mA} \left(1 \text{ k}\Omega \middle\| 2 \text{ k}\Omega \right)$$
$$= 3 \text{ mA} \times 667 \text{ }\Omega$$
$$= 2 \text{ V}$$

The net output voltage is the sum or superposition of the two responses.

$$V_0 = V_0' + V_0''$$
$$= 6 \text{ V} + 2 \text{ V}$$
$$= 8 \text{ V}$$

Capacitors, Inductors, and Complex Numbers

Unlike resistors, capacitors and inductors are circuit elements that store energy. Capacitors store energy in the form of an electrostatic field, while inductors store energy in a magnetic field. When we analyze circuits containing R, L, and C elements, we can look at transient behavior, steady-state behavior, or both.

This section is very brief and cursory. I considered putting a tutorial on phasor algebra in here, but it would increase the size of the appendix like crazy, and we still couldn't do the topic justice. I recommend consulting a basic circuit analysis text. There are also some very good videos on basic circuit analysis online as well.

Transient Behavior

By definition, transient behavior only lasts for a short time. When we first apply a voltage (power or a signal), capacitors and inductors must reach equilibrium; capacitors charge and inductors build magnetic flux. Both charging events usually happen in an exponential manner.

Figure A.6 shows the charging voltage and current for an RC circuit. If the switch is thrown from position 1 to position 2 at $t = 0$, the capacitor voltage will increase exponentially toward the input voltage. The current will jump initially to $I_{max} = V_{in}/R$ and then decrease exponentially toward zero. The voltage and current equations for the charging capacitor are

$$v_C(t) = V_{in} \left(1 - e^{-t/RC} \right)$$

$$i(t) = \frac{V_{in}}{R} e^{-t/RC}$$

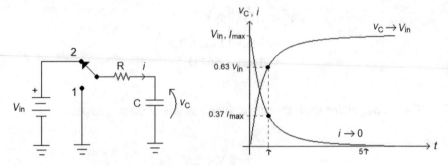

Fig. A.6 Capacitor charging voltage and current

where the product of R and C is the time constant.

$$\tau = RC \text{ (seconds)}$$

After 5τ, the capacitor is 99% charged, and the circuit is generally considered to have reached equilibrium.

If we move the switch from position 2 back to position 1, the capacitor will discharge through the resistor. The capacitor voltage and current equations for discharge are

$$v_C(t) = V_o e^{-t/RC}$$

$$i(t) = \frac{-V_o}{R} e^{-t/RC}$$

where V_o is the voltage across the capacitor at the instant the switch is thrown. There is a negative sign in the $i(t)$ equation because the capacitor discharges back through the resistor, opposite to the indicated direction of current in the schematic diagram.

Other examples of places where we use exponential functions are when examining transistor behavior, signal envelopes, and signal decay properties. Capacitor charging and discharging behavior would be used in more detailed analysis of the relaxation oscillator. That is beyond the scope of this book, but it's interesting stuff, and I recommend taking a deeper look into these topics if you get the chance.

Steady-State Behavior

Steady-state behavior or steady-state response is more applicable to the analyses that we are performing in this book. When we apply a sinusoidal signal to the input of an amplifier and look at various points with an oscilloscope, we are viewing steady-state voltages. These signals persist and are relatively constant over the time we are looking at things.

   Resistance is the opposition to the flow of current through a circuit element that does not store energy (e.g., a resistor). The opposition to the flow of current through a capacitor is called *capacitive reactance*. There are two equivalent ways to express capacitive reactance.

$$X_C = \frac{1}{2\pi fC} \angle -90° \,(\text{Polar Form})$$

$$X_C = \frac{-j}{2\pi fC} \,(\text{Rectangular Form})$$

   This notation is sometimes called the *phasor form*. In the *rectangular form* equation, $j = \sqrt{-1}$. Multiplying by $-j$ is equivalent to phase-shifting by $-90°$. The rectangular form is usually somewhat difficult to visualize, so throughout the book, the polar form is used when describing signals and impedances. Often we don't need to worry about the phase angle, and we can just use the magnitude of the reactance.
   Inductive reactance is the opposition to the flow of sinusoidal current through an inductor. The formulas for inductive reactance are

$$X_L = 2\pi fL \angle 90° \,(\text{polar form})$$

$$X_L = j2\pi fL \,(\text{rectangular form})$$

   Inductors have the opposite phase shift from capacitors. As with capacitors, often we don't care about the phase angle and may simply use the magnitude of inductive reactance.
   Capacitive reactance decreases with frequency, while inductive reactance increases. This is shown in Fig. A.7. This characteristic is what allows us to use capacitors for coupling AC signals while blocking DC levels. At very low

Fig. A.7 Variation of capacitive and inductive reactance with frequency

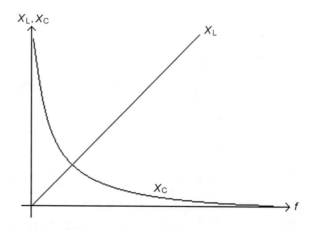

frequencies, the capacitor behaves like an open circuit. At high frequencies, the capacitor can be treated as a short circuit.

Combining resistance and reactance results in impedance Z. That is,

$$Z = R \pm jX$$

The positive sign is used with inductors and negative for capacitance. The magnitude of the impedance is found using the Pythagorean theorem.

$$|Z| = \sqrt{R^2 + X^2}$$

Conductance G is the reciprocal of resistance, and the reciprocal of impedance is *admittance Y.*

$$Y = \frac{1}{Z}$$

Summary of Useful Formulas

Ohm's Law

$$V = IR \qquad I = \frac{V}{R} \qquad R = \frac{V}{I}$$

$$P = IV \qquad P = I^2 R \qquad P = \frac{V^2}{R}$$

n Series Resistances

$$R_{eq} = R_1 + R_2 + \ldots + R_n$$

n Parallel Resistances

$$R_{eq} = \frac{1}{\frac{1}{R_1} + \frac{1}{R_2} + \ldots + \frac{1}{R_n}}$$

Frequency and Period

$$f = \frac{1}{T} \qquad T = \frac{1}{f} \qquad \omega = 2\pi f = \frac{2\pi}{T} \qquad f = \frac{\omega}{2\pi}$$

Charging and Discharging Capacitor

$$v_C(t) = V_{in}\left(1 - e^{-t/RC}\right) \qquad i(t) = \frac{V_{in}}{R} e^{-t/RC} \text{ (charging)}$$

$$v(t) = V_o e^{-t/RC} \qquad\qquad i(t) = \frac{-V_o}{R} e^{-t/RC} \text{ (discharging)}$$

Inductive and Capacitive Reactance $\left(j = \sqrt{-1}\right)$

$$X_C = \frac{1}{2\pi fC} \angle -90° \qquad X_C = \frac{-j}{2\pi fC} \qquad X_C = \frac{1}{j2\pi fC} \qquad |X_C| = \frac{1}{2\pi fC}$$

$$X_L = 2\pi fL \angle 90° \qquad X_L = j2\pi fL \qquad |X_L| = 2\pi fL \qquad |Z| = \sqrt{R^2 + X^2}$$

Bipolar Transistor Relationships

$$I_C = \beta I_B \qquad I_E = I_C + I_B \qquad r_e = \frac{V_T}{I_{CQ}} \qquad I_C \cong I_S e^{V_{BE}/\eta V_T}$$

Triode Relationships

$$g_m = \mu r_P \qquad \mu = \frac{g_m}{r_P} \qquad r_P = \frac{\mu}{g_m} \qquad I_P = k(V_P + \mu V_G)^{3/2}$$

JFET Relationships

$$g_{m0} = \frac{2I_{DSS}}{V_P} \qquad I_D = I_{DSS}\left(1 - \frac{V_{GS}}{-V_P}\right)^2$$

Appendix B: Selected Tube Characteristic Curves

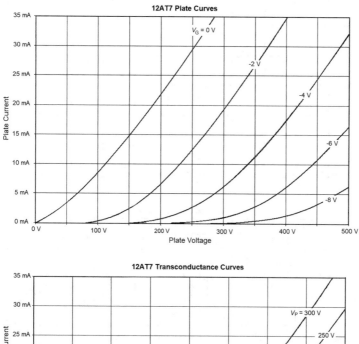

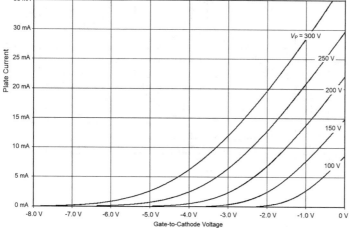

Fig. B.1   12AT7 plate and transconductance curves

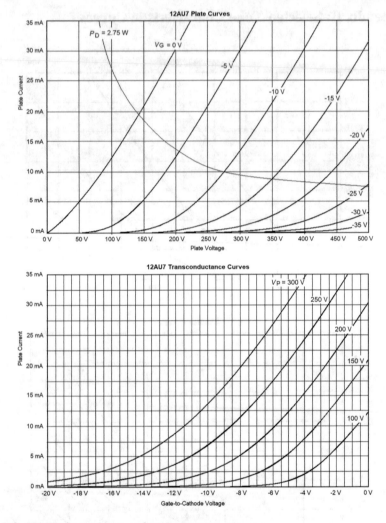

Fig. B.2 12AU7 plate and transconductance curves

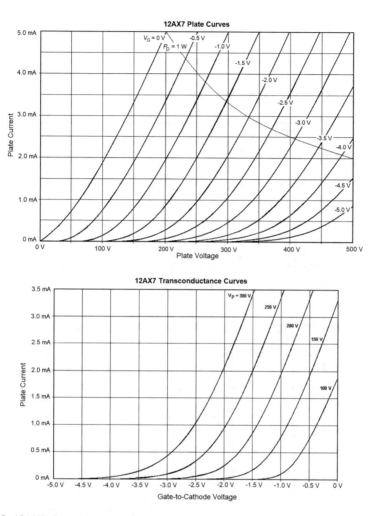

Fig. B.3 12AX7 plate and transconductance curves

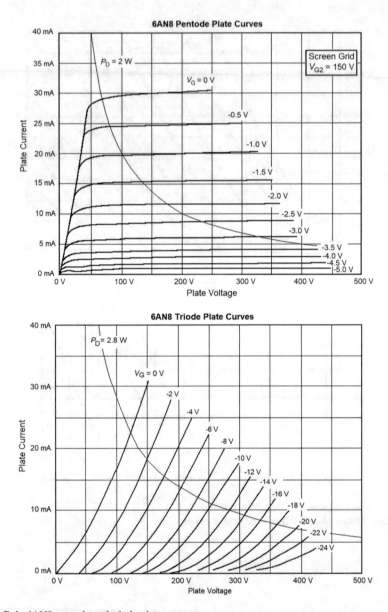

Fig. B.4 6AN8 pentode and triode plate curves

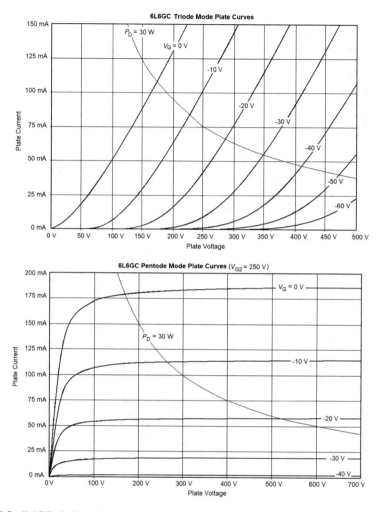

Fig. B.5 6L6GC triode- and pentode-mode plate curves

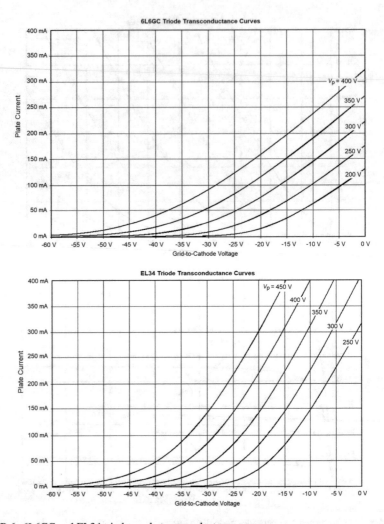

Fig. B.6 6L6GC and EL34 triode-mode transconductance curves

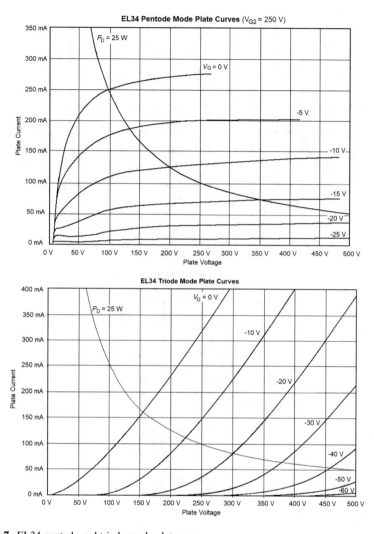

Fig. B.7 EL34 pentode and triode-mode plate curves

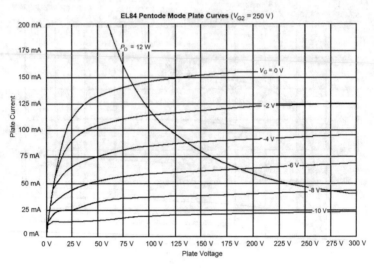

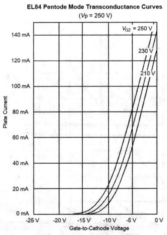

Fig. B.8 EL84 pentode-mode plate and transconductance curves

Appendix C: Basic Vacuum Tube Operating Principles

Diodes

The diode is the simplest vacuum tube. Recall that the vacuum tube diode operates by thermionic emission of electrons from a heated cathode within the evacuated glass envelope of the tube, as shown in Fig. C.1.

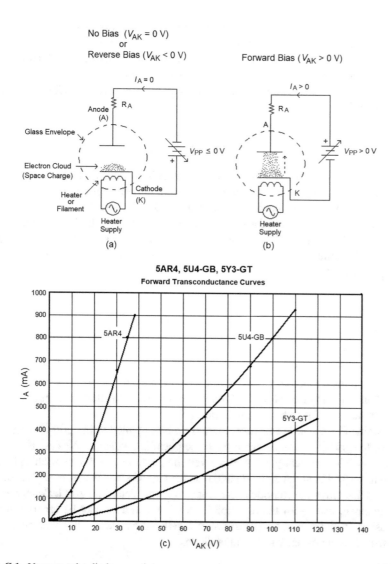

Fig. C.1 Vacuum tube diode operation

Reverse Bias

When the anode is at a negative potential with respect to the cathode $V_{AK} \leq 0$ V, the diode is said to be reverse biased, and the current flow through the tube is approximately zero. This occurs because the free electrons around the cathode are repelled by the negative potential at the anode, as shown in Fig. C.1a.

Forward Bias

If the anode terminal is made positive with respect to the cathode $V_{AK} > 0$ V, electrons that are attracted from the cathode are free to travel through the vacuum to the anode, as shown in Fig. C.1b. This is condition is referred to as forward bias.

   The current that flows in a forward-biased vacuum tube diode is not a linear function of the applied voltage, but rather obeys what is called the Child-Langmuir law, which is

$$I_A \cong kV_P^{3/2} \tag{C.1}$$

   The constant k is the *perveance* of the diode. A high value of perveance results in a diode that drops less voltage at a given forward current. For typical vacuum tube diodes, k ranges from approximately 0.003–0.00035. The forward bias current versus voltage relationships for three common vacuum tube diodes, the 5AR4, the 5U4-GB, and the 5Y3-GT, are shown in Fig. C.1c. All three diodes closely follow the Child-Langmuir law, with the value of k varying from 0.003 for the 5AR4 to 0.00035 for the 5Y3-GT.

Triodes

If we take a vacuum tube diode and place a fine wire *control grid*, in between the cathode and the anode (which we will now refer to as the *plate*), a triode is formed. A triode is shown in Fig. C.2. The function of the grid is to allow control of the flow of electrons from the cathode to the plate.

   Examine Fig. C.2a. In this case, there is no bias applied to the grid of the triode. That is, the grid and cathode are at the same potential, which in this case is ground, and $V_{GK} = 0$ V. Physically, the grid has very little cross-sectional area. This plus the fact that the grid is at the same potential as the cathode means that nearly all of the electrons that leave the cathode make it across to the plate. The grid leakage current is quite small, usually of the order of 1 μA or less, and so may be ignored in this discussion. With no bias, the plate current I_P will increase with plate voltage V_P just as it would for a normal vacuum tube diode.

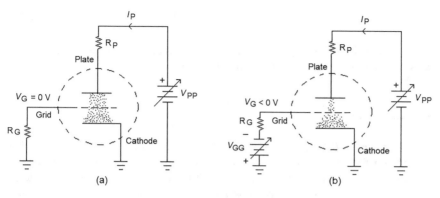

Fig. C.2 The addition of the grid creates a triode

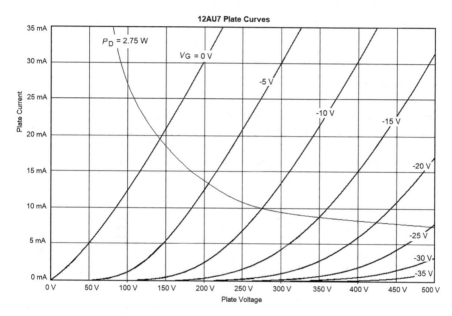

Fig. C.3 Plate characteristic curves for the 12AU7

In Fig. C.2b, the grid is biased to a more negative potential than the cathode. Now, the positive potential of the plate exerts less attractive force on the electrons surrounding the cathode, and the plate current will be reduced relative to that which would flow with no bias on the grid. If we step the grid bias voltage from zero to some maximum value, a family of plate characteristic curves is generated. Figure C.3 shows the plate characteristic curves for the 12AU7 triode. The equation that describes these curves is

$$i_P = k(V_P + \mu v_G)^{3/2} \tag{C.2}$$

When there is no bias applied to the triode, the term μv_G is zero, and (C.2) reduces to the standard Child-Langmuir equation for a diode (C.1).

Amplification Factor

The parameter μ (Greek mu) is called the amplification factor. The value of μ depends on the construction of the tube. For example, the 6AS7-G is a low-mu triode with $\mu = 2$, while the 12AX7 is a high-mu triode with $\mu = 100$. The actual value of μ varies somewhat from one tube to another of the same type, but it is generally close enough to the published value that in many cases we can consider it to be the same for every tube of the same type. This is in sharp contrast to bipolar transistors where, for example, beta may vary wildly from one device to another with the same part number.

Transconductance

The parameter that relates plate current I_P to grid bias voltage V_{GK} is called *transconductance*, which is designated as g_m. We can plot plate current as a function of grid voltage at various plate voltage levels. This produces a family of transconductance curves, such as those shown for the 12AU7 in Fig. C.4.

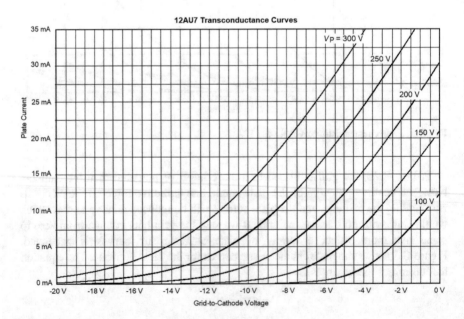

Fig. C.4 Transconductance curves for the 12AU7

Transconductance allows us to view the triode as a voltage-controlled current source. Looking at the triode from an AC signal standpoint, we have

$$i_P = g_m v_G \tag{C.3}$$

where the AC plate current i_P is controlled by the AC grid voltage v_G. We are taking some liberties here, assuming that $v_G = v_{GK}$, which is true when there is no cathode resistance in the AC equivalent circuit.

Dynamic Plate Resistance

The effective internal resistance of the plate r_P is another basic tube parameter that is very important. For example, maximum power transfer from the tube to a load that is driven by the plate occurs when $r_P = R_L$. Although it is generally not necessary to have a precise match between r_P and R_L, triodes with low plate resistance are better suited for driving low-resistance loads than triodes with high plate resistance.

Plate resistance, amplification factor, and transconductance are related by the following equation:

$$\mu = g_m r_P \tag{C.4}$$

Tetrodes

A tetrode is a four-terminal tube that is created with the addition of a *screen grid* between the plate and the control grid. The function of the screen grid is to shield the control grid from the plate, which reduces the effective input capacitance of the tube. The screen grid is typically held at a slightly less positive voltage than the plate and is often bypassed to ground to help keep the suppressor grid voltage steady and noise-free. The schematic symbol for a tetrode is shown in Fig. C.5.

The presence of the screen grid has some negative effects on tube characteristics as well. For example, since the screen grid is held at a relatively high positive potential, it will accelerate electrons as they move from the cathode toward the plate. The increased energy of the electrons striking the plate causes secondary electrons to be emitted from the plate back toward the screen grid. Many of these secondary electrons are captured by the screen grid, resulting in high screen grid current and power dissipation which is undesirable. The creation of secondary emission current can also cause undesirable nonlinear behavior of the tetrode, even causing negative resistance under some conditions.

Fig. C.5 Schematic symbol
for the tetrode

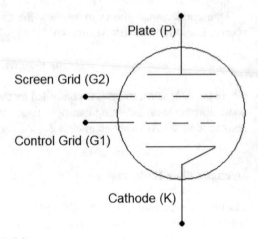

Fig. C.6 Schematic symbol
for the pentode

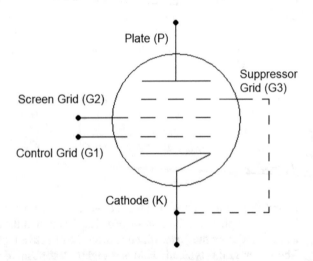

Pentodes

The undesirable characteristics of the tetrode are improved by the addition of a fifth
element, called a *suppressor grid*. This forms a *pentode*. The schematic symbol for a
pentode is shown in Fig. C.6. The suppressor grid G3 is located between the screen
grid and the plate inside the tube. Most often, the suppressor grid is tied directly to
the cathode, which holds it at a very negative potential relative to the plate. The
function of the suppressor grid is to repel secondary emission electrons back to the
plate, preventing them from causing excessive current flow and power dissipation in
the screen grid.

Pentodes generally have much higher effective plate resistance r_P, and amplification factor μ, than triodes. The plate characteristic curves for pentodes are very
similar to those of bipolar transistors and field effect transistors. The plate curves for
the EL34 pentode are shown in Fig. C.7.

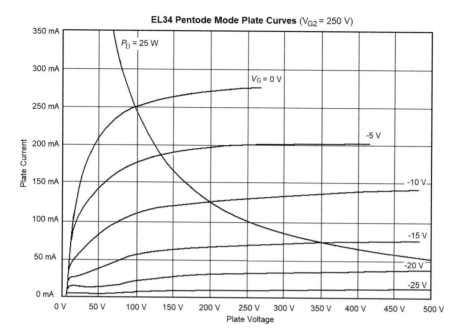

Fig. C.7 Plate characteristic curves for the EL34 pentode

Even though the pentode has a suppressor grid, the secondary emission electrons still affect the plate characteristics to some extent. Notice in Fig. C.7 how the low current plate curves dip or kink at low plate voltages. This effect is caused by secondary emission, and it is even more pronounced in tetrode characteristic curves. For this reason, tetrodes were rapidly superseded by pentodes.

The pentode can also be configured to operate as a triode simply by connecting the screen grid directly to the plate. Operation as a triode significantly reduces r_P, μ, and maximum power output but is often used to obtain particularly desirable overdrive and distortion characteristics.

Index

© The Editor(s) (if applicable) and The Author(s), under exclusive license to
Springer Nature Switzerland AG 2022
D. J. Dailey, *Electronics for Guitarists*, https://doi.org/10.1007/978-3-031-10758-0

Printed in the United States
by Baker & Taylor Publisher Services